THE CHEMISTRY OF NONAQUEOUS SOLVENTS

Edited by J. J. LAGOWSKI

DEPARTMENT OF CHEMISTRY
THE UNIVERSITY OF TEXAS AT AUSTIN
AUSTIN, TEXAS

Volume VB

ACIDIC AND APROTIC SOLVENTS

1978

ACADEMIC PRESS New York San Francisco London
A Subsidiary of Harcourt Brace Jovanovich, Publishers

ACADEMIC PRESS, INC.
111 Fifth Avenue, New York, New York 10003

United Kingdom Edition published by
ACADEMIC PRESS, INC. (LONDON) LTD.
24/28 Oval Road, London NW1 7DX

Library of Congress Cataloging in Publication Data

Lagowski, J J ed.
The chemistry of non-aqueous solvents.

Includes bibliographies.
CONTENTS: v. 1. Principles and techniques.–
v. 2. Acidic and basic solvents.--v. 3. Inert,
aprotic, and acidic solvents. [etc.]
1. Nonaqueous solvents.
TP247.5.L3 660.2'9'482 66-16441
ISBN 0-12-433841-0 (v. 5B)

PRINTED IN THE UNITED STATES OF AMERICA

Contents

List of Contributors

Numbers in parentheses indicate the pages on which the authors' contributions begin.

ALBERT W. JACHE, Department of Chemistry, Marquette University, Milwaukee, Wisconsin (53)

D. H. KERRIDGE, Department of Chemistry, The University, Southampton, England (269)

DOMINIQUE MARTIN, Division de Chimie, Centre D'Études Nucléaires de Saclay, Gif-sur-Yvette, France (157)

JOHN B. MILNE, Department of Chemistry, University of Ottawa, Ottawa, Ontario, Canada (1)

SARASWATHI NATARAJAN, Department of Chemistry, Marquette University, Milwaukee, Wisconsin (53)

RAM CHAND PAUL, Department of Chemistry, Panjab University, Chandigarh, India (197)

ROGER ROUSSON, Division de Chimie, Centre D'Études Nucléaires de Saclay, Gif-sur-Yvette, France (157)

GURDEV SINGH, Department of Chemistry, Panjab University, Chandigarh, India (197)

JEAN-MARC WEULERSSE, Division de Chimie, Centre D'Études Nucléaires de Saclay, Gif-sur-Yvette, France (157)

Preface

Volume V of this treatise completes the original plan established in 1965 to attempt a critical analysis of the subject from two points of view, viz., (1) a discussion of the theoretical aspects of nonaqueous solution chemistry independent of solvent and (2) a consideration of individual solvents or solvent types for which reasonably comprehensive information has been gathered to this point in time. Chapters 1–4 in this volume contribute to the first point of view whereas the remaining chapters, for the most part, contain information on individual solvent systems; there is, of course, some discussion of special aspects of theory in the latter class of chapters.

Taken as a whole, the 38 chapters in the 5 volumes of this treatise would have been organized according to the following outline, if the Editor had been able to overcome the logistic problems associated with the personal work schedules of the contributors.

The Chemistry of Nonaqueous Solvents

Part I. Practical Aspects

1. Experimental Techniques for Low-Boiling Solvents, Jindřich Nassler, Ch. 6, Vol. I
2. Experimental Techniques in the Study of Fused Salts, R. A. Bailey and G. J. Jany, Ch. 7, Vol. I
3. Ion-Selective Electrodes in Nonaqueous Solvents, E. Pungor and K. Tóth, Ch. 4, Vol. VA
4. Electrochemical Methods, P. Zuman and W. Wawzonek, Ch. 3, Vol. VA
5. Conductivity in Nonaqueous Solvents, John H. Roberts, Ch. 1, Vol. IV
6. Solvent Extraction of Inorganic Species, Leonard T. Katzen, Ch. 5, Vol. I

Part II. Theoretical Aspects

7. Lewis Acid–Base Interactions in Polar Nonaqueous Solvents, Devon W. Meek, Ch. 1, Vol. I
8. Brønsted Acid–Base Behavior in "Inert" Organic Solvents, Marion Maclean Davis, Ch. 1, Vol. III

9. Solvation of Electrolytes and Solution Equilibria, Elton Price, Ch. 2, Vol. I
10. Solvation and Complex Formation in Protic and Aprotic Solvents, Sten Ahrland, Ch. 1, Vol. VA
11. Solvent Basicity, Robert L. Benoit and Christen Louis, Ch. 2, Vol. VA
12. Electrode Potentials in Nonaqueous Solvents, H. Strehlow, Ch. 4, Vol. I
13. Redox Systems in Nonaqueous Solvents, Michel Rumeau, Ch. 3, Vol. IV
14. Acidity Functions for Amphiprotic Media, Roger G. Bates, Ch. 3, Vol. I
15. Hydrogen Bonding Phenomena, Ann T. Lemley, Ch. 2, Vol. IV

Part III. Acidic Solvents

16. Liquid Hydrogen Chloride, Hydrogen Bromide, and Hydrogen Iodide, Frank Klanberg, Ch. 1, Vol. II
17. Anhydrous Hydrogen Fluoride as a Solvent and a Medium for Chemical Reactions, Martin Kilpatrick and John G. Jones, Ch. 2, Vol. II
18. Liquid Hydrogen Sulfide, F. Fehér, Ch. 4, Vol. III
19. Sulfuric Acid, Witt Lee, Ch. 3, Vol. II
20. Halosulfuric Acids, S. Natarajan and A. W. Jache, Ch. 2, Vol. VB
21. Nitric Acid, W. H. Lee, Ch. 4, Vol. II
22. Anhydrous Acetic Acid as Nonaqueous Solvent, Alexander I. Popov, Ch. 5, Vol. III
23. Other Carboxylic Acids, Alexander I. Popov, Ch. 6, Vol. III
24. Trifluoroacetic Acid, John B. Milne, Ch. 1, Vol. VB

Part IV. Basic Solvents

25. Liquid Ammonia, J. J. Lagowski and G. A. Moczygemba, Ch. 7, Vol. II
26. The Physical Properties of Metal Solutions in Nonaqueous Solvents, J. C. Thompson, Ch. 6, Vol. II
27. Hydrazine, D. Bauer and Ph. Gaillochet, Ch. 6, Vol. VA
28. Pyridine, J.-M. Nigretto and M. Jozefowicz, Ch. 5, Vol. VA
29. Tetramethylurea, Barbara J. Barker and Joseph A. Caruso, Ch. 4, Vol. IV
30. Amides, Joe W. Vaughn, Ch. 5, Vol. II

Part V. Aprotic Solvents

31. Liquid Sulfur Dioxide, D. F. Burow, Ch. 2, Vol. III
32. Sulfolane, Jukka Martinmaa, Ch. 2, Vol. IV
33. Inorganic Acid Chlorides of High Dielectric Constant (with Special Reference to Antimony Trichloride), E. C. Baughan, Ch. 5, Vol. IV
34. Inorganic Halides and Oxyhalides, R. C. Paul and G. Singh, Ch. 4, Vol. VB
35. Acyl Halides as Nonaqueous Solvents, Ram Chand Paul and Sarjit Singh Sandhu, Ch. 3, Vol. III
36. Cyclic Carbonates, W. H. Lee, Ch. 6, Vol. IV
37. The Interhalogens, D. Martin, R. Rousson, and J.-M. Weulersee, Ch. 3, Vol. VB
38. Molten Salts as Nonaqueous Solvents, D. H. Kerridge, Ch. 5, Vol. VB

The division of some chapters between Parts I and II in the outline is somewhat arbitrary since many chapters in Part I are organized about well-established or evolving principles.

I should like to acknowledge the help of Ms. R. Schall who assisted in numerous ways during the preparation of Volume V. The cooperation of the staff of Academic Press in many ways since the inception of this treatise has been outstanding. Finally, the contributions of the numerous authors, both in terms of the manuscripts they produced and their numerous suggestions are gratefully appreciated. It is apparent that a very large number of persons contributed materially to the success of this effort since its inception in 1965.

J. J. LAGOWSKI

Contents of Other Volumes

VOLUME I PRINCIPLES AND TECHNIQUES

VOLUME II ACIDIC AND BASIC SOLVENTS

VOLUME III INERT, APROTIC, AND ACIDIC SOLVENTS

VOLUME IV SOLUTION PHENOMENA AND APROTIC SOLVENTS

VOLUME VA PRINCIPLES AND BASIC SOLVENTS

THE CHEMISTRY OF NONAQUEOUS SOLVENTS

Volume VB

ACIDIC AND APROTIC SOLVENTS

~ 1 ~

Trifluoroacetic Acid

JOHN B. MILNE

Department of Chemistry, University of Ottawa
Ottawa, Canada

I. INTRODUCTION

Since its initial preparation by F. Swarts in 1922,[1] there has been a continuing interest in trifluoroacetic acid (HOTFA). The solvent has attracted researchers because of its excellent solvent properties for both inorganic[2] and organic solutes,[3–5] as well as for many polymers,[6,7] its weak nucleophilicity and tendency to promote ionization in organic reactions[8,9] and its quite high resistance to chemical attack.[2] Trifluoroacetic acid is prepared commercially by the electrolysis of acetic acid, acetyl halides, or acetic anhydride in HF.[10–13] While the higher molecular weight perfluorocarboxylic acids find commercial use as surfactants and anti-wetting agents,[14] no extensive commercial use has been found for HOTFA.[15] Several patents are, however, held for its use as a solvent for processing linear terphthalates[7] and as a catalyst for the esterification of cellulose[16,17] and the synthesis of conjugated diene polymers with high cis content.[18]

Trifluoroacetic acid is a very hygroscopic liquid which fumes profusely in air. No special precautions are required for handling apart from avoiding moisture when working with the anhydrous material. In contrast to the high toxicity of monofluoroacetic acid, trifluoroacetic acid is said to be nontoxic.[19] The acid does, however, cause severe burns and care should be taken to avoid inhalation and contact with skin and eyes.[5] Some toxicity testing of the acid has been done with mammals and insects and the toxicity of sodium trifluoroacetate for bacteria has been investigated.[20] The compounds were found to be of low toxicity.

The chemistry of HOTFA as a nonaqueous solvent has been the subject of a short review by A. I. Popov, covering the literature up to 1969.[21] While it is now very much out of date, Ref. 5 contains useful information concerning the acid. More recently, the manufacture and uses of HOTFA have been reviewed.[22] This present review will concern itself principally with the chemistry of the bulk solvent. However, topics which have a bearing on this are also included, such as in the case of the dimerization equilibrium in aprotic solvents, acid strength in water, and the preparation of trifluoroacetates.

II. Solvent Purification

Trifluoroacetic acid can be obtained commercially in a highly purified state with small quantities of water the usual principal contaminant. Trifluoroacetic acid forms an azeotrope with water (79.4% acid by weight) which boils at 105.46°C.[23] Pure acid may be obtained by fractional distillation at atmospheric pressure. Removal of water is accomplished by distillation from concentrated sulfuric acid[23] or boric oxide[24] or by addition of trifluoroacetic anhydride [bp 40°C[5]] and separating the initial low-boiling fraction.[25] Distillation from phosphorus pentoxide has also been used,[26–28] but this procedure is not recommended since trifluoroacetic anhydride is produced and, unless the pentoxide has been completely consumed, fraction selection may not ensure a product free of trifluoroacetic anhydride. This drawback does not exist with sulfuric acid. Eaborn *et al.* purified their acid by distillation from sulfuric acid followed by a distillation from silver trifluoroacetate, which was used to remove HCl present as an impurity in their starting acid.[29]

Simons and Lorentzen, in their electrical conductivity work, used a recrystallization procedure to purify their acid. They accepted as final a product acid freezing at $-15.25°C$.[30,31]

III. Physical Properties and Solvent Structure

A. Physical Properties

Trifluoroacetic acid has a convenient and reasonably broad liquid range. Its relatively low boiling point is convenient for easy solvent removal in preparative chemistry. Important physical constants are listed in Table I.[5,10,26,30–47] The variation of several of these physical properties with temperature has been studied. The variation of density with temperature was reported by Fialkov and Zhikharev,[36] Swarts,[23] Kreglewski,[33] and Jasper and Wedlick.[37] The last authors give the expression:

$$d = 1.5375 - 0.002346t \text{ (°C) g/ml} \tag{1}$$

Kreglewski[33] gives for the molar volume:

$$V = 72.2625 + 0.10635t + 3.253 \times 10^{-4}t^2 \text{ (°C) ml} \tag{2}$$

Viscosity change with temperature has been investigated by Swarts,[40] Kauck and Diesslin,[10] and Fialkov and Zhikharev.[36] Such information is also given in Ref. 5. Many of the measurements in this reference exactly duplicate those of Kauck and Diesslin and, since the measurements were made for the same industrial firm, they are no doubt identical. For this reason the measurements given in Ref. 5, which are presented graphically there, are not given here. The data are given in Table II.

Jasper and Wedlick[37] have studied the variation in surface tension with temperature and give the general equation

$$\gamma = 15.638 - 0.8444t \text{ (°C) dyn/cm} \tag{3}$$

DeWith[48] has calculated from theory the dipole moment of the HOTFA monomer. He gives the values 2.15 D for the structure with an internal hydrogen bond to oxygen and 2.12 D for the open structure. The experimental value is 2.28 D.[41]

TABLE I

PHYSICAL PROPERTIES OF TRIFLUOROACETIC ACID

Property	Value	Reference
Melting point	−15.22°C	30
	−15.25°C	31
	−15.20 ± 0.05°C	32, 33
	−15.216°C	34
Boiling point (760 mm pressure)	71.78°C	33
	71.8°C	35
Density (25°C)	1.4776 g/ml	5
	1.4785 g/ml	33
	1.477 g/ml	36
	1.4788 g/ml	37, 38
Molar volume (25°C)	77.125 ml/mole	33
Adiabatic compressibility (25°C)	1.463 atm	39
Viscosity (25°C)	0.807 cP	40
	0.854 cP[a]	10
	0.813 cP	36
Surface tension (25°C)	13.527 dyn/cm	37
(20°C)	15.0 dyn/cm	5
Dipole moment (monomer vapor)	2.28 D	41
Refractive index (25°C)	1.283	42
(20°C)	1.285	5
Dielectric constant (25°C)	8.32[b]	26
	8.40[b]	32
	8.2	43
	8.25	44
Specific conductance (25°C)	0.0064×10^{-6} ohm^{-1} cm^{-1}	45
Enthalpy of vaporization (71.8°C)	8.30 kcal/mole	10
	7.949 kcal/mole	35
	8.126 kcal/mole	33
Entropy of vaporization (71.8°C)	24.1 cal deg^{-1} mole^{-1}	10
	23.55 cal deg^{-1} mole^{-1}	33
Cryoscopic constant	6.458 deg kg mole^{-1}	34
Enthalpy of fusion	2.33 kcal/mole	34
Enthalpy of formation (25°C)	−253 ± 0.8 kcal/mole	46
Heat capacity at constant pressure (25°C)	0.464 cal deg^{-1} g^{-1}	39
Critical temperature	218.13°C	47
Critical pressure	32.15 atm	47
Critical volume	0.204 liter/mole	47

[a] Calculated by the present author from values in centistokes using $\rho_{25} = 1.478$ g/ml.
[b] Interpolated.

TABLE II

VARIATION OF VISCOSITY WITH TEMPERATURE

Temperature (°C)	η^a (cP)	Temperature (°C)	η^b (cP)	η^c (cP)
19.75	0.867	20	0.919	—
—	—	25	0.854	0.813
30.05	0.750	30	0.803	—
39.92	0.653	40	0.704	—
50.35	0.569	50	0.631	0.576
60.40	0.502	58	0.576	—
65.42	0.473	—	—	—

[a] Swarts.[40]

[b] Calculated by the present author from values in centistokes using $\rho_{25} = 1.478$ g/ml.[10]

[c] Fialkov and Zhikharev.[36]

The measurement of the dielectric constant of HOTFA has been the subject of some confusion. Simons and Lorentzen reported a value of 40.2 at 21.5°C with a positive temperature coefficient[49] and Maryott and Smith gave a similar value of 39.5 at 25°C.[50] The abnormal positive temperature coefficient for a liquid with such a high dielectric constant lead Dannhauser and Cole to repeat the measurements.[26] They showed that, using higher frequencies than those used by Simons and Lorentzen, the dielectric constant was 8.42 at 20°C. Dannhauser and Cole used a cell made of stainless steel which is known to be attacked by HOTFA.[5] Moreover, they purified their acid by distillation from P_2O_5 which could result in contamination by $(CF_3CO)_2O$. The best dielectric constant values appear to be those given by Harris and O'Konski.[32] These authors employed a platinum cell and observed no electrode polarization down to a frequency of 100 cps. Their values at various temperatures are given in Table III. The low dielectric constant for HOTFA has been confirmed by two other groups of workers[43,44] and accounts for the electrical conductivity behavior of trifluoroacetates in the acid.[25] However, the early values near 40 have led to confusion even in the more recent literature.[51-54] Borovikov and Fialkov have reported the dielectric constant of mixtures of HOTFA and H_2SO_4.[44]

Vapor pressure studies have been carried out by Kauck and Diesslin[10] and Taylor and Templeman.[35] The measurements from both studies, which are in excellent agreement with each other, are given in Table IV. Taylor and Templeman also determined the vapor pressure curve for deuterotrifluoroacetic acid[35] and observed the boiling point to be 71.2°C and the heat of vaporization at the boiling point to be 8.150 kcal/mole.

TABLE III

VARIATION OF DIELECTRIC CONSTANT WITH TEMPERATURE[a]

Temperature (°C)	ε
−10	9.54
0	9.21
10	8.90
20	8.55
30	8.26

[a] Harris and O'Konski.[32]

TABLE IV

VAPOR PRESSURES OF TRIFLUOROACETIC ACID

Temperature (°C)	Pressure (mm Hg)[a]	Temperature (°C)	Pressure (mm Hg)[b]
37.0	191	−9.1	15.48
43.1	250	−4.7	21.24
47.3	298	−4.3	21.30
49.5	326	0.0	28.75
60.8	503	9.0	50.63
66.7	625	19.5	84.08
71.1	734	29.9	135.6
—	—	40.3	224.1
—	—	50.4	353.0
—	—	60.9	516.7
—	—	65.3	599.6
—	—	65.4	601.1
—	—	71.5	749.0

[a] Kauck and Diesslin.[10]
[b] Taylor and Templeman.[35]

B. Acid Strength

HOTFA is a very strong acid relative to other carboxylic acids, but compared to other mineral acids it is of medium strength. In aqueous solution it is a weaker acid than nitric acid,[55] although a considerable spread of values for the K_a of HOTFA have been reported. The various K_a values are listed in Table V. Henne and Fox determined K_a from electrical conductivity data using the Ostwald dilution law.[30] Their value agrees closely with the estimate made originally by Swarts from conductivity studies.[23] Redlich

TABLE V

ACID DISSOCIATION CONSTANTS IN WATER

K_a	Method	Reference
0.58	Conductivity	23
0.588	Conductivity	30
1.8	NMR	59
1.4	H_0	60
0.93	Raman	55
1.1	Refractometry	62
4–8	Raman and NMR	63

rejects this value[56,57] because of the poor applicability of this method to moderately dissociated electrolytes. However, treatment of conductivity measurements[30] using the 1965 Fuoss equation[58] leads to $K_a = 0.67$,[34] a value still at variance with those determined by other methods. Hood *et al.*[59] used both ^{1}H- and ^{19}F-NMR and arrived at a much higher value. A somewhat arbitrary choice of the shift of undissociated HOTFA was necessary to determine K_a. The concentration range studied was considerably higher than that studied by Henne and Fox. Henne and Fox studied concentrations below 0.09 moles/liter while Hood *et al.* used solutions of concentration 2.1 to 13.06 moles/liter. Högfeldt[60] used Randles and Tedder's H_0 values[61] to determine K_a. The agreement of his value with that from the NMR data[59] is reasonable. Both studies cover the same concentration region. Ignoring activity coefficients, a preliminary estimate of K_a from Raman intensity measurements on the asymmetric CO stretching mode at 1435 cm^{-1} of the $OFTA^-$ ion gave a value for K_a of 0.93 moles/liter.[55] Grunwald and Haley[62] used refractometry in a lower concentration region (0.0–0.6 *M*) and found the value 1.1 moles/liter for K_a at 22°C. Covington *et al.*[63] have, more recently, employed the Raman and ^{1}H-NMR techniques in conjunction with activity coefficients for HCl and CsCl to estimate a value of K_a between 4 and 9 moles/liter. Their choice for NMR shift of undissociated HOTFA in aqueous solution as that of pure liquid HOTFA is open to question. Furthermore, the assumption that the Raman band at 1435 cm^{-1} is due only to the $OTFA^-$ ion, has been questioned.[64,65] Certainly the anion, the acid,[66] the acid dimer,[66] and the hydrogen-bridged dianion $[H(OTFA)_2{}_-]$[67] all have bands in this region. Covington *et al.* suggest that the great disparity between their values and those measured by conductivity[30] and indicator methods[60] is a result of the presence of both ionization (K_i) and dissociation (K_d) equilibria.

$$\mathrm{HOTFA + H_2O \underset{}{\overset{K_i}{\rightleftharpoons}} [H_3O^+OTFA^-] \overset{K_d}{\rightleftharpoons} H_3O^+ + OTFA^-} \qquad (4)$$

The Raman and NMR methods would include both the ion pair and the dissociated ion concentrations in the determination of the degree of dissociation, while the conductivity and indicator methods would measure only the dissociated ions.[63]

The large discrepancy in K_a values for aqueous HOTFA has yet to be completely accounted for. Other acids of moderate strength have been studied by a wide variety of methods and good agreement is observed between the K_a values determined.[68] According to a correlation of ν_{OH} and pK_a in aqueous solution for a series of acids, the pK_a of trifluoroacetic acid should be near zero.[69] The enthalpy of ionic dissociation of HOTFA in the gas phase has been determined by mass spectrometric methods and found to be 320.2 kcal/mole.[70]

The relative acid strength of HOTFA has been measured in chloroform using indicator methods.[71] It is a stronger acid than picric acid but weaker than hydrogen chloride in this solvent. According to an ^{1}H-NMR study of acids in acetic acid as solvent,[72] HOTFA is stronger than trichloroacetic acid but weaker than nitric acid. However, on the basis of electrical conductivity studies in acetic acid, HOTFA and HNO_3 are of comparable strength, with HOTFA marginally stronger.[73]

There is a large range of values for the Hammett acidity function, H_0, for the pure acid, as Table VI shows.[29,74–76] The disparity between values may well be a result of the variation in the degree of dissociation of the protonated indicator ion pair with concentration. The Hammett acidity function in its simplest form is given by:

$$H_0 = pK_{IH^+} - \log \frac{[IH^+]}{[I]} \tag{5}$$

where K_{IH^+} is the dissociation constant of the protonated indicator, I, in aqueous solution. In a solvent of low dielectric constant, some IH^+ will be present as the ion pair IH^+OTFA^-. As the stoichiometric indicator concentration is decreased, the ion pair will be dissociated to a greater extent and

TABLE VI

HAMMETT ACIDITY FUNCTION

$-H_0$	Indicator	Reference
4.4	Benzalacetophenone	74
3.1	*p*-Nitrodiphenylamine	75
3.03	2,4-Dichloro-6-nitroaniline	76
2.77	2,4-Dichloro-6-nitroaniline	29

the relative concentration of IH^+ present as both the free ion and in the ion pair will be greater, giving an apparently lower H_0 value. Moreover, a change of indicator with a different ion-pair dissociation constant may result in a different H_0 value. Eaborn *et al.*[29] have suggested that the difference between their H_0 value and those of other authors may be due to the presence of HCl in the acid used by other authors.

The H_0 values of several mixtures of trifluoroacetic acid with other solutes have been measured and Mackor *et al.*[3] state that it is a versatile solvent in that it can provide a wide range of acidity. Three groups of workers have determined the H_0 for H_2O–HOTFA mixtures.[29,62,77] There is quite good agreement between the sets of results over the region common to the three studies. Eaborn *et al.*[29] find a maximum H_0 near the HOTFA mole fraction $N_{HOTFA} = 0.9$. A similar maximum was found in the rate of detritiration and desilylation by these authors. They attributed the maximum to a balance between increasing acidity with increasing N_{HOTFA} and decreasing solvation of ions as the mole fraction of water, N_{H_2O}, decreases.

Sulfuric acid in HOTFA has been studied by Kilpatrick and Hyman,[75] Hyman and Garber,[76] and Mackor *et al.*[3] There is good agreement between the results of Mackor *et al.* and those of Hyman and Garber in the region of overlap (0–1 M H_2SO_4) where conventional Hammett indicators were used. However, the measurements of Kilpatrick and Hyman are generally lower by as much as 1.4 H_0 units, which is probably a result of the use of hexamethylbenzene as the indicator base. Other acids for which H_0 has been determined in trifluoroacetic acid are HF,[76] $HClO_4$, HBF_3OH, HPO_2F_2, and $HPO_2(CF_3)_2$.[78] In using this acidity function information, it should be remembered that H_0 values in low dielectric constant solvents are dependent not only on the overall dissociation constant of the acid but also on the ion-pair dissociation constants of the indicator base salts.[79] Bessière and Petit[80] have measured the H_0 for solutions of *n*-butylamine in trifluoroacetic acid containing 0.5 M tetraethylammonium perchlorate. The H_0 values indicate a greater acidity for these mixtures than for comparable concentrations of H_2SO_4 in HOTFA. This anomaly is very likely a result of triple ion formation, which has been shown to take place in this solvent at solute concentrations above 0.02 M.[81] Spurious H_0 values have also been shown to arise at high concentrations in acetic acid due to triple ion formation.[79] The high concentration of tetraethylammonium perchlorate would favor the formation of ion triples, $[(IH)_2ClO_4^+]$ and $[IH(ClO_4)_2^-]$, increasing the apparent IH^+ content of the solutions. Bessière and Petit[80] have also evaluated $R_0(H)$[82] in trifluoroacetic acid but, until the effect of triple ion formation on electrode potentials has been investigated, little can be said about these values.

C. Solvent Structure

1. Vapor Phase

HOTFA exists in the vapor phase in monomeric, dimeric, and polymeric forms, depending on temperature and pressure. The equilibrium between monomer and dimer has been extensively studied using vapor density and infrared techniques. Values for the association constant, K_{assoc}, corresponding temperature, and the enthalpy of association are given in Table VII.[35,83–89] Karle and Brockway[83] measured vapor pressures at room temperature and, assuming the enthalpy of dimerization to be the same as that for acetic acid, were able to set the temperature and pressure conditions for their electron diffraction study of the monomer and dimer. Taylor and Templeman[35] give for the variation of the dimer dissociation constant with temperature:

$$\log K_{dissoc}\ (\text{in mm}) = -3071/T + 10.869\ (T \text{ in } ^\circ\text{K}) \tag{6}$$

Both measurements of K_{assoc} by the IR method have used the OH stretching region where a clear distinction between the monomer (3589 cm^{-1}) and dimer (~3100 cm^{-1}) bands can be made.[85,88] The photoelectron spectrum of HOTFA vapor shows changes as the temperature is varied and this has been attributed to the monomer–dimer equilibrium.[90] The presence of small amounts of polymers of a greater state of association than the dimer (at room temperature) has been indicated by deviations from ideality in the vapor pressure[84] and from the appearance of bands in the IR spectrum in addition to those of the monomer and dimer in matrix isolation studies.[91] Trifluoroacetic acid is similar to acetic acid in this regard.[92]

TABLE VII

VAPOR PHASE ASSOCIATION CONSTANTS, K_{assoc}, AND ENTHALPY OF ASSOCIATION, ΔH

K_{assoc} (mm^{-1})	$-\Delta H_{assoc}$ (kcal/mole)	Method	Reference
0.2 (32°C)	—	Vapor pressure	83
1.8×10^{-4} (160°C)	14.0	Vapor pressure	84
0.273 (25°C)	14.05	Vapor pressure	35
—	13.7	IR	85
0.46 (25°C)	—	Vapor pressure	86, 87
0.2 (25°C)	13.4	IR	88
—	17.0	IR	89

The magnitude of the enthalpy of dimerization in conjunction with the conclusions of some spectroscopic studies[52,91] indicate that the dimer is cyclic.

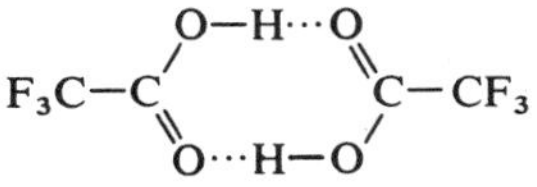

```
          O—H···O
         /       \\
F3C—C             C—CF3
     \\           /
      O···H—O
```

(I)

Karle and Brockway[83] fit their electron diffraction scattering curves to a cyclic dimer model with the parameters given in Table VIII which includes *d*(CF) and ∠(FCF) for the monomer as determined by the same authors.

TABLE VIII

ELECTRON DIFFRACTION PARAMETERS FOR TRIFLUOROACETIC ACID[a]

Parameter	Monomer	Dimer
d(C–F)	1.36 ± 0.05 Å	1.36 ± 0.03 Å
∠(FCF)	110 ± 4°	109 ± 2°
d(C–C)	—	1.47 ± 0.03 Å
d(OH ··· O)	—	2.76 ± 0.06 Å
d(CO) (avg)	—	1.30 ± 0.03 Å
∠(OCO)	—	130 ± 3°

[a] Karle and Brockway.[83]

Several authors have made assignments of the vibrational spectra. Bands belonging to the monomer and dimer have been identified by variations in temperature[85,88,93] and pressure.[66] A brief report on the Raman spectrum of the vapor has appeared[94] and the matrix isolation technique has been employed.[91,95] As would be expected, the monomer bands particularly altered by dimerization involve those groups directly affected by ring formation, ν_{CO} (monomer, 1826 cm^{-1}; dimer, 1788 cm^{-1}) and ν_{OH} (monomer, 3587 cm^{-1}; dimer, ~3000 cm^{-1}, broad). The breadth and complexity of the OH-stretching region in the spectrum of the dimer has been accounted for by anharmonic coupling of ν_{OH}, and the stretching mode resulting from ring formation, ν_{O-HO}, as well as by CO stretching overtones which fall in this region of the spectrum.[95] Such a proposal has accounted for the breadth and complexity of the ν_{OH} stretching region in the spectrum of the formic acid dimer as well.[96] Fuson and Josien have given an alternative explanation.[97] These authors propose that the position of the hydrogen in the dimer ring is quantized and this leads to a series of OH-stretching bands. Bratoz *et al.* have given a critical discussion of several of the theories which have been proposed to account for this region of the spectrum of carboxylic acid dimers.[98]

The far IR region of the spectrum of HOTFA vapor has been studied by Clague and Novak.[99] They assign $\nu_{O \cdots HO}$ at 158 cm^{-1}

2. In Aprotic Solvents

There is abundant evidence that the monomer and cyclic forms of HOTFA are present in solutions in aprotic solvents.[66,88,93,100,101] The existence of the open dimer **(II)** in the same solvents[54,100,102,103] is still not certain.

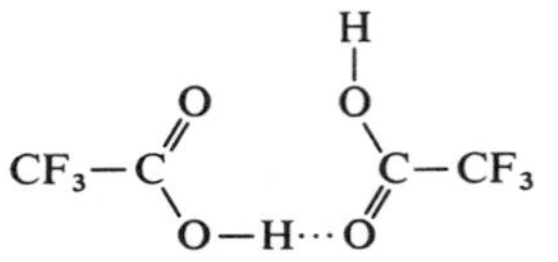

(II)

Where the solvent is a strong proton acceptor, ionization may occur giving
HOTFA are present in solutions in aprotic solvents.[66,88,93,100,101] The
on the dielectric constant.[100,104,105] In addition, there is evidence for polymer formation in aprotic solvents.[88,100]

In the first study of a solution of trifluoroacetic acid in carbon tetrachloride, Fuson *et al.*,[66] using IR and Raman techniques, noted that the $\nu_{C=O}$ assigned to the dimer in the vapor phase at 1788 cm^{-1} was virtually unshifted upon solution in CCl_4, while the $\nu_{C=O}$ of the monomer in the vapor phase at 1826 cm^{-1} was shifted to 1810 cm^{-1} in solution. Taken with other spectroscopic evidence, this led them to conclude that the cyclic dimer and monomer were present in solution. This interpretation for CCl_4 solutions was substantiated by Bellanto and Barcelo[106] and Barcelo and Otero.[93] Statz and Lippert[107] studied the low wave number IR spectrum of solutions in CCl_4 and identified $\nu_{O \cdots HO}$ at 93 cm^{-1}. Solutions in 1,2-dichloroethane have also been studied.[100,101,104] Reeves[100] concluded that the monomer HOTFA interacted much more strongly with 1,2-dichloroethane than with CCl_4 on the basis of the shift in ν_{OH} in going from vapor phase to solution ($\Delta\nu = -190$ cm^{-1} for 1,2-dichloroethane; -80 cm^{-1} for CCl_4). Kirszenbaum *et al.*[101,104] similarly report that only the CO double bond stretching frequency of the monomer at 1804 cm^{-1} is observed in a 0.02 moles/liter solution of HOTFA in 1,2-dichloroethane. The $\nu_{C=O}$ of the dimer (1782 cm^{-1}) is only observed at 0.04 moles/liter HOTFA. The dimer is first observed in CCl_4 at a concentration two orders of magnitude lower than this.[101] The monomer–dimer equilibrium in 1,2-dichloroethane must be quite temperature sensitive inasmuch as Christian and Stevens observed both CO stretches at 0.02 moles/liter at 32.8°C[88] and Reeves[100] gives a room-temperature IR spectrum of the OH stretching region which shows the presence of both the monomer and dimer. The monomer is the favored species relative to the dimer in solvents more basic than, and of greater polarity than, CCl_4. Only the monomer has been observed at comparable concentrations in benzene,[101] nitromethane and acetonitrile.[104] Acetonitrile undergoes a slow reaction with HOTFA[104] which has been attributed to the protonation of acetonitrile, giving trifluoroacetate ion in solution.

While most researchers favor the interpretation of vibrational spectra in terms of the cyclic dimer–monomer equilibrium, Murty and Pitzer[54] found evidence for a linear dimer in addition to the cyclic dimer and monomer in the solvents CCl_4 and benzene. They point out that solvation favors stabilization of structure (II) by bonding to the terminal OH group. Their experimental evidence consists of three parts: (1) the breadth of the ν_{OH} band for the bridging OH group indicates both dimers are present; (2) the observation in benzene of two sharp ν_{OH} bands for terminal OH groups, one for the monomer and one for the open dimer; and (3) the observation of three ν_{CO} bands, one each for monomer, cyclic dimer, and open dimer, which show concentration dependence. These findings are substantiated by other workers. Reeves[100] finds two sharp ν_{OH} bands in the spectrum of HOTFA in 1,2-dichloroethane. Perelygin and Afanas'eva[102] find, in addition to these bands, three ν_{CO} bands as found by Murty and Pitzer. On the other hand, the broad OH-stretching band of the cyclic dimer in the vapor phase has found other explanations.[95,97] Other authors have been unable to find a third CO stretching band nor any concentration dependence with solutions in several solvents.[88,101] In addition, with carefully purified materials only a single terminal OH-stretching band is found.[88,101] The additional ν_{OH} observed by Murty and Pitzer has been attributed to the presence of water.[86,88] Where benzene was used as solvent, the second ν_{OH} observed arose from imprecise solvent compensation.[101] The weight of evidence against the open dimer seems convincing but the question of whether there are two or three CO stretching bands remains open.[103]

Nuclear magnetic resonance spectroscopy has been used to study the HOTFA species present in various solvents.[100,105,108,109] The initial shift of the proton resonance to low field with increasing HOTFA mole fraction in 1,2-dichloroethane[100] and CCl_4[109] is accounted for by dimer formation with a greater proportion of hydrogens involved in hydrogen bonding. The subsequent shift again to high field at a high HOTFA mole fraction is attributed to polymer formation with non-hydrogen-bonded terminal OH groups making up an increasing proportion of the proton population.

Some quantitative studies of the monomer–dimer equilibrium in aprotic solvents have been carried out.[88,110–112] The values of K_{assoc} and the enthalpy of dimerization as well as the solvent and method used for the determination are given in Table IX. Higazy and Taha[111] found that the inclusion of a second polymerized species in their equilibrium model gave no improvement in the fit to their vapor pressure data. This is further evidence against the presence of both open and ring dimers in solution at low concentrations.[54,103] Christian and Stevens[88] have developed an IR spectral partition method for determining K_{assoc} in cases where the solvent bands interfere with quantitative detection of the monomer and dimer. They measure the IR spectrum of the vapor and obtain sufficient information to evaluate the equilibrium in solution. A slope–intercept plot of the data permits an evaluation of the Henry's Law constant for the HOTFA monomer and the dimerization constant, K_{assoc}. The agreement between K_{assoc} in the solvent, determined in this way, and that determined directly in the solvent is good, as the results in Table IX show. Christian and Stevens[88] show from deviations of the slope–intercept plot from linearity, that association beyond the dimer takes place in benzene solutions.

TABLE IX

K_{assoc} AND ENTHALPY OF DIMERIZATION FOR HOTFA IN VARIOUS APROTIC SOLVENTS

K_{assoc} (liter/mole)	$-\Delta H$ (kcal/mole)	Solvent	Method	Reference
320 (25°C)	—	Cyclohexane	Dielectric constant	110
377 (40°C)	18.097	Tetradecane	Vapor pressure	111
2.84 (40°C)	7.447	Diphenylmethane	Vapor pressure	111
2.64 (40°C)	—	Diphenylmethane	Vapor pressure	112
192 (25°C)	11.7	Cyclohexane	IR (spectral partition)	88
149 (25°C)	9.2	Carbon tetrachloride	IR (spectral partition)	88
128 (25°C)	—	Carbon tetrachloride	IR (direct in solvent)	88
2.6 (25°C)	7.4	Benzene	IR (spectral partition)	88
1.5 (25°C)	7.0	1,2-Dichloroethane	IR (spectral partition)	88

Thyrion and Decroocq[110] measured the variation in dielectric constant with the concentration of HOTFA in cyclohexane and from this determined K_{assoc} ($=320 \pm 29$ liters/mole). The agreement between their value and that determined by Christian and Stevens (192 ± 36 liters/mole)[88] is not unreasonable in the light of the assumptions made by Thyrion and Decroocq that (1) the molar volume of HOTFA dimer is double that of the monomer, and (2) that their simplified expression is valid for such a low dimerization constant.

3. Solid Phase

The structure of solid HOTFA has not been definitely established. No x-ray crystal structure determination has been reported. Solid acetic[113] and formic[114] acids have chain structures while trichloroacetic acid has a cyclic dimer structure.[115] Clague and Novak[52] have demonstrated the mutual exclusion of the Raman and IR spectra which could be evidence for a structure consisting of cyclic dimers located on C_i lattice sites. Berney has shown, however, that the spectra may be accounted for equally well by a unit cell having an inversion center[51] and he proposes two chain structures having this requirement. Berney's preference for chains is based in part, however, on erroneous dielectric constant measurements[49,50] and the existence of the open dimer[54] which is in some doubt (see Section III,C,2). Dunnel *et al.*[116] have provided perhaps the best evidence thus far for the chain structure. Their calculation of the second moment for protons from the wide-line ^{1}H-NMR of solid HOTFA shows that the cyclic dimer model with a single close H–H interaction does not account for the second moment in HOTFA. Both Berney and Dunnel *et al.* report that a phase change occurs in solid HOTFA but the transition temperatures given are not the same: Berney gives 220.5°K while Dunnel *et al.* report that the change occurs between 77°K and 185°K. Additional evidence for chains comes from the matrix isolation work of Redington and Lin,[91,95] which indicates that species other than the monomer and cyclic dimer, which are likely to be higher polymers, are formed when the sample is annealed. The balance of the evidence favors a chain structure like that of acetic and formic acids, but only a crystal structure determination will provide a definite answer.

4. Liquid Phase

The exact nature of the structure of liquid HOTFA is not known. Early studies[66,93] of the vibrational spectrum of the liquid were interpreted in terms of the cyclic dimer known to exist in the vapor phase and in aprotic solvents. The absence of certain IR bands, assigned to the HOTFA monomer in the liquid state, indicated that the monomer was not present.[66] An estimation of polymer chain length from the molecular volume determined by sound velocity experiments suggests that the bulk liquid contains mostly the dimer. Ultrasonic sound absorption measurements[38] have been interpreted in terms of an equilibrium between cyclic and open dimers:

$$K = [\text{open dimer}]/[\text{cyclic dimer}] = 1.7 \times 10^{-2} \qquad (7)$$

The enthalpy of ring opening reported is 2.9 kcal/mole. Thyrion and Decroocq[110] estimated the dielectric constant of the dimer to be 1.62 from studies of solutions of HOFTA in cyclohexane. The IR evidence[66] that no monomer is present, taken together with this low dielectric constant for the dimer and the reported value of a dielectric constant of 8.40 for the liquid,[32] suggests that some polymeric chain structure must be present. The neutron inelastic scattering pattern for the pure liquid has been shown to be unchanged upon freezing; this suggests that the structure of the solid and the liquid are similar.[51,117] Moreover, ^{1}H-NMR[100] and IR[102] studies have been interpreted to show that polymerization beyond the dimer stage occurs in liquid HOTFA.

IV. Electrochemistry

Because of the low dielectric constant of HOTFA (8.40 at 25°C), ionophores give dissociated ions only at very low concentrations. At higher concentrations, ion pairs and higher ion aggregates are formed.[25,81] Both the Bjerrum and the Fuoss equations predict, for an ionophore with a cation to anion distance of closest approach of 5 Å in the ion pair, that the ion pair dissociation constant will be $\sim 10^{-5}$/mole.[118] An expression derived by Fuoss[118,119] permits one to predict the concentration at which triple ion behavior will become apparent:

$$C = 3.2 \times 10^{-7} \varepsilon^3 \text{ (at 25°C)} \tag{8}$$

This expression leads to a concentration of 1.9×10^{-4} moles/liter for HOTFA. Experimental studies have shown, however, that electrical conductivity is not affected by triple ion formation until a concentration of 6×10^{-4} moles/liter is reached.[25] These low concentrations indicate how critical it is to remove trace impurities such as water when doing studies on this solvent in the ion-pair region of concentration ($C < 6 \times 10^{-4}$ moles/liter).

Simons and Lorentzen[49] were the first to carry out conductivity studies in HOTFA. They extrapolated their data by the method of Fuoss[119] but used an erroneous value for the dielectric constant. Bessière,[120] using the method of Shedlovsky[121] and a dielectric constant of 7.8, extrapolated the results of Simons and Lorentzen to establish a reference system for his potentiometric studies. For low dielectric constant solvents the Fuoss and Shedlovsky plots have been shown to yield nearly identical values for the limiting equivalent conductivity, Λ^0, and the dissociation constant, K_D.[25,118,122] Harriss and Milne[25] studied the electrical conductivity of alkali metal and ammonium trifluoroacetates, which are strong bases in the solvent, in the ion-pair region of concentration. Their results are given in Table X. Comparison with the values obtained by extrapolation of the results of Simons and Lorentzen in the ion-pair region of concentration, using $\varepsilon = 8.4$, shows good agreement

TABLE X

LIMITING EQUIVALENT CONDUCTIVITIES (Λ^0), DISSOCIATION CONSTANTS (K_D), AND TRIPLE-ION DISSOCIATION CONSTANTS (k) OF SOLUTES IN ANHYDROUS HOTFA[a,b,c]

Solute	Λ^0 (ohm^{-1} cm^2 equiv.$^{-1}$)	$10^5 K_D$ (moles/liter)	k (moles/liter)
$[H_2OTFA]OTFA$	26.2 ± 4.1[d]	—	—
LiOTFA	112[e]	0.0098	0.0563
NaOTFA	73.8 ± 1.3	0.0398	0.060
KOTFA	63.7 ± 0.5	0.219	0.033
KOTFA[f]	61.2 ± 5.7	0.38	—
RbOTFA	58.6 ± 0.2	0.422	0.023
CsOTFA	52.4 ± 0.2	1.01	0.017
NH_4OTFA	54.7 ± 0.4	0.455	0.029
$Cs[B(OTFA)_4]$	48.6 ± 3.2	1.72	0.023
$B(OTFA)_3$	22.3 ± 0.7	0.12	—
SbF_5	18.8 ± 1.0	1.75	—

[a] Harriss and Milne.[25]
[b] Harriss and Milne.[45]
[c] Harriss and Milne.[81]
[d] Hypothetical completely dissociated HOTFA calculated from Λ^0s of $Cs[B(OTFA)_4]$, $B(OTFA)_3$, and CsOTFA.
[e] Data evaluated from triple-ion plot, Fig. 1.[81]
[f] From conductivity data of Simons and Lorentzen[49] extrapolated by Hariss and Milne.[25]

(Table X). It has been suggested[25] that Bessière's much higher value for the dissociation constant of KOTFA ($10^{-4.6}$)[120] is due to the inclusion of points in the triple-ion region in his extrapolation. Bessière's value for NaOTFA ($10^{-4.5}$) is also high. Comparison of the K_D for the trifluoroacetates shows that the ion-pair with the smaller cation is the least dissociated, indicating that contact pairs are formed. Apparently the cations are not strongly solvated and the anion may readily displace solvent molecules in the solvation sheath to give the contact pair. This behavior parallels that in nitrobenzene,[123] trifluoroethanol,[124] and acetone,[125] among other solvents. The limiting equivalent conductivity for the trifluoroacetates shows the effects of the weak solvating power of this solvent also. The smaller ion has greater mobility except in the case of the proton which is strongly solvated. Related behavior has been observed in dimethylacetamide,[126] but normally such behavior is only observed for large cations.[125,127] The poor cation solvating properties of HOTFA are reflected in its very weak basic properties.[128,129] Other solutes which are bases in HOTFA and for which electrical conductivity studies have been made are: tri-*n*-butylamine,[49] *N*,*N*-diethylaniline,[130] amino acids,[131] polyvinyl pyrrolidone, and polypeptides.[132] The data in the last two studies are in the triple-ion region of concentration.

Several nonelectrolytes have been studied in HOTFA by electrical conductivity. Simons and Lorentzen[49] showed that $(C_4F_9)_3N$, $(C_4H_9)_2O$, and water were virtually nonelectrolytes. Harriss and Milne[25] studied water and trifluoroacetic anhydride conductivities up to 0.1 M and Randles *et al.*[133] studied the system at much higher concentrations. Fialkov and Zhikharev[134] investigated water–HOTFA mixtures over the whole composition range. These results confirm that water is virtually a nonelectrolyte with a specific conductivity 1/1000th of that of potassium trifluoroacetate at comparable concentrations. If one assumes Λ^0 for $H_3O^+TFA^-$ to be the same as that of KOTFA, then free ion concentrations for H_2O would be 1/1000th that for potassium trifluoroacetate and $K_{D(H_2O)}$ would be 10^{-6} $K_{D(KOTFA)} \cong 10^{-11}$ moles/liter. The behavior of water in HOTFA is in striking contrast to the strongly basic character of bulk water, reflected in the moderately strong acid behavior of HOTFA in water, and parallels the behavior of H_2O in nitric acid.[135] The proton affinities of molecular H_2O and HOTFA have been evaluated by mass spectrometric methods and were found to be very similar, HOTFA being the stronger proton acceptor (H_2O: 165 ± 3 kcal/mole; HOTFA: 167 ± 3 kcal/mole).[136] Fialkov and coworkers have studied the electrical conductivity of several solutes which behave as acids in water but which, on the basis of studies over the complete composition range, are nonelectrolytes in HOTFA. These acids are: HNO_3,[134] H_2SO_4,[36,137] trichloro- and monochloroacetic acids.[138] Quarterman *et al.*[139] have reported the specific conductivity of mixtures of HOTFA and HF over the entire composition range.

Acidic behavior may occur (1) by dissociation of protonic acids, as in the case of HSO_3F:

$$HSO_3F + HOTFA \rightleftharpoons H_2OTFA^+ + SO_3F^- \tag{9}$$

or (2) by anion abstraction, as with SbF_5:

$$2HOTFA + SbF_5 \rightleftharpoons H_2OTFA^+ + SbF_5(OTFA)^- \tag{10}$$

Electrolyte behavior has been demonstrated for several acids which are normally considered to be strong protic acids. Acid dissociation constants have been determined from electrical conductivity data for $H_2S_2O_7$,[137,140] $HClO_4$,[140,141] HSO_3CF_3,[142] and HSO_3F.[143] The dissociation constants were determined by the antiquated Kraus–Bray method,[144] which yields only K_D (and not Λ^0). Unfortunately, since the actual specific conductivities are not given, extrapolation by modern methods[118] is not possible. Bessière[120] and Harriss and Milne[25] report the presence of traces of water (0.003–0.005 moles/liter) even in acid purified by careful methods normally adequate for nonaqueous solvent work. Comparison of the specific conductivity reported by Fialkov and co-workers[36,130] with that reported by Harriss and Milne[25] suggests that Fialkov's acid is no better. However, even these traces of water are serious if one wishes to study solutes in the ion-pair region of concentration ($<6 \times 10^{-4}$). This is especially so for the study of acids where, in all likelihood, the electrolyte present in solutions of an acid HA would be $H_3O^+A^-$ rather than $H_2OTFA^+A^-$. For this reason Harriss and Milne carried out their studies in the presence of 0.1% anhydride which would have minimal effects on the results.[25,45,81] Only two acids were found which did not react with anhydride[45,145]: $B(OTFA)_3$ and SbF_5, which both act as acids by accepting trifluoroacetate ion from the solvent:

$$B(OTFA)_3 + 2HOTFA \rightleftharpoons H_2OTFA^+ + B(OTFA)_4^- \qquad (11)$$

Limiting equivalent conductivities and dissociation constants for these are given in Table X. The conductimetric tritation of $B(OTFA)_3$ with CsOTFA is accounted for quantitatively by the reaction

$$B(OTFA)_3 + CsOTFA \rightarrow Cs[B(OTFA)_4] \qquad (12)$$

and the salt, $Cs[B(OTFA)_4]$, has been isolated.[45] Fluorine-19 NMR has been used to demonstrate the presence of $HSbF_5(OTFA)$ in solution.[145]

The solvent self-dissociation constant of HOTFA, K_{sd}, has been determined from the lowest specific conductivity recorded for the acid ($K_{sp} = 0.0064\ \mu$mho/cm)[45] and a knowledge of Λ^0 for the hypothetical 100% dissociated H^+OTFA^- which is given in Table X.

$$2HOTFA \rightleftharpoons H_2OTFA^+ + OTFA^- \qquad (13)$$

$$K_{sd} = [H_2OTFA^+][OTFA^-] = 4 \times 10^{-14}\ \text{moles}^2/\text{liter}^2 \qquad (14)$$

Harris and Alder[146] have attempted a theoretical treatment of dielectric constant measurements[26] assuming that only cyclic dimers are present and that these are in equilibrium with charge separated cyclic dimers

```
         OH···O
        /      \
CF3—C         C—CF3
        \+    –/
         OH···O
```

(III)

The ionization constant

$$K_i = \frac{[\text{ion-pair}]}{[\text{dimer}]} \qquad (15)$$

calculated from their thermodynamic data is 0.012. The ion-pair dissociation constant calculated from K_i and K_{sd} is 2.6×10^{-13}, which is far smaller than the observed K_D values for ionophores (Table X) and suggests that the dielectric constant is better interpreted in terms of the presence of polymeric chains.

If conductivity measurements in HOTFA are carried out at higher concentrations, Λ is seen to pass through a minimum and rise again.[81,132,147] This kind of behavior is observed in many low dielectric constant solvents and has been attributed to the formation of triple ions

$$3A^+ + 3B^- \rightleftharpoons 3AB \rightleftharpoons A_2B^+ + AB_2^- \tag{16}$$

Triple ion dissociation constants, k, are defined by:

$$k_+ = [A^+][AB]/[A_2B^+] \qquad \text{and} \qquad k_- = [B^-][AB]/[AB_2^-] \tag{17}$$

If one assumes $k_+ = k_- = k$, $K_D = c\alpha^2$, where c is the stoichiometric concentration (moles/liter) of solute and α is the degree of dissociation and $\Lambda_{A_2B^+} = \Lambda_{AB_2^-}$, then it is possible to derive[118,148] the expression

$$C^{1/2}\Lambda = K_D\Lambda^0 + \frac{K_D\Lambda_T^0}{k}C \tag{18}$$

where Λ_T^0 is the limiting equivalent conductivity of the triple ion electrolyte $[A_2B^+][AB_2^-]$. A plot of $C^{1/2}\Lambda$ vs. c yields a straight line with slope $K_D\Lambda_T^0/k$ and intercept $K_D\Lambda^0$. A more exact expression which makes a blanket correction for long-range interionic effects is given by:

$$\frac{c^{1/2}f\Lambda}{1 - \dfrac{(B_1\Lambda^0 + B_2)(c\Lambda)^{1/2}}{(\Lambda^0)^{3/2}}(1 - \Lambda/\Lambda^0)^{1/2}} = K_D\Lambda^0 + \frac{K_D\Lambda_T^0}{k}(1 - \Lambda/\Lambda^0)c \tag{19}$$

where Λ_T^0 is the limiting equivalent conductivity of the triple ion electrolyte tion and B_1 and B_2 are functions of the dielectric constant, viscosity and temperature, which are defined in Robinson and Stokes[148]: A plot of Eq. 19 for the various electrolytes studied by Harriss and Milne[81] appears in Fig. 1. The plots are reasonably straight lines except for cases where Λ^0 is small and in these cases the departures from theory have been accounted for.[81] Harriss and Milne have shown that in cases where reliable conductivity data cannot be obtained over a broad range of the ion-pair region of concentration, a triple-ion plot in conjunction with a single conductivity measurement at the high concentration end of the ion-pair region can yield reasonable estimates of Λ^0 and K_D.[81]

Petit and Bessière[149–151] have investigated several indicator and reference electrodes for use in potentiometric studies in HOTFA. The reference elec-

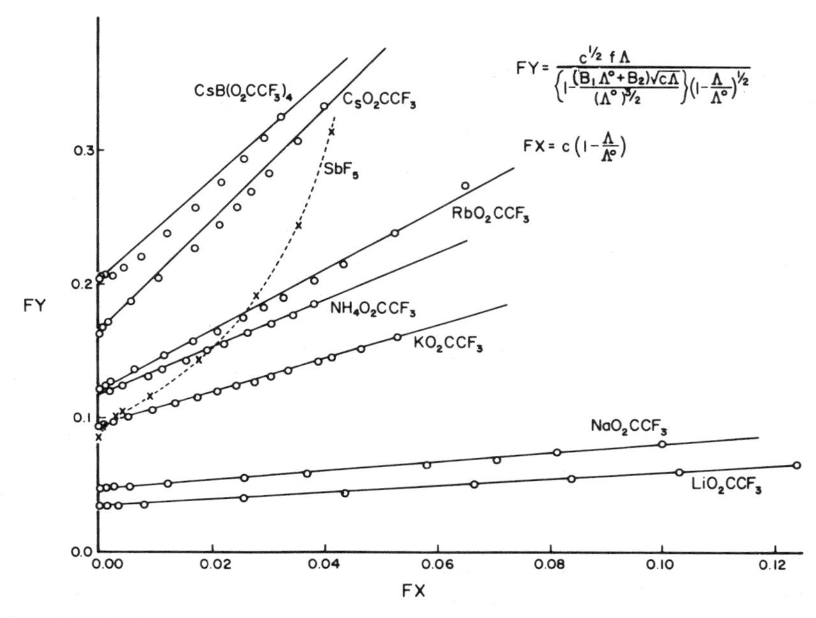

FIG. 1. Triple-ion plots using Eq. 19. Reproduced with permission of the National Research Council of Canada from Harriss and Milne.[81]

trodes used were $Ag(s)/AgClO_4(s)/Et_4NClO_4$[120] and $Ag(s)/AgOOCCF_3$, $NaOOCCF_3$[150] and the indicator electrodes studied were the silver[120,150,151] and glass electrodes[120]; rotating platinum and glossy carbon[149,150]; bright and platinized platinum, gold, and mercury[150]; hydrogen at platinum and ferrocene/ferrocinium at mercury.[151] Standard potentials have been determined for several of these. Bessière[120] reports that the potentials are not reproducible in studies on HOTFA where electrolyte concentrations are low. The later work of Petit and Bessière used HOTFA containing 0.5 moles/liter Et_4NClO_4. The presence of such a large concentration of ionophore will lead to high concentrations of triple ions and, very likely, quadruple ions.[118] However, the interpretation given is in terms of simple ions and ion pairs.

Conway and Dzieciuch[27] and Conway and Vijh[24] have studied the electrochemical decomposition of HOTFA containing sodium trifluoroacetate, the so-called Kolbe reaction:

$$2CF_3COO^- \rightarrow C_2F_6 + 2CO_2 + 2e^- \tag{20}$$

This produces high yields, and no peroxide, $(CF_3CO)OO(COCF_3)$, ester, $(CF_3CO)OCF_3$, or fluoroform are formed.

V. Solubilities

Trifluoroacetic acid is a good solvent for organic compounds[5] and polymers[6,7,132] and a reasonable solvent for many inorganic compounds.[2,152] Trifluoroacetic acid is miscible with ether, acetone, methanol, ethanol, isopropyl alcohol, *n*-butyl alcohol, benzene, carbon tetrachloride, xylene, and water.[5] It is an excellent solvent for aromatic and alphatic hydrocarbons.[3] It is not soluble in carbon disulfide,[5] and 85% phosphoric acid and HOTFA are immiscible.[153] It is useful as a solvent for processing terphthalate resins[7] and it dissolves polymers with basic functional groups such as proteins,[6,154,155] polypeptides, and polyvinyl pyrrolidone.[132] Trifluoroacetic acid is also a very good solvent for amino acids.[156,157] Fujioka and Cady studied the solubility of several gaseous inorganic compounds.[2] The results are listed in Table XI and, when compared with those in water, show that HOTFA is an excellent solvent for gases. The few exceptions, such as HCl and HBr, are more soluble in water because of reaction with the solvent. Hara and Cady[152] determined the approximate solubilities for a variety of salts and these are given in Table XII. Noteworthy for their exceptionally low solubility in HOTFA compared to water are the dipositive metal sulfates and sodium and zinc perchlorates. Lithium perchlorate, on the other hand, has a much greater solubility (11.6 g/100 g HOTFA)[158] while silver perchlorate, for which the solubility product has been determined potentiometrically ($K_{sp} = 10^{-7.6}$

TABLE XI

Solubilities of Gases in Trifluoroacetic Acid[a]

Gas	Temperature (°C)	Partial pressure (cm)	ml of gas dissolved in 1 ml of CF_3COOH
CO_2	27	65.2	3.5
CO	26	64.6	0.0
Cl_2	25.5	65.6	9.3
HBr	26	65.7	6.6
HCl	26	65.2	4.1
H_2S	26	65.7	8.6
N_2	26	63.6	0.1
N_2O	24.5	66.5	4.3
O_2	27	64.2	0.2
PH_3	26	65.3	15.9
SO_2	26	66.7	23.4

[a] Reproduced with permission of Academic Press. Data from Fujioka and Cady.[2]

TABLE XII

APPROXIMATE SOLUBILITIES OF SALTS IN TRIFLUOROACETIC ACID AT ABOUT 25°C[a]

Amount dissolved greater than 10 g anhydrous salt per 100 g CF_3COOH,
- Acetates: Na, K, NH_4, Sr, Ba, Cd, Hg(II), Cu(II), Co(II), Ni, Pb(II)
- Other salts: CsCl, NH_4NO_3, K_2SO_4, K_3PO_4

Amount dissolved from 1 to 10 g anhydrous salt per 100 g CF_3COOH
- Fluorides: Na, K, Ba, Cu(II), Ag(I), Cd
- Chlorides: Na, K, Mg, Ca, Fe(III)
- Bromides and iodides: Na, K, Ca
- Sulfates: Na, NH_4
- Nitrates, nitrites, oxalates, chlorates, bromates, iodates, acid sulfates, and dichromates of Na and K
- Thiocyanates: Na, K, NH_4
- Phosphates: Na, Ca, Co, Zn, HPO_4^{2-} and $H_2PO_4^-$ salts of Na, K, Ca
- Chromates: Na, K, Ca, Ba, Pb, Co, Zn
- Others: $KMnO_4$, $K_4Fe(CN)_6$, CrO_3, $(NH_4)_6Mo_7O_{24} \cdot 4H_2O$, $[Co(NH_3)_6]Cl_3$, $[Co(NH_3)_6](NO_3)_3$, $[Co(NH_3)_5Cl]Cl_2$, $[Co(NH_3)_5H_2O](NO_3)_3$, *trans*-$[Co(NH_3)_4(NO_2)_2]Cl$, *trans*-$[Co(NH_3)_4NO_2Cl]Cl$, $[Co(NH_3)_3(NO_3)_3]$

Amount dissolved from 0.1–1 g/100 g CF_3COOH
- Halogenides: $BeCl_2$, $SrCl_2$, $BaCl_2$, $CdCl_2$, $CoCl_2$, $CrCl_3$, $CdBr_2$, CdI_2, MgF_2, NiF_2
- Others: $Ca(NO_3)_2$, $Ba(NO_3)_2$, $Cr_2(SO_4)_3$

Amount dissolved less than 0.1 g/100 g CF_3COOH
- Halogenides: CaF_2, PbF_2, MnF_2, FeF_3, AgCl, $ZnCl_2$, $AlCl_3$, $PbCl_2$, $HgCl_2$
- Perchlorates: Na, K, Zn
- Periodates: Na, K
- Sulfates: Be, Mg, Ca, Ba, Zn, Ni, Co, Fe(II), Cu(II)
- Others: $Ce_2(C_2O_4)_3 \cdot 9H_2O$, $BaS_2O_6 \cdot 2H_2O$

[a] Reproduced with permission of the American Chemical Society from Hara and Cady.[152]

moles^2/liter^2), has a solubility comparable to that of $KClO_4$ and $NaClO_4$. Iodine is of low solubility in HOTFA (0.002 *M*).[149]

Hara and Cady[152] determined in addition the solubility of a series of metal trifluoroacetates and established the composition of the solid phase in equilibrium with the solution. Their data are listed in Table XIII. Later measurements by Jones and Dyer[159] are essentially in agreement with those of Hara and Cady. The solubilities are in general greater than the comparable acetates in acetic acid[21] and hydroxides in water,[42] the exceptions being the alkali metal trifluoroacetates. In most instances the solid phase is a solvate.

TABLE XIII

SOLUBILITIES OF TRIFLUOROACETATES[a,b]

Cation of trifluoroacetate solute	Solubility (g anhydrous salt/100 g HOTFA)[c]	Solid phase, apparent number of moles HOTFA/mole of salt
Li^+	21.8†	—
Na^+	13.1	2.1
	19.5†	—
K^+	50.13	1.0
	44.0†	—
Rb^+	59.6†	—
Cs^+	101.8†	—
$(CH_3)_4N^+$	153.9†	—
$C_6H_5(CH_3)_3N^+$	608.9†	—
Mg^{2+}	0.570	1.9
Ca^{2+}	6.260	3.8
Sr^{2+}	23.19	2.2
Ba^{2+}	41.66	3.5
Zn^{2+}	0.87	0.2
Cd^{2+}	7.61	0.1
Hg^{2+}	49.72	0.4
Ag^+	15.15	0.1
Cu^{2+}	19.73	0.1
Al^{3+}	0.01	0.0
La^{3+}	0.142	3.1
Ce^{3+}	0.101	—
Pr^{3+}	0.084	3.0
Nd^{3+}	0.059	2.6
Th(IV)	0.016	0.4
UO_2^{2+}	0.035	—
Co^{2+}	16.65	1.7
N_i^{2+}	16.48	1.7

[a] Hara and Cady.[152]
[b] Jones and Dyer.[159]
[c] At $T = 29.8°C$[152] except where † is indicated for which no exact temperature is specified.[159]

VI. REACTIONS IN TRIFLUOROACETIC ACID

A. Acid–Base Reactions, Solvates, and Adducts

The known adducts and solvates for trifluoroacetic acid are listed in Table XIV[36,86,87,108,112,130,134,138,141–143,152,159–166] along with the method used to characterize them.

1. H_2O ADDUCTS

The H_2O–HOTFA system has been studied by a variety of methods. Fialkov and Zhikharev[134] reported the viscosity, density, and specific con-

TABLE XIV

SOLVATES AND MOLECULAR ADDITION COMPOUNDS

Compound	Method of characterization	Reference
$H_2O \cdot 2HOTFA$	IR in CCl_4 solution	160
$H_2O \cdot HOTFA$	IR in CCl_4 solution	160
	Freezing point	161
$2H_2O \cdot HOTFA$	Vapor pressure	86
$4H_2O \cdot HOTFA$	Freezing point	161
$CH_3OH \cdot HOTFA$	IR in CCl_4 solution	160
HOAc · HOTFA	Vapor pressure	86, 87
	Microwave spectroscopy	162
	Physical methods[a]	138
2HOAc · HOTFA	Physical methods[a]	138
HCOOH · HOTFA	Microwave spectroscopy	162, 163
$ClCH_2COOH \cdot HOTFA$	Physical methods[a]	138
$FCH_2COOH \cdot HOTFA$	Microwave spectroscopy	162
$H_2SO_4 \cdot 2HOTFA$	Physical methods,[a] NMR	164
$H_2SO_4 \cdot HOTFA$	Physical methods,[a] NMR	36, 164
$H_2S_2O_7 \cdot HOTFA$	Physical methods[a]	134
$2H_2S_2O_7 \cdot HOTFA$	Physical methods[a]	134
$HClO_4 \cdot 2HOTFA$	Physical methods[a]	141
$HClO_4 \cdot HOTFA$	Physical methods[a]	141
$HSO_3CF_3 \cdot HOTFA$	Physical methods[a]	142
$HSO_3F \cdot HOTFA$	Physical methods[a]	143
$(C_2H_5)_2O \cdot HOTFA$	IR in CCl_4 solution	160
$C_6H_5NO_2 \cdot 2HOTFA$	NMR	108
$C_6H_5NO_2 \cdot HOTFA$	NMR	108
$2C_6H_5NO_2 \cdot HOTFA$	NMR	108
Dioxane · HOTFA	Vapor pressure	86
$(C_6H_5)_2CO \cdot 2HOTFA$	Vapor pressure	112
$(C_6H_5)_2CO \cdot HOTFA$	Vapor pressure	112
NaOTFA · 2HOTFA	HOTFA vapor uptake	165
	Chemical analysis of solid	152
NaOTFA · HOTFA	HOTFA vapor uptake	165
	NMR	159
KOTFA · HOTFA	Chemical analysis of solid	152
	Crystal structure determination	166
	NMR	159
RbOTFA · HOTFA	Crystal structure determination	166
	NMR	159
CsOTFA · HOTFA	Crystal structure determination	166
	NMR	159
$(CH_3)_4NOTFA \cdot HOTFA$	NMR	159
$(C_6H_5)(CH_3)_3NOTFA \cdot HOTFA$	NMR	159
$(C_6H_5)(C_2H_5)_2N \cdot HOTFA$	Physical methods[a]	130

[a] Viscosity, electrical conductivity, and density.

ductivity over the entire composition range. Although deviations from ideal mixing occur, it was not possible to select any definite stoichiometries for the H_2O–HOTFA adducts. Cady and Cady[161] determined the freezing point composition curve of H_2O–HOTFA, which is given in Fig. 2. These authors also list freezing points for the range of acid molality 0–6.5 moles/kg, from which activities for HOTFA have been calculated.[59] Such a calculation requires an assumption about the dissociation constant of HOTFA which is not definitely established (Section III,B). Maximum freezing points occur at compositions 1 : 1 and 4 : 1 H_2O/HOTFA. It would be of interest to know if these solids contain the known H_3O^+ and $H_9O_4^+$ cations. Water is known to be a nonelectrolyte in HNO_3[135] but the 1 : 1 adduct has an ionic structure, $H_3O^+NO_3^-$.[167] From a study of the three vibrational modes of H_2O and ν_{CO} of HOTFA, de Villepin *et al.*[160] were able to demonstrate the existence of homoassociated H_2O and HOTFA aggregates as well as the 1 : 1 and 1 : 2 H_2O/HOTFA adducts in CCl_4 solution. From the small shift in the symmetric stretching mode of H_2O upon complex formation, it was concluded that the 1 : 1 adduct has the form

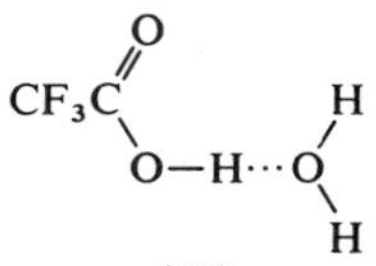

$CF_3C(=O)$–O–H···OH_2

(IV)

No decision could be made between two possible structures for the 1 : 2 adduct:

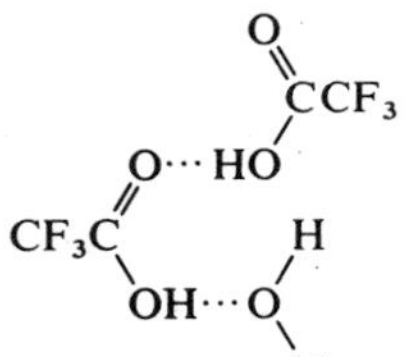

(V) 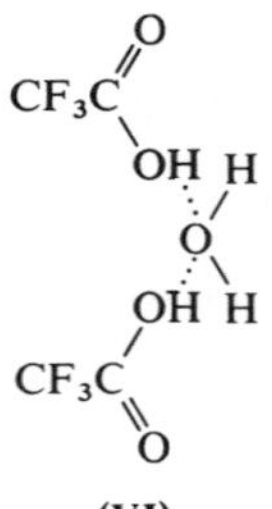 **(VI)**

Denisov *et al.*[168] used spectroscopic methods to measure the heat of formation of the H_2O–HOTFA complex in CCl_4 solution. They find the value −8.4 kcal/mole. Christian *et al.*[86] have shown that the 2 : 1 H_2O/HOTFA complex is the stable species in the vapor phase. They find

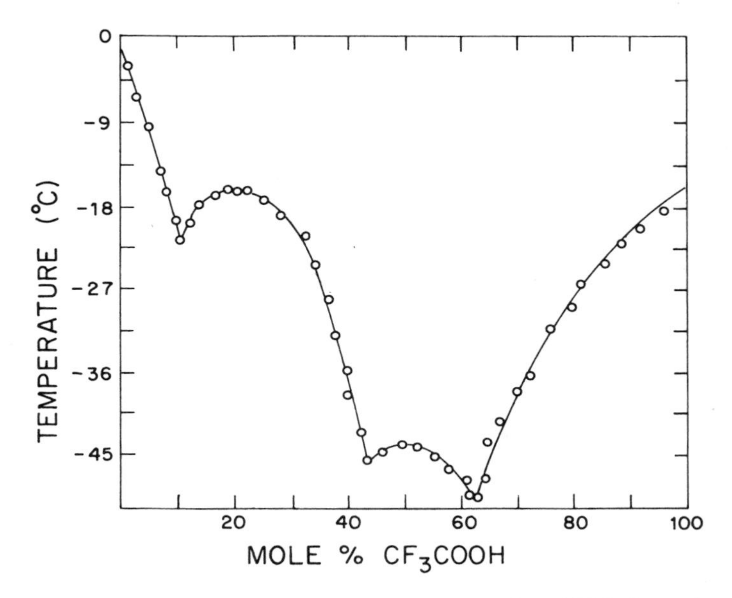

FIG. 2. Freezing point–composition plot for the system H_2O–HOTFA. Reproduced with permission of the American Chemical Society from Cady and Cady.[161]

$K_{assoc} = 0.10/\text{mm}^2$ and they propose a structure with an eight-membered ring:

```
            H
           /
     O···H—O
    //      :
CF3C        H
    \      /
     OH·····O
            \
             H
```

(VII)

N. G. Zarakhani and N. P. Vorob'eva,[169] on the other hand, find no evidence for H_2O–HOTFA association in vapors over aqueous HOTFA solutions. From band intensities of HOTFA monomer and dimer OH-stretching bands, they have evaluated K_{assoc} as 0.015/mm (25°C), considerably at variance with reported values (Table VII). These authors have used their data to evaluate monomer and dimer activities in the aqueous solutions, taking 100% HOTFA as the standard state. They have extended their study[170] using the Gibbs–Duhem equation[148] to determine the water activities as well. However, they have not considered ionic dissociation in these calculations and they ignore dimer concentration.

2. Carboxylic Acid Adducts

The evidence for heteroassociation of formic, acetic, and fluoroacetic acids with HOTFA in 1 : 1 adducts is very good.[87,162] Only 1 : 1 adducts are found, and these are considerably more stable than the homoassociated dimers: $(HOTFA)_2$, $\Delta H_{assoc} = -14$ kcal/mole; HOAc–HOTFA, $\Delta H_{assoc} = -17$ kcal/mole.

The photoelectron spectrum of mixed dimers in the vapor phase has been obtained.[90] The shift of the bands assigned to the nonbonding σ orbital on the carbonyl oxygen and the nonbonding π orbital on the hydroxyl oxygen for both acids upon adduct formation, indicates that the hydroxyl group of HOTFA bonded to the acetic acid carbonyl oxygen is the stronger of the two hydrogen bonds. Ling *et al.*[87] had suggested this to be the case from their measurement of heats of association for adducts with acetic acid and HOTFA. The infrared spectrum of the HOTFA adducts with formic and acetic acid has been measured and assigned in the region 300–20/cm,[99] and microwave studies have also been done.[162,163]

The case for adducts in liquid phase is equivocal. The variation of density over the entire composition range shows no evidence for compound formation, while viscosity and viscosity-corrected electrical conductivity indicate 1 : 1 and 2 : 1 HOAc/HOTFA adducts.[138] Randles *et al.*[133] reported similar findings but they point out that relative to the $HOAc/H_2SO_4$ system, the changes in specific conductivity are small. Emeléus *et al.*[73] measured the electrical conductivity of dilute solutions of HOTFA in acetic acid and found it to be a very weak electrolyte compared to HCl and H_2SO_4 in this solvent, but marginally stronger than HNO_3. Fialkov and Zhikharev[138] estimated the pK_a of HOTFA in HOAc to be 9.38 from the data of Emeléus *et al.*[73] Rode *et al.*[72] determined a value of 11.4 from NMR measurements. Monochloroacetic acid is said to form a 1 : 1 adduct with HOTFA, as indicated by deviation of molar volume from additivity.[138] However, the viscosity–composition plot shows a minimum rather than a maximum and the electrical conductivity–composition plot gives no evidence for compound formation. Physical measurements on trichloroacetic acid–HOTFA mixtures[138] and nitric acid–HOTFA mixtures[134] give no evidence for adduct formation, although the proton shift for the latter mixture shows a slight inflection at 1 : 1 composition.[171]

3. Strong Acid Adducts

Fialkov and co-workers have used the departure from additivity of viscosity, electrical conductivity, and molar volume to predict the existence of adducts of some strong acids with HOTFA. The adducts formed are listed in Table XIV. The evidence for these adducts is, in most cases, tenuous inasmuch as the effects are small and all the parameters measured do not necessarily indicate adduct formation. Where there is a strong interaction in the adduct, all of the physical methods predict the same stoichiometry as in the case of $(C_6H_5)N(C_2H_5)_2$–HOTFA.[130] Yet, where adduct formation is shown to occur by different methods, as in the case of H_2O–HOTFA, the departures from additivity do not predict the same stoichiometry.[134] In the case of sulfuric acid, Fialkov and Zhikharev claim a 1 : 1 adduct exists on the basis of a maximum decrease in molar volume from additivity (2.4%).[36] This stoichiometry as well as that of 1 : 2 H_2SO_4/HOTFA has been confirmed by NMR measurements.[164] At low concentrations of H_2SO_4 in HOTFA, there are conflicting results. Bessière[120] used the indicator method of Kolthoff and Bruckenstein[172] to determine the dissociation constant of H_2SO_4 which he reports as $10^{-6.3}$. However, Fialkov and Zhikharev[36] claim that H_2SO_4 is a nonelectrolyte in HOTFA. Barr *et al.*[128] have shown that HOTFA is a nonelectrolyte in 100% H_2SO_4. Kachurin[173] has titrated 97.5% sulfuric acid with oleum in HOTFA and he finds a minimum in conductivity at the composition 1 : 1 H_2O/SO_3. He claims, however, that his results are not consistent with the formation of H_2OTFA^+ and may be accounted for by ionic self-dissociation of H_2SO_4, as well as H_3O^+ and HSO_4^- ions on the water-rich side of the end point. Dissociation constants for some of the other strong acids in HOTFA have been reported. Bessière[120] gives $10^{-4.5}$ for perchloric acid, although it is not clear whether this is for anhydrous acid or 70% aqueous perchloric acid. Indicator studies on perchloric acid in HOTFA by the same author[174] were done in the presence of 0.02 M H_2O. Fialkov and Borovikov[137,140] report the dissociation constant of $H_2S_2O_7$ to be $10^{-3.1}$. However, Gillespie and Malhotra find that HOTFA is a nonelectrolyte in 100% $H_2S_2O_7$.[129] Fialkov *et al.*[143] claim the dissociation constant of HSO_3F to be $10^{-3.85}$. The dissociation constants reported by Fialkov and co-workers for $H_2S_2O_7$ and HSO_3F were determined by the outdated Kraus–Bray method.[144] Unfortunately, the raw conductivity data are not given.

4. Adducts with Aprotic Donors

Proton and ^{19}F-NMR spectroscopy and dielectric constant measurements have been used to establish the stoichiometry of adducts formed in CCl_4 solution with nitrobenzene.[108] The interaction of diethyl ether and HOTFA has been studied in CCl_4 solution by infrared methods.[160] Christian *et al.*[86] have shown that dioxane forms a 1 : 1 complex with HOTFA in the vapor phase for which $K_{assoc} = 0.12$/mm at 40°C. Taha and Christian[112] studied the adduct formation between benzophenone and HOTFA in diphenylmethane as solvent. They measured the vapor pressure of HOTFA over the solution and, from a knowledge of Henry's Law Constants and K_{assoc} of HOTFA, established that 1 : 1 and 1 : 2 benzophenone/HOTFA adducts were formed with K_{assoc} 40 liters/mole and 275 liters^2/mole^2, respectively, at 40°C. They prefer **VIII** for the 1 : 2 adduct structure rather than **IX**.

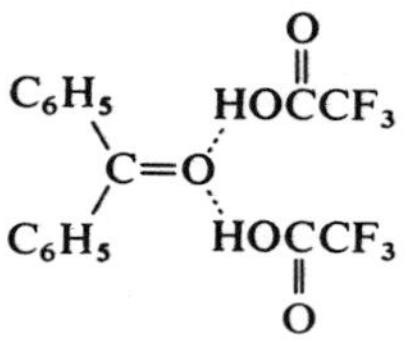

```
                  O
                  ||
C6H5          HOCCF3
    \          :
     C=O
    /          :
C6H5          HOCCF3
                  ||
                  O

        (VIII)
```

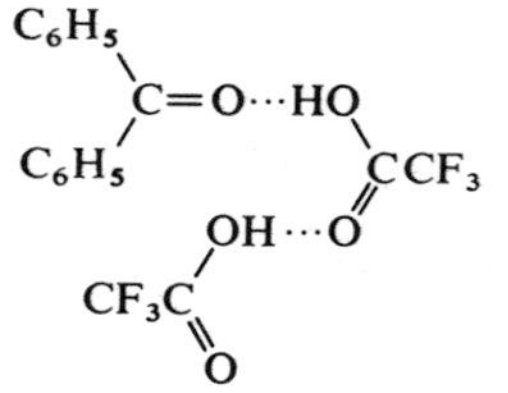

```
C6H5
    \
     C=O···HO
    /         \
C6H5           CCF3
              //
       OH···O
      /
 CF3C
     \\
      O

        (IX)
```

They argue that K_{assoc} for the homoassociation of HOTFA (Table IX), where two hydrogen bonds are involved, is so much smaller than that of the 1 : 1 benzophenone/HOTFA adduct that structure **(VIII)** is favored. Kirszenbaum *et al.*[104] have studied complexes of HOTFA with tetrahydrofuran (THF) and acetonitrile using infrared spectroscopy. They find that as THF is added to a 0.02 *M* solution of HOTFA in 1,2-dichloroethane, the HOTFA monomer peak at 1804 cm^{-1} weakens while the peak at 1781 cm^{-1} and the shoulder at lower wave numbers grow in intensity. These new bands are due to the THF–HOTFA adduct and the ion pair $THFH^+ \cdots OTFA^-$. The solvent and THF are nearly isodielectric and, as the concentration is changed, the intensity ratio of the new peaks due to adduct and ion-pair formation remains constant. However, if the same experiment is carried out in nitromethane as solvent, the ion-pair peak (1762 cm^{-1}) is relatively more intense than that of the adduct (1781 cm^{-1}) and their intensity ratio changes with concentration, reflecting the change in dielectric constant. A 0.02 *M* solution of HOTFA in acetonitrile initially shows a peak at 1792 cm^{-1} due to the CH_3CN–HOTFA adduct which is gradually replaced by a band at 1708 cm^{-1}. The trifluoroacetate ion has a peak at 1696 cm^{-1} in acetonitrile and for this reason it is thought that the adduct is gradually changed into an ionic form. Forcier and Olver[175] claim that HOTFA acts as an acid in acetonitrile with a dissociation constant of 4×10^{-8} moles/liter. Hauptschein and Grosse[176] have studied mixtures of HOTFA with pyridine, acetonitrile, tertiary amines, and ethers. They used the constant boiling point of a mixture as proof of compound formation, but this is not a reliable criterion, and their measurements in all likelihood represent boiling points for azeotropic mixtures. In some instances, however, crystalline compounds were formed, as in the case of $(C_6H_5)N(CH_3)_2$–HOTFA. Water also forms an azeotrope with HOTFA.[23]

5. Solvates of Ionic Trifluoroacetates

In view of the weak donor strength of HOTFA and its strong tendency to form hydrogen bonds, it is expected that in all of the solvates of ionic trifluoroacetates listed in Table XIV the solvent molecule is bound to the anion. In all of the solvates listed except two, the ratio of trifluoroacetate to solvent is 1 : 1, indicating that the $H(OTFA)_2^-$ ion is present. Several of the species which remained as the solid phase in the solubility studies should be considered along with the solvates listed in Table XIV[152]; these are given in Table XIII. For instance, the compounds $Pr[H(OTFA)_2]_3$, $La[H(OTFA)_2]_3$, and $Mg[H(OTFA)_2]_2$ may be formulated for the solid phase from the data. Jones and Dyer[159] have provided ^{1}H-NMR evidence for the existence of the $H(OTFA)_2^-$ ion in solution. They found that, as trimethylphenylammonium trifluoroacetate is added to HOTFA, the ^{1}H resonance shifts to lower field until an equimolar ratio of salt and solvent is reached at which point the shift remains constant with further additions of salt. Golic and Speakman[166] have determined the crystal structure of the potassium, rubidium, and cesium trifluoroacetate solvates. The potassium and rubidium salts are isomorphous with $d(O \cdots H \cdots O) = 2.435$ Å; the cesium salt has $d(O \cdots H \cdots O) = 2.38$ Å. The hydrogen bond has been shown to be symmetric by neutron diffraction studies[177] and from the infrared spectrum.[67,177] There is a disagreement concerning the assignment of the symmetric ν_{OHO} mode. Miller *et al.*[67] have assigned the symmetric ν_{OHO} band, which is only Raman active and not shifted upon deuteration, at 792 cm^{-1} in the potassium compound, while Hadzi *et al.*[178] assign this mode at 130 cm^{-1}.

The existence of the 1 : 2 NaOTFA/HOTFA solvate has been confirmed by two groups.[152,165] Klemperer and Pimentel[165] evaluated heats of dissociation of both the 1 : 1 and 1 : 2 solvates from vapor pressure–temperature studies. They find ΔH_{dissoc} is 16 kcal/mole for the 1 : 1 solvate and 14 kcal/mole of HOTFA for the 1 : 2 solvate.

Fialkov and Kholodnikova[130] have shown that diethylphenylamine forms a 1 : 1 adduct with HOTFA. The departure of viscosity, conductivity, molar volume, and temperature coefficients for viscosity and conductivity, from additivity are large and the compound is very likely ionic. Tsarevskaya[179] has used cryoscopy to establish the stoichiometry of the association equilibria of carboxylic acid–tertiary amine adducts in benzene. Her results for halocarboxylic acids are accounted for by the equilibria:

$$2A + B \rightleftharpoons A_2B \tag{21}$$

$$2A_2B \rightleftharpoons (A_2B)_2 \tag{22}$$

where A represents the acid and B the base. However, in the case of HOTFA adducts, association of the adduct, A_2B, proceeds beyond the dimer stage and it can only be concluded that Eq. 22 goes to completion. The extensive aggregation in this case suggests that the adducts are ionic.

6. Protonation of Amides

There have been several studies of solutions of amides in HOTFA. The interest here lies in the mechanism whereby helix–coil transformations in polypeptides take place in solutions in or containing HOTFA.[147] Amides serve as model compounds to study this mechanism. While the protonation of amides on oxygen has been clearly established for media more strongly acidic than HOTFA, such as aqueous $HClO_4$[180] and HSO_3F,[181] the evidence for protonation by HOTFA is still disputed.[182] Reeves[100] provided NMR evidence for the protonation of *N*-methylformamide (NMF) and dimethylformamide (DMF) by HOTFA. The proton chemical shift moves to low field as the amide is added to HOTFA, indicating hydrogen bond formation. The shift is much greater than that observed in acetic acid and is greater for DMF than for NMF, reflecting the greater basicity of DMF. At higher amide mole fraction the chemical shift moves again to high field, indicating ionization. Hanlon *et al.*[183] demonstrated that the infrared band at 1.51 μm (6623 cm^{-1}) in solutions of the model compound *N*-methylacetamide (NMA) as well as polypeptides in HOTFA is due to $2\nu_{NH}$ of the protonated amide with the structure

OH
C
‖
N
H +

(X)

Furthermore, mixtures of NMA in HOTFA have molar volumes less than those predicted from additivity and this has been attributed to electrostriction resulting from the formation of ions. Klotz *et al.*[147] measured the electrical conductivity of solutions of NMA in HOTFA and, from the high equivalent conductivities observed, showed that protonation of amides does take place. Stoke and Klotz extended the evidence for protonation of amides in HOTFA and HOTFA–$CHCl_3$ mixtures to include polypeptides.[132] Purkina *et al.*[184] studied six amides in HOTFA and HOTFA–$CHCl_3$ mixtures by ^{19}F- and ^1H-NMR and infrared spectroscopy and showed that there is an equilibrium between free, hydrogen-bonded, and protonated amide. As infrared criteria for protonation they used the appearance of $\nu_{C=N}$ and bands characteristic of $OTFA^-$ and $H(OTFA)_2^-$ ions. These authors point out that dimeric HOTFA is a stronger acid than the monomer. Tam and Klotz[105] have provided additional infrared evidence to show that amides are protonated in 100% HOTFA or in its solutions in $CHCl_3$ or CCl_4. They show that, upon addition of NMA to HOTFA, the acid monomer and dimer peaks at 3510, 3120, and 2940 cm^{-1} are replaced by a broad band at 2300–2800 cm^{-1} and that a band at 1620 cm^{-1} attributed to the $OTFA^-$ ion, appears in the spectra. They give NMR results similar to those of Reeves[100] to support their conclusions, although Combelas and Garrigou-LaGrange[182] suggest that the results of Tam and Klotz may be interpreted in terms of a hydrogen-bonded complex and dispute whether amides are protonated by HOTFA. The evidence falls clearly on the side of protonation.

7. Acid–Base Scales

Bessière has used potentiometric measurements and indicator methods to establish scales of acid and base strengths in HOTFA.[120,174,185] He finds the relative order of acid strength to be[185]:

$$HClO_4 > H_2SO_4 > HSO_3Cl > HSO_3CH_3 >$$

$$p\text{-toluenesulfonic} > HBr > HBF_4 > HI > HCl > HNO_3$$

The scales established for HOTFA are compared to those found in aqueous H_2SO_4 and glacial acetic acid.[185] Basicity scales for solutions of $HClO_4$ (in the presence of 0.05 *M* H_2O), H_2SO_4, HSO_3Cl, and HBr in HOTFA and HOTFA itself show the same relative order of basic strengths.[174] The basic dissociation constants for NaOTFA and KOTFA and the acid dissociation constants for H_2SO_4 are, however, in conflict with other reported values.[25,36,173,174]

8. Other Acid–Base Interactions

Zinc trifluoroacetate may be the only example of an ampholyte in HOTFA. Hara and Cady[152] showed that the concentration of dissolved $Zn(OTFA)_2$ increased from 0.87 g/100 g to 13 g/100 g HOTFA upon simultaneous saturation with NaOTFA, which increased in concentration from 13.1 g/100 g to 23 g/100 g HOTFA. The mole ratio of trifluoroacetate to zinc ion in the saturated solution is 5.78, suggesting that the complex in solution may be $Na_4Zn(OTFA)_6$. Surprisingly, $Al(OTFA)_3$ shows only a slight tendency to dissolve in NaOTFA solutions in HOTFA. The same is true of $Mg(OTFA)_2$ and $Nd(OTFA)_3$.

Trifluoroacetic acid has been used as a medium for acid-base titrations. Petit and Bessière[150] titrated H_2O (0.28 *M*) with HSO_3F (17.4 *M*) using anthraquinone as indicator. The titration was carried out in 0.05 *M*

Et_4NClO_4 in HOTFA and was followed potentiometrically using a hydrogen electrode with $Ag(s)|AgClO_4(s)$ reference electrode. It is interesting that in HOTFA no hydrolysis of HSO_3F takes place as has been observed for solutions in 100% HSO_3F[186]:

$$H_2O + HSO_3F \rightarrow H_3O^+ + SO_3F^- \rightleftharpoons HF + H_2SO_4 \qquad (23)$$

Kresze and Schmidt[131] have titrated amino acids and polypeptides conductometrically with anhydrous $HClO_4$ in HOTFA. These compounds have been shown to be very soluble in HOTFA.[157] Such titrations are less successful in acetic acid due to the low solubility of amino acids in this solvent. Kresze and Schmidt state that the titration may be carried out in the presence of some water. They[187] have also discussed the general utility of HOTFA as a solvent and reactant for characterizing polypeptides. Similarly de Vries *et al.*[188] have titrated weak bases such as thiourea and nitroguanidine potentiometrically in HOTFA with 0.1 M $HClO_4$, which may be standardized against NaOTFA. Aqueous $HClO_4$ was used to make up the titrant and apparently the water introduced does not affect the titrations. A Pt–Pt electrode system was found to be most effective. The weak bases thiourea and nitroguanidine cannot be titrated in acetic acid. Inasmuch as ionic perchlorates are virtually insoluble in HOTFA,[152] the reactions being followed are probably precipitation reactions. Harriss and Milne[45] titrated the acid $[H_2OTFA][B(OTFA)_4]$ with CsOTFA in the ion-pair region of concentration conductometrically and from a knowledge of Λ and K_D for the acid, base, and salt they were able to calculate the observed titration curve. The end-point inflection point is very slight because the Λ and K_D of all three solutes are very similar:

$$CsOTFA + [H_2OTFA][B(OTFA)_4] \rightarrow Cs[B(OTFA)_4] + 2HOTFA \qquad (24)$$

Trifluoroacetic acid[189,190] and its mixtures with BF_3–H_2O[191] have been used as media for studying the protonation of arenechromium tricarbonyls, arenechromium dicarbonyl triphenylphosphines, and manganese cyclopentadienyl carbonyl complexes. The introduction of a triphenylphosphine group in the chromium complexes increases the basicity sufficiently to permit protonation in HOTFA.[189]

B. Reactions of Inorganic Compounds

1. Preparation of Trifluoroacetates

In this section, the main methods of preparation of trifluoroacetates are discussed, but a complete survey of trifluoroacetates, which would be extensive, is not given. Garner and Hughes have recently reviewed this subject.[191a]

Ionic trifluoroacetates are readily prepared by the action of aqueous HOTFA on the metal oxide, hydroxide, or carbonate.[23,152,192,193] Care must be taken to avoid excess HOTFA which may lead to crystallization of an acid salt.[166] For easily oxidized elements, the metal itself may be used.[152,153,194] Some metal trifluoroacetates crystallize in the anhydrous state from aqueous solution (Na, K, Rb, Cs, Ag, Hg^{2+}), while others require special dehydration treatment. Swarts[192] effected crystallization at room temperature by desiccating over concentrated sulfuric acid. Hara and Cady[152] dried many of their salts under vacuum at 100°C but, in some cases (Al^{3+}, $UO_2{}^{2+}$, Th^{4+}), the glassy hydrate obtained was dehydrated by using trifluoroacetic acid anhydride. This method was also used by Norman *et al.*[195] to prepare anhydrous $Ce(OTFA)_4$. Swarts[192] noted that some anhydrous salts could be further purified by sublimation under vacuum (Ag^+, Pb^{2+}, Cu^{2+}) although the procedure was often not efficient.

In some instances the acetate is used as a starting point.[152,196]

$$UO_2(OAc)_2 + HOTFA \rightarrow UO_2(OTFA)_2 + HOAc \quad (25)$$

For trifluoroacetates of the non-metals and complex trifluoroacetates other routes are used. Sartori and co-workers[194,197] reacted chlorides with HOTFA:

$$ZrCl_4 + 4HOTFA \xrightarrow[40^\circ C]{} Zr(OTFA)_4 + 4HCl \quad (26)$$

It is also possible to convert metal aryls[198] and hydrides[199] into trifluoroacetates with HOTFA:

$$(C_6H_5)BCl_2 + 3HOTFA \xrightarrow[n\text{-pentane}]{20^\circ C} B(OTFA)_3 + 2HCl + C_6H_6 \quad (27)$$

$$\pi\text{-}Cp_2ZrH_2 + 2HOTFA \xrightarrow[HOTFA]{} \pi\text{-}Cp_2Zr(OTFA)_2 + H_2 \quad (28)$$

A metathetic reaction between silver or mercuric trifluoroacetate and a chloride in an inert solvent or HOTFA has been used by several workers.[197,200–202]

$$CrCl_3 + 3AgOTFA \xrightarrow[CH_3NO_2]{} Cr(OTFA)_3 + 3AgCl \quad (29)$$

Iodine(I) trifluoroacetate complexes are prepared by metathesis of AgOTFA and iodine.[203–205]

$$2AgOTFA + I_2 \xrightarrow[C_6H_5NO_2]{} AgI(OTFA)_2 + AgI \quad (30)$$

$$AgOTFA + C_5H_5N + I_2 \xrightarrow[C_6H_5]{} I(C_5H_5N)(OTFA) + AgI \quad (31)$$

A series of complex trifluoroacetates have been made using the action of trifluoroacetic anhydride on oxometallates[206–208]:

$$Na_2TeO_3 + 3(TFA)_2O \rightarrow Na_2Te(OTFA)_6 \quad (32)$$

Fluorides also react with anhydride to give trifluoroacetates[205,209–211]; the driving force of the reaction being the volatility of CF_3COF (bp = −59°C)[5]:

$$IF_3 + 3(CF_3CO)_2O \rightarrow I(OOCCF_3)_3 + CF_3COF \quad (33)$$

Reaction of the anhydride with a chloride has also been used.[45]

Peroxotrifluoroacetic acid has been employed for the preparation of some trifluoroacetates of iodine.[212,213]

$$3HOOTFA + 3(TFA)_2O + I_2 \rightarrow 2I(OTFA)_3 + 3HOTFA \quad (34)$$

$$5HOOTFA + (TFA)_2O + I_2 \rightarrow 2IO_2OTFA + 5HOTFA \quad (35)$$

$$H(CF_2CF_2)_nCH_2I + HOOTFA + HOTFA \rightarrow H(CF_2CF_2)_nCH_2I(OTFA)_2 + H_2O \quad (36)$$

$$n = 1, 2, 3$$

2. Survey of Reactions of Inorganic Compounds with HOTFA

Fujioka and Cady[2] made a very extensive survey of the reactions of metal trifluoroacetates dissolved in HOTFA [alkali metals, alkaline earths, ammonium, Cr(III), Mn(II), Fe(III), Co(II), Ni(II), Cu(II), Ag(I), Zn(II), Cd(II), Hg(I), Hg(II), As(III), Sb(III), Bi(III), Sn(II), Sn(IV)] with a variety of inorganic reagents: HCl, H_2S, $HClO_4$, H_2SO_4, SO_3, H_3PO_4, HPO_3, Cl_2, CrO_3, and $KMnO_4$. The reactions were carried out under anhydrous conditions with the addition of trifluoroacetic acid anhydride. They found that for reactions of the protonic acids, precipitation was common. Insoluble chlorides, sulfides, and perchlorates were formed with varying degrees of completion. An exception was the reaction of Pb(II) with H_2S, which did not give a precipitate. Anhydrous sulfates [Ca, Sr, Ba, Co(II), Zn] and hydrogen sulfates were formed on reaction with H_2SO_4. Sulfur trioxide produced sulfates and pyrosulfates. Orthophosphoric acid gave mono- and dihydrogen phosphates; metaphosphoric acid produced metaphosphates. Antimonyl metaphosphate ($SbOPO_3$) was formed from $Sb(OTFA)_3$, while $Bi(OTFA)_3$ gave $Bi(PO_3)_3$. Chromic oxide and potassium permanganate failed to produce insoluble chromates and permanganates. Lee and Johnson[214] observed that a Mn(VII) and Cr(VI) species came over with HOTFA when solutions of $KMnO_4$ and CrO_3 in HOTFA were distilled. They suggested that $KMnO_4$ and H_2CrO_4 may codistill with HOTFA. Hara and Cady found that both reagents acted as oxidizing agents in HOTFA. Trifluoroacetic acid itself was slowly attacked by $KMnO_4$ to give CO_2, CF_3COF, and COF_2. Manganese heptoxide, Mn_2O_7, was claimed to be present in the $KMnO_4$ solutions. Manganese(III) was usually produced in the oxidations by $KMnO_4$; Cr(III) resulted from CrO_3 reactions. Chromic oxide, potassium permanganate, and chlorine all oxidized Hg(I), As(III), and Sb(III) to their higher oxidation states. Sulfur dioxide was oxidized to sulfuric acid by CrO_3 and $KMnO_4$. Chromium trioxide yielded a peroxochromium complex with H_2O_2 and CrO_2Cl_2 with HCl.

Several attempts have been made to prepare xenon trifluoroacetates.[215-217]

$$XeF_2 + HOTFA \underset{-24^\circ C}{\rightleftarrows} XeF(OTFA) + HF \quad (37)$$

$$XeF(OTFA) + HOTFA \rightarrow Xe(OTFA)_2 + HF \quad (38)$$

The reactions have been carried out at low temperatures (−24°C) in HOTFA, $(TFA)_2O$, and HF. Both the mono- and bistrifluoroacetates are obtained, depending upon stoichiometry. They decompose explosively at room temperature:

$$2XeF(OTFA) \xrightarrow[20^\circ C]{} Xe + XeF_2 + CO_2 + C_2F_6 \quad (39)$$

$$Xe(OTFA)_2 \xrightarrow[20^\circ C]{} Xe + CO_2 + C_2F_6 \quad (40)$$

Buckles and Mills[218] found that the frequency of maximum optical density in the visible spectrum of I_2 dissolved in HOTFA ($\lambda_{max} = 515$ nm) was comparable to that of I_2 dissolved in weak donor solvents such as H_2SO_4 ($\lambda_{max} = 502$ nm) and CCl_4 ($\lambda_{max} = 517$ nm). The much greater donor strength of acetic acid is reflected in the lower λ_{max} (478 nm). Although HOTFA is a very weak donor solvent, studies of organic reaction kinetics[8,219] indicate that it solvates anions strongly. This would also account for the finding of Buckles and Mills[220,221] that polyhalide ions are unstable with respect to the halogen or interhalogen and the smallest halide anion in the polyhalide. This is illustrated by the following dissociations which go to completion:

$$ICl_2^- \rightarrow ICl + Cl^- \quad (41)$$

$$IBrCl^- \rightarrow IBr + Cl^- \quad (42)$$

Alcais *et al.*[222] evaluated the association constant K_{assoc} for the tribromide ion by spectrophotometry:

$$Br_2 + Br^- \rightleftharpoons Br_3^- \quad (43)$$

They report the value, $K_{assoc} < 0.91$ liter/mole. Petit and Bessière have studied the redox behavior of iodine in HOTFA containing Et_4NClO_4 or NaOTFA by voltammetry using glassy carbon and rotating shiny platinum indicator electrodes.[149] They showed that iodine is directly reduced to iodide, with no intermediate triodide formation, but that iodide cannot be oxidized electrochemically in dry HOTFA to which a small amount of anhydride has been added. Iodine is shown to be oxidized to the ^+I state

$$I_2 + 2HOTFA + 2ClO_4^- - 2e^- \rightarrow 2IOTFA + 2HClO_4 \quad (44)$$

but the reduction of I ($^+$I) cannot be accounted for quantitatively due to the decomposition of IOTFA:

$$IOOCCF_3 \rightarrow ICF_3 + CO_2 \quad (45)$$

Iodine ($^+$I) may be further oxidized to I ($^+$III), no doubt present as I(OTFA)$_3$, but this reaction is also not reversible.

If SO_3 is dissolved in an equimolar quantity of HOTFA and the mixture distilled, no separation takes place and the compound CF_3COHSO_4 distills over.[223] The product reacts with HOTFA to give anhydride:

$$CF_3COHSO_4 + CF_3COOH \rightarrow (CF_3CO)_2O + H_2SO_4. \quad (46)$$

Subsequent to this work, Harriss and Milne[145] showed by ^{19}F-NMR studies that H_2SO_4 is extensively dehydrated by $(TFA)_2O$ to give SO_3 or possibly CF_3COHSO_4. The reaction of SF_4 with HOTFA has been studied[224] and shown to yield CF_3COF rather than the perfluorinated product produced with other carboxylic acids.

Apart from the preparation of AsO(OTFA), Sb(OTFA)$_3$, and Bi(OTFA)$_3$ reported by Hara and Cady,[152] little work has been done on the chemistry of elements of group V in HOTFA. Antimony pentafluoride has been shown to behave as an acid in the solvent (Eq. 10), but attempts to titrate it with KOTFA result in immediate precipitation of $KSbF_6$.[45] Apparently there is a disproportionation reaction but the nature of the other product in solution is unknown.

The tetrachlorides of Si, Zr, Hf, and Rh all react at moderate temperatures with HOTFA to give the tetrakistrifluoroacetates (Eq. 26).[197,225] Titanium tetrachloride reacts with HOTFA to give the oxide bistrifluoroacetate, 3TiO(OTFA)$_2$–2HOTFA, which loses HOTFA at 60°C under vacuum to yield [TiO(OTFA)$_2$]$_n$. The germanium and tin tetrakistrifluoroacetates have been made from the tetrachloride or bromide by metathesis in HOTFA–(TFA)$_2$O, using AgOTFA or Hg(OTFA)$_2$ and the tetraphenyl compounds undergo reaction with HOTFA.[197,225]

$$Sn(C_6H_5)_4 + 4HOTFA \xrightarrow{40°C} Sn(OTFA)_4 + 4C_6H_6 \text{ (7\% yield)} \quad (47)$$

$$Ge(C_6H_5)_4 + HOTFA \rightarrow (C_6H_5)_3GeOTFA + C_6H_6 \text{ (quantitative)} \quad (48)$$

The lead compound is made from Pb_3O_4, HOTFA, and (TFA)$_2$O. All of the Group IV trifluoroacetates lose (TFA)$_2$O upon careful thermal decomposition as, for example, with Si(OTFA)$_4$:

$$Si(OTFA)_4 \xrightarrow[in\ vacuo]{200°C} [(TFAO)_3Si]_2O + TFA_2O \quad (49)$$

Under forcing conditions, decomposition also occurs to CO, CO_2, COF_2, and CF_3COF.[197] Sara and Taugbol[226] have investigated the reaction of

$SnCl_4$ with HOTFA in benzene. They find that a disubstituted product, $SnCl_2(OTFA)_2$, can be obtained from the dried reaction mixture. On the basis of its insolubility in weakly polar or nonpolar solvents such as dichloromethane and benzene, as well as from its infrared spectrum, $SnCl_2(OTFA)_2$ is claimed to have a structure with trifluoroacetate bridges.

The rather unusual trifluoroacetate of graphite may be prepared by electrolytic or chemical oxidation of graphite in dry HOTFA.[227] Trifluoroacetate to carbon ratios up to 1 : 30 are reached.

Except for the Tl(III) compound, all of the tristrifluoroacetates of the elements of Group III are known. Aluminum tristrifluoroacetate has been prepared from aqueous solution followed by treatment with $(TFA)_2O$[152] and from anhydrous aluminum trichloride and HOTFA.[194] This compound is unique in not dissolving in HOTFA to any appreciable extent upon addition of NaOTFA.[152] Indium tristrifluoroacetate has been prepared from the metal while Ga(OTFA) and $B(OTFA)_3$ have been made from their respective chlorides.[194] There is some controversy concerning the existence of $B(OTFA)_3$.[45] This compound was first prepared by Muetterties[228] from B_2O_3 and $(TFA)_2O$ but the products were impure, apparently containing considerable $B_2O(OTFA)_4$. Gerrard *et al.*[198] prepared $B(OTFA)_3$ from BCl_3 and HOTFA in *n*-pentane at room temperature and, according to Eq. 27, Harriss and Milne[45] were unable to successfully prepare pure $B(OTFA)_3$ by this or other methods. Their products always contained some $B_2O(OTFA)_4$ formed by the reaction

$$2B(OTFA)_3 \rightarrow B_2O(OTFA)_4 + (TFA)_2O \quad (50)$$

Gerrard *et al.* reported that $B_2O(OTFA)_4$ was formed from $B(OTFA)_3$ by sublimation under vacuum, although Sartori *et al.*[194] found that the compound was much less stable. The existence in HOTFA solution of $B(OTFA)_3$ or its solvated analog, $H_2OTFA^+B(OTFA)_4^-$, accounts quantitatively for the conductometric titration of dilute solutions with CsOTFA.[45] Adducts of the group III tristrifluoroacetates with tetrahydrofuran, pyridine, dimethyl sulfoxide, and dimethylformamide have been prepared as well as several complex compounds of the formula, $Cs[M^{III}(OTFA)_4]$ (M^{III} = B, Al, Ga, In).[45,194] Should $B(OTFA)_3$ be stable with respect to $B_2O(TFA)_4$ at room temperature, it is the only $B(OOCR)_3$ compound known which behaves in this way.[229]

HOTFA is an excellent solvent for organic compounds and would be a good solvent for organometallic compounds, yet little work has been done in HOTFA with this class of compounds. Wailes and Weigold[199] showed that bis(π-cyclopentadienyl)zirconium dihydride is converted to the trifluoroacetate in the presence of excess HOTFA (Eq. 28). If the zirconium compound is in excess, then bridged trifluoroacetate complexes are formed. Petit and Bessière[151] found that ferrocene is slowly oxidized by HOTFA to ferricinium ion. Castagnola *et al.*[230] report similar findings for solutions in benzene and give the reaction

$$Cp_2Fe + 3CF_3COOH \xrightarrow[C_6H_6]{} 2Cp_2Fe^+CF_3COO^- + CF_3CHO + H_2O \quad (51)$$

Blake and Pritchard[231] find that HOTFA is pyrolyzed at 300°–390°C to yield as principal products CO, CO_2, CF_3COF, and CF_3COOCH_2F. The reaction takes place on the walls of the silica and iron containers.

Uranium hexafluoride reacts with dry HOTFA on standing according to the reaction[232]

$$2UF_6 + 2CF_3COOH \rightarrow 2UF_5 \cdot HF + CF_3COF + COF_2 + CO_2 \quad (52)$$

C. Organic Reactions in HOTFA

Trifluoroacetic acid undergoes all of the common reactions of carboxylic acids. It may be readily converted to the anhydride, the nitrile (CF_3CN), the aldehyde (CF_3CHO), and the amine, $CF_3CH_2NH_2$, as well as to derivative acid halides, esters, amides, and alcohols by standard methods.[5]

Trifluoroacetic acid is stable to oxidation by all but the most powerful oxidizing agents[152] and has been used as the solvent for oxidation of organic compounds. Lee and Johnson[214] found that the mechanism of oxidation of 2-propanol by $KMnO_4$ or CrO_3 in aqueous HOTFA resembled that in other aqueous strong acid solutions. Norman *et al.* have studied the oxidation of methyl-substituted benzenes by $Ce(OTFA)_4$[195] and $Pb(OOCCH_3)_4$[233,234] in anhydrous HOTFA. Mesitylene reacted instantaneously at 25°C to give principally the biaryl **(XI)** with smaller amounts of diarylmethane **(XII)** and benzyltrifluoroacetate **(XIII)**

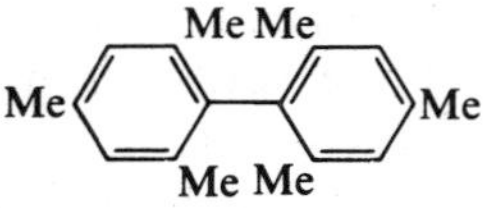

(XI) 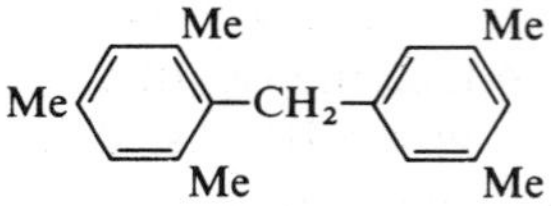 **(XII)** 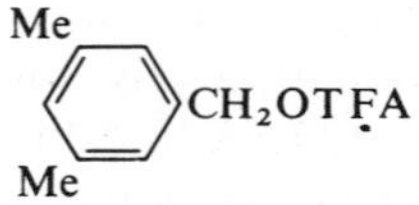 **(XIII)**

Addition of LiOTFA was found to enhance the yield of the diarylmethane and benzyltrifluoroacetate. Toluene, xylene, and durene also reacted rapidly, but benzene reacted slowly even at 70°C. Mixtures of Fe(II)–$(C_2H_5)_2NOH$ or Fe(II)–$(C_2H_5)_3NO$ in 100% HOTFA have been used by Deno and Pohl[235] to oxidize cyclohexane, heptane, decane, and 1-octyl-trifluoroacetate. Trifluoroacetates are formed with no evidence for further oxidation. High selectivity (72%) was found for the 7 position of 1-octyltrifluoroacetate to produce the bistrifluoroacetate of 1,7-octanediol.

Electrophilic attack on aromatic compounds in HOTFA has been studied by several groups. Andrews and Keefer investigated the kinetics of iodination with ICl[219] and chlorination using Cl_2[236] in CCl_4 and HOTFA. The aromatics studied were mesitylene and pentamethylbenzene. They found that halogenation was rapid and followed kinetics, first-order in halogen, with HOTFA as solvent. In the halogenations in CCl_4, iodine monochloride catalyzes the reaction by acting as an acceptor for the leaving halide ion in the transition state. This accounts for the kinetics of the iodination with ICl in CCl_4 which are higher than first-order. Trifluoroacetic acid can fill this role of halide acceptor in the reactions in HOTFA with its hydrogen bonding capability which accounts for the first-order kinetics in halogen and the rapid reactions in this case. Brown and Wirkkala[8,237,238] found that bromination, nitration, and mercuration of aromatics all occurred at far faster rates in HOTFA than in HOAc. In the case of mercuration, the rate was 690,000 times faster. In addition, the reactions did not produce side products, as when acetic acid was used, and the kinetics were simple. The bromination and mercuration of toluene were highly selective for the para position. It was suggested that HOTFA will be a useful medium for studying the kinetics and isomer distributions of less reactive aromatics. Himoe and Stock have carried out related kinetic studies on the chlorination of aromatics in HOTFA.[239] The tribromide ion has been shown to be of minor importance in the bromination of aromatics in HOTFA.[222]

Detritiation and desilylation of the aromatic compounds in HOTFA has been studied by Eaborn *et al.*[29] These reactions proceed via a protonated transition state. They find that the addition of water accelerates the rate of these reactions up to a point and then causes a slowing down. They attribute this to an initial favorable solvation by water followed by a deceleration due to reduced acidity. Lithium bromide, chloride, and perchlorate were found to greatly accelerate the reactions, while lithium trifluoroacetate had little effect. The authors suggest that the large increases in rate are due to the formation of free HCl, HBr, and $HClO_4$, which would result in more effective proton transfer. Dedeuteration of alkyl benzenes in HOTFA containing H_2SO_4, $HClO_4$, and HBF_3OH has been investigated by Dallinga and ter Maten[78] and Mackor *et al.*[3] Exchange was found to occur on the ring as well as in the side chain. Exchange has been observed to occur in tertiary alkyl trifluoroacetates also.[240]

Trifluoroacetic acid is a very weak nucleophile that is able to promote ionization by hydrogen bonding to the leaving anion. For these reasons solvolytic displacement reactions proceed predominantly by the S_N1 mechanism. The solvent is more "limiting" in this sense[241] than formic or acetic acids which impart an S_N2 character to the solvolysis. Bentley *et al.*[242] have measured the relative nucleophilicities for various media based on the solvolysis of methyl tosylate. They find that HOTFA has a nucleophilicity comparable to that of fluorosulfuric acid, HSO_3F, but much lower than that of formic and acetic acids. The considerable literature on trifluoroacetolysis deserves a review in itself and only a few "lead-in" references will be discussed here. Peterson[243] showed that solvolysis of 2-hexyltosylate in HOTFA, containing 25% NaOTFA, proceeded four times faster than the same reaction in formic acid. The reaction in HOTFA yielded 20% 3-hexyltrifluoroacetate while in formic acid, hitherto considered to be the best solvent for rearrangements, only 6% of the comparable product was obtained. These rearranged products arise from hydride transfer facilitated by the S_N1 mechanism. Peterson also found that solvolysis of 1-hexene and *cis*- and *trans*-2-hexene took place much more rapidly in HOTFA–NaOTFA than in formic acid–sodium formate. Moreover, hydrocarbons were much less soluble in the latter solvent. Lee and Chwang[244] found that trifluoroacetolysis of 1-propyltosylate led to 1- and 2-propyltrifluoroacetates in 100% HOTFA but that the rearranged product was produced in lower yield if NaOTFA was added. Apparently NaOTFA increased the contribution of the S_N2 mechanism to the reaction.

Nordlander and Kelly found that in the trifluoroacetolysis of 1-phenyl-2-propyl-*p*-toluenesulfonate[245] there was complete retention of configuration, indicating that anchimeric assistance by the phenyl group takes place. Such anchimeric assistance is enhanced in a weakly nucleophilic solvent such as HOTFA. Nordlander and Deadman studied the trifluoroacetolysis of 2-phenylethyl-*p*-toluenesulfonate[246] and found the reaction to be rapid, indicating anchimeric assistance. Deuterium labeling permitted these authors to show that equal amounts of the two possible trifluoroacetates were found. These reactions occur via the ethylene–phenonium cation intermediate:

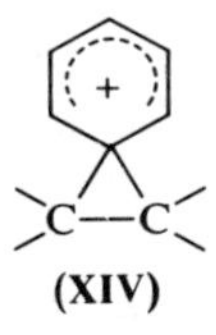

(XIV)

Anchimeric effects involving acetylene groups have been shown to be greatest in HOTFA, compared to acetic and formic acids as the reaction medium.[247] Winstein and co-workers have shown that the anchimerically assisted mechanism, k_Δ, for phenylpropyltosylates[248] and simple alkyl tosylates[249] predominates over the unassisted route, k_s, in HOTFA. For 1-phenyl-2-propyltosylate the ratio of k_Δ to k_s is $36{,}800 \times 10^{-6}$ to 6.6×10^{-6}/sec. Halogen participation accompanied by 1,2, 1,4, 1,5, and 1,6 shifts, has been found to occur in the trifluoroacetolysis of the halogen derivatives of primary alkyl *p*-nitrobenzenesulfonates, RONs.[250]

$$\text{(Cl-substituted alkyl)-ONs} \xrightarrow[\text{HOFTA}]{-\text{ONs}} \left[\text{cyclic chloronium ion}\right] \xrightarrow[\text{HOFTA}]{+\text{OTFA}} \text{TFAO-alkyl-Cl} \qquad (53)$$

Such reactions may only be studied in weakly nucleophilic solvents.

Solvolysis of olefins has also been studied, using HOTFA as solvent. Peterson[243] found that 1-hexene reacted to give 76% 2-hexyltrifluoroacetate and 24% 3-hexyltrifluoroacetate. Mason and Norman[251] also found evidence for hydride transfer with an S_N1 mechanism in the reactions of olefins they studied. They found, however, that where ring formation was stereochemically favorable, tetralins were formed.

$$C_6H_5(CH_2)_3CH{=}CH_2 \xrightarrow{\text{HOFTA}} \underset{74\%}{\text{1-methyltetralin } (CH_3)} + C_6H_5(CH_2)_3\underset{\text{OTFA}}{\underset{|}{C}}HCH_3 \qquad (54)$$

Propene reacts within one minute with *p*-bromobenzenesulfonic acid (HOBs) in HOTFA to give isopropyl-*p*-bromobenzenesulfonate while the half-life of isopropyl-*p*-bromobenzenesulfonate in HOTFA is 189 min.[252] The initial "hidden return" reaction takes place via the tight ion-pair, $C_3H_7{}^+OBs^-$. Isopropyl alcohol is, on the other hand, immediately converted to the isopropyltrifluoroacetate via an H_2O-separated ion-pair, $C_3H_7{}^+\text{–}H_2O\text{–}OBs^-$, which reacts rapidly with HOTFA.

Halomethyl ethers undergo 1,2 and 1,4 halogen shifts with dimethyl ether as leaving group upon reaction with $(CH_3)_3O^+BF_4^-$ in HOTFA[253]:

$$CH_3CHX(CH_2)_nCH_2OCH_3 + (CH_3)_3O^+BF_4^- \rightarrow CH_3CHX(CH_2)_nCH_2O(CH_3)_2{}^+BF_4^- + (CH_3)_2O + BF_4^- \qquad (55)$$

$$CH_3\text{–}CH(X)\text{–}(CH_2)_n\text{–}CH_2\text{–}\overset{+}{O}(CH_3)_2 \longrightarrow CH_3\text{–}\overset{+}{CH}\text{–}(CH_2)_n\text{–}CH_2 \text{ (bridged by X)} + (CH_3)_2O \qquad (56)$$

$$CH_3\text{–}\overset{+}{CH}\text{–}(CH_2)_n\text{–}CH_2 \text{ (bridged by X)} \xrightarrow{\text{HOFTA}} CH_3\underset{\text{OTFA}}{\underset{|}{C}}H(CH_2)_nCH_2X \qquad (57)$$

$n = 0, 2;\quad X = Cl, Br, I$

The reaction is possible because of the stability of the oxonium tetrafluoroborate in HOTFA.

Catt and Matier[254] have developed a new synthesis for primary sulfamides using the cleavage of *N-tert*-butyl amides in HOTFA:

$$t\text{-BuNHSO}_2\text{Cl} \xrightarrow{R_3N} [t\text{-BuN}{=}\text{SO}_2] \xrightarrow{RNH_2} t\text{-BuNHSO}_2\text{NHR} \tag{58}$$

$$t\text{-BuNHSO}_2\text{NHR} \xrightarrow[\text{HOTFA}]{} t\text{-BuOTFA} + \text{RNHSO}_2\text{NH}_2 \tag{59}$$

Trimethylsilylation of amino acids, amino alcohols, and nucleic acid bases with *N*-methyl-*N*-trimethylsilyl trifluoroacetamide in HOTFA has been used to produce compounds suitable for chromatographic purposes.[4]

D. Other Reactions in HOTFA

1. The Trifluoroacetylium Cation, CF_3CO^+

While the acetylium cation, CH_3CO^+, is well characterized, little work has been done on the trifluoroacetylium cation, CF_3CO^+. Lindner and Kranz[255,256] prepared the solid adducts $CF_3COF\text{–}MF_5$ (M = P, As, Sb) by the reaction:

$$\underset{(X = \text{Cl, Br})}{CF_3COX} + \underset{(M = \text{P, As, Sb})}{AgMF_6} \xrightarrow{\text{P, As, } -50°\text{; Sb, } -25°\text{C}} AgX\downarrow + CF_3COF\text{–}MF_5 \tag{60}$$

The reaction was not quantitative with $AgPF_6$[256] and all products decomposed to CF_3COF and the pentafluoride at 0°C. The electrical conductivity of $CF_3COF\text{–}SbF_5$ in liquid SO_2 at −25°C showed electrolyte behavior with $\Lambda = 83.6$ ohm^{-1} cm^2 mol^{-1} at 0.0075 *M*.[255] The infrared spectrum at low temperatures shows a characteristic $\nu_{C\equiv O}$ at 2371 cm^{-1} for the antimony compound, which may be compared with $\nu_{C\equiv O}$ of CH_3CO^+ at 2294 cm^{-1}. The infrared spectrum changes above −57°C and is more consistent with that of a donor adduct structure which the authors formulate as

```
       O···SbF5
      //
  CF3C
      \
       F
```

(XV)

2. Reactions of Trifluoroacetic Anhydride

Trifluoroacetic anhydride reacts with many acids to give mixed anhydrides. Randles *et al.*[133] have studied the complete composition diagram for $(CH_3CO)_2O$, $(TFA)_2O$, and H_2O. They showed that the mixed anhydride, $CH_3CO(OTFA)$, is formed and they attribute the increased conductivity, resulting from the removal of H_2O from the $HOOCCH_3\text{–}HOTFA$ mixture, to ionization to acetylium and trifluoroacetate ions. Cyanic acid forms a mixed anhydride as well[257]:

$$HNCO + (TFA)_2O \rightarrow \underset{(bp\ 35°C)}{TFANCO} + HOTFA \tag{61}$$

Trifluoroacetylnitrite is formed in the reaction of nitrosyl chloride with the anhydride[258]:

$$NOCl + (TFA)_2O \rightarrow TFAONO + TFACl \tag{62}$$

Thermal decomposition of the mixed anhydride at 200°C yields trifluoronitrosomethane, CF_3NO, in 85% yield. Nitric acid reacts with $(TFA)_2O$ at 100°C to give HOTFA, CO_2, and a 7% yield of nitrotrifluoromethane, CF_3NO_2.[259] Boric acid reacts with $(TFA)_2O$ to give an unstable boron tristrifluoroacetate which decomposes readily to $B_2O(OTFA)_4$[45]:

$$B(OH)_3 + 3(TFA)_2O \rightarrow B(OTFA)_3 + 3HOTFA \tag{63}$$

Harriss and Milne[145] have established the stoichiometry of the reactions of $(TFA)_2O$ with several strong acids dissolved in HOTFA using ^{19}F-NMR techniques. They showed that $HClO_4$, HSO_3F, and HNO_3 all react to give mixed anhydrides:

$$HClO_4 + (TFA)_2O \xrightarrow[HOTFA]{} TFAClO_4 + HOTFA \tag{64}$$

$$HSO_3F + (TFA)_2O \xrightarrow[HOTFA]{} TFASO_3F + HOTFA \tag{65}$$

$$HNO_3 + (TFA)_2O \xrightarrow[HOTFA]{} TFANO_3 + HOTFA \tag{66}$$

While separate NMR shifts are observed for $TFAClO_4$ and $TFASO_3F$, $TFANO_3$ undergoes exchange with the solvent. Tetrafluoroboric acid reacts with $(TFA)_2O$, but the trifluoroacetylium tetrafluoroborate formed is apparently unstable at room temperature and dissociates to CF_3COF and BF_3:

$$HBF_4 + (TFA)_2O \xrightarrow[HOTFA]{} [TFABF_4] + HOTFA \tag{67}$$

$$[TFABF_4] \xrightarrow[HOTFA]{} TFAF + BF_3 \tag{68}$$

Lindner and Kranz found similar behavior for trifluoroacetylium hexafluorophosphate, -arsenate, and -antimonate.[255,256]

3. Reactions of Proteins

Trifluoroacetic acid is a good solvent for proteins.[6] The conclusions of an earlier report,[6] indicating that the proteins are recovered unchanged from the solvent, were later shown by the authors to be erroneous.[154] The process whereby helix–coil transformations occur in HOTFA solutions has been the subject of some dispute.[147] The ^1H-NMR spectrum of polypeptides dissolved in acidic media shows two resonances which have been assigned to the protons on the α-carbon in helix and coil environments.[260] Tam and Klotz[261] have given convincing evidence, however, that the two signals in HOTFA solutions are due to protonated and unprotonated polypeptides. They showed that both poly-DL- and poly-L-alanine dissolved in $CDCl_3$ containing 1% HOTFA give the two ^1H signals for the α-CH group. The two resonances cannot arise from helix and coil forms, since there is no helix form possible in poly-DL-alanine. Moreover, as acid strength is increased the low field signal at $\tau \sim 4.26$ increases in strength at the expense of the high field signal at $\tau \sim 3.9$, and shifts to low field. It was concluded that the two signals arise from protonated and unprotonated polyalanine. Further evidence was provided by the observation of only the low field peak in 100% HOTFA.

Note Added in Proof

Since the preparation of this manuscript the results of a study of cryoscopy in 100% HOTFA have appeared.[262] This work casts considerable doubt on the existence of triple ions in this solvent, at least at concentrations below about 0.2 molal for 1 : 1 electrolytes. The authors calculated theoretical osmotic coefficients, using the expressions derived by Fowler and Guggenheim[263] as described by Robinson and Stokes[148] (p. 233) for two cases: (1) assuming ion triples and pairs in equilibrium with simple ions and using ion-pair[25] and triple ion[81] dissociation constants as determined from electrical conductivity studies, and (2) assuming only ion-pairs and simple ions to be present and using only the ion-pair dissociation constants,[25] which are considered to be of great accuracy, since they are determined by extrapolation of results measured at low concentrations. The experimental osmotic coefficients for the electrolytes, NaOTFA, KOTFA, and CsOTFA show good agreement with those calculated for case (2) but differ significantly from those calculated for case (1), especially for the CsOTFA results. Thus, the explanation of the minimum in the equivalent conductance, Λ, vs. concentration plots in terms of triple ions, as discussed in Section IV of this chapter, is doubtful and the minimum must be accounted for in another way. Perhaps the use of a more precise expression than the extended Debye–Hückel equation for the ionic activity coefficient would account for these electrical conductivity results.

A more accurate value of the cryoscopic constant than that given in Table I was also determined,[262] $k_f = 6.25$ deg kg mole^{-1}. In addition to this cryoscopic study on anhydrous HOTFA, a report on some cryoscopy in eutectic HOTFA (10.6 mole % H_2O) has appeared.[264]

References

1. F. Swarts, *Bull. Cl. Sci., Acad. R. Belg.* **8**, 331 (1922).
2. G. S. Fujioka and G. H. Cady, *J. Am. Chem. Soc.* **79**, 2451 (1957).
3. E. L. Mackor, P. J. Smit, and J. H. van der Waals, *Trans. Faraday Soc.* **53**, 1309 (1957).
4. M. Donike, *J. Chromatogr.* **85**, 1 (1973).
5. "3M Brand Perfluorocarboxylic Acid." 3M Co., St. Paul, Minnesota, 1960.
6. J. J. Katz, *Nature (London)* **174**, 509 (1954).
7. S. S. Sweet, M. H. van Horn, and P. T. Newsome, U.S. Patent 2,710,848 (1955); *C. A.* **49**, 12850g (1955).

8. H. C. Brown and R. A. Wirkkala, *J. Am. Chem. Soc.* **88**, 1447 (1966).
9. J. E. Nordlander, R. R. Gruetzmacher, W. J. Kelly, and S. P. Sindal, *J. Am. Chem. Soc.* **96**, 181 (1974).
10. E. A. Kauck and A. R. Diesslin, *Ind. Eng. Chem.* **43**, 2332 (1951).
11. H. M. Scholberg and H. G. Brice, U.S. Patent 2,717,871 (1955); *C. A.* **49**, 15572h (1955).
12. J. Burdon and J. C. Tatlow, *Adv. Fluorine Chem.* **1**, 129 (1960).
13. G. Fernschild, H. Paricksoh, and J. Massonne, French Patent 2,166,358 (1973); *C. A.* **80**, 70339t (1973).
14. H. G. Bryce, *Fluorine Chem.* **5**, 295 (1964).
15. K. G. Schoenroth, 3M Company, personal communication (1975).
16. P. W. Morgan, U.S. Patent 2,629,716 (1953); *C. A.* **48**, 715h (1954).
17. P. W. Morgan, *Ind. Eng. Chem.* **43**, 2575 (1951).
18. T. Yamawaki, M. Usami, T. Suzuki, and T. Uematsu, Japanese Patent 73/12,194 (1973); *C. A.* **80**, 71871x (1973).
19. B. C. Saunders, *Adv. Fluorine Chem.* **2**, 197 (1961).
20. H. C. Hodge, F. A. Smith, and P. S. Chen, *Fluorine Chem.* **3**, 1 (1963).
21. A. I. Popov, *in* "The Chemistry of Nonaqueous Solvents" (J. J. Lagowski, ed.), Vol. 3, p. 366. Academic Press, New York, 1970.
22. N. Ishikawa, *Yuki Gosei Kagaku Kyokai Shi* **29**, 83 (1971); *C. A.* **74**, 124670r (1952).
23. F. Swarts, *Bull. Cl. Sci., Acad. R. Belg.* **8**, 343 (1922).
24. B. E. Conway and A. K. Vijh, *J. Phys. Chem.* **71**, 3637 (1967).
25. M. G. Harriss and J. B. Milne, *Can. J. Chem.* **49**, 1888 (1971).
26. W. Dannhauser and R. H. Cole, *J. Am. Chem. Soc.* **74**, 6105 (1952).
27. B. E. Conway and M. Dzieciuch, *Can. J. Chem.* **41**, 38 (1963).
28. J. H. Bradbury and M. D. Fenn, *J. Mol. Biol.* **36**, 231 (1968).
29. C. Eaborn, P. M. Jackson, and R. Taylor, *J. Chem. Soc. B* p. 613 (1966).
30. A. L. Henne and C. J. Fox, *J. Am. Chem. Soc.* **73**, 2323 (1951).
31. J. H. Simons and K. E. Lorentzen, *J. Am. Chem. Soc.* **74**, 4746 (1952).
32. F. E. Harris and C. T. O'Konski, *J. Am. Chem. Soc.* **76**, 4317 (1954).
33. A. Kreglewski, *Bull. Acad. Pol. Sci., Ser. Sci. Chim.* **10**, 629 (1962).
34. M. G. Harriss, Ph.D. Dissertation, University of Ottawa (1971).
35. M. D. Taylor and M. B. Templeman, *J. Am. Chem. Soc.* **78**, 2950 (1956).
36. Y. Y. Fialkov and V. S. Zhikharev, *J. Gen. Chem. USSR (Engl. Transl.)* **33**, 3397 (1963).
37. J. J. Jasper and H. Wedlick, *J. Chem. Eng. Data* **9**, 446 (1964).
38. T. Sano, N. Tatsumato, Y. Mende, and T. Yasunaga, *Bull. Chem. Soc. Jpn.* **45**, 2673 (1972).
39. G. W. Marks, *J. Acoust. Soc. Am.* **27**, 680 (1955).
40. F. Swarts, *J. Chim. Phys.* **28**, 622 (1931).
41. J. H. Gibbs and C. P. Smyth, *J. Am. Chem. Soc.* **73**, 5115 (1951).
42. R. C. Weast, ed., "The Handbook of Chemistry and Physics," 50th ed., p. E230. Chem. Rubber Publ. Co., Cleveland, Ohio, 1969.
43. J. M. Tedder, *J. Chem. Soc.* p. 2646 (1954).
44. Yu. Ya. Borovikov and Yu. Ya. Fialkov, *Sov. Electrochem. (Engl. Transl.)* **1**, 989 (1965).
45. M. G. Harriss and J. B. Milne, *Can. J. Chem.* **49**, 3612 (1971).
46. V. P. Kolesov and I. D. Zenkov, *Russ. J. Phys. Chem. (Engl. Transl.)* **40**, 743 (1966).
47. A. C. Zawisza, *Bull. Acad. Pol. Sci., Ser. Sci. Chim.* **15**, 307 (1967).
48. G. Dewith, *J. Mol. Struct.* **18**, 241 (1973).
49. J. H. Simons and K. E. Lorentzen, *J. Am. Chem. Soc.* **72**, 1426 (1950).
50. A. A. Maryott and E. R. Smith, *Natl. Bur. Stand. (U.S.), Circ.* **514** (1951).
51. C. V. Berney, *J. Am. Chem. Soc.* **95**, 708 (1973).
52. D. Clague and A. Novak, *J. Chim. Phys.* **67**, 1126 (1970).

53. H. P. Marshall, F. G. Borgardt, P. Noble, Jr., and N. S. Bhacca, *J. Phys. Chem.* **75**, 499 (1971).
54. T. S. S. R. Murty and K. S. Pitzer, *J. Phys. Chem.* **73**, 1426 (1969).
55. A. K. Covington, M. J. Tait, and Lord Wynne-Jones, *Discuss. Faraday Soc.* **39**, 172 (1965).
56. O. Redlich, *Chem. Rev.* **39**, 333 (1946).
57. O. Redlich and G. C. Hood, *Discuss. Faraday Soc.* **24**, 87 (1957).
58. R. M. Fuoss, L. Onsager, and J. F. Skinner, *J. Phys. Chem.* **69**, 2581 (1965).
59. G. C. Hood, O. Redlich, and C. A. Reilly, *J. Chem. Phys.* **23**, 2229 (1955).
60. E. Högfeldt, *J. Inorg. Nucl. Chem.* **17**, 302 (1961).
61. J. E. B. Randles and J. M. Tedder, *J. Chem. Soc.* p. 1218 (1955).
62. E. Grunwald and J. F. Haley, *J. Phys. Chem.* **72**, 1944 (1968).
63. A. K. Covington, J. G. Freeman, and T. H. Lilley, *J. Phys. Chem.* **74**, 3773 (1970).
64. A. K. Covington, *Trans. Faraday Soc.* **39**, 176 (1965).
65. M. M. Kreevoy and C. A. Mead, *Trans. Faraday Soc.* **39**, 171 (1965).
66. N. Fuson, M. L. Josien, E. A. Jones, and J. R. Lawson, *J. Chem. Phys.* **20**, 1627 (1952).
67. P. J. Miller, R. A. Butler, and E. R. Lippincott, *J. Chem. Phys.* **57**, 5451 (1972).
68. A. D. Pethybridge and J. E. Prue, *Trans. Faraday Soc.* **63**, 2019 (1967).
69. J. D. S. Poulden, *Spectrochim. Acta* **6**, 129 (1954).
70. K. Hiraoka, R. Yamdagni, and P. Kebarle, *J. Am. Chem. Soc.* **95**, 6833 (1973).
71. M. Rumeau, *Ann. Chim.* (*Paris*) [14] **8**, 131 (1973).
72. B. M. Rode, A. Englebrecht, and J. Schantl, *Z. Phys. Chem.* (*Leipzig*) **253**, 17 (1973).
73. H. J. Emeléus, R. Haszeldine, and R. Paul, *J. Chem. Soc.* p. 881 (1954).
74. G. Van Dyke Tiers, *J. Am. Chem. Soc.* **78**, 4165 (1956).
75. M. Kilpatrick and H. H. Hyman, *J. Am. Chem. Soc.* **80**, 77 (1958).
76. H. H. Hyman and R. A. Garber, *J. Am. Chem. Soc.* **81**, 1847 (1959).
77. D. P. N. Satchell, *J. Chem. Soc.* p. 3904 (1958).
78. G. Dallinga and G. ter Maten, *Recl. Trav. Chim. Pays-Bas* **79**, 737 (1960).
79. S. Bruckenstein, *J. Am. Chem. Soc.* **82**, 307 (1960).
80. J. Bessière and G. Petit, *Anal. Chim. Acta* **54**, 360 (1971).
81. M. G. Harriss and J. B. Milne, *Can. J. Chem.* **50**, 3789 (1972).
82. H. Strehlow, *in* "The Chemistry of Nonaqueous Solvents" (J. J. Lagowski, ed.), Vol. 1, p. 157. Academic Press, New York, 1966.
83. J. Karle and L. O. Brockway, *J. Am. Chem. Soc.* **66**, 574 (1944).
84. R. E. Lundin, F. E. Harris, and L. K. Nash, *J. Am. Chem. Soc.* **74**, 4654 (1952).
85. R. E. Kagarise, *J. Chem. Phys.* **27**, 519 (1957).
86. S. D. Christian, H. E. Affsprung, and Chii Ling, *J. Chem. Soc.* p. 2378 (1965).
87. C. Ling, S. D. Christian, H. E. Affsprung, and R. W. Gray, *J. Chem. Soc. A* p. 293 (1966).
88. S. D. Christian and T. L. Stevens, *J. Phys. Chem.* **76**, 2039 (1972).
89. H. Dunken and G. Marx, *Abh. Dtsch. Akad. Wiss. Berlin, Kl. Math., Phys. Tech.* No. 6, p. 101 (1964).
90. R. K. Thomas, *Proc. R. Soc. London, Ser. A* **331**, 249 (1972).
91. R. L. Redington and K. C. Lin, *Spectrochim. Acta, Part A* **27**, 2445 (1971).
92. H. L. Ritter and J. H. Simons, *J. Am. Chem. Soc.* **67**, 757 (1945).
93. J. R. Barcelo and C. Otero, *Spectrochim. Acta* **18**, 1231 (1962).
94. M. L. Josien, N. Fuson, J. R. Lawson, and E. A. Jones, *C. R. Hebd. Seances Acad. Sci.*, Ser. B, **234**, 1163 (1952).
95. R. L. Redington and K. C. Lin, *J. Chem. Phys.* **54**, 4111 (1971).
96. P. Excoffon and Y. Marechal, *Spectrochim. Acta, Part A* **28**, 269 (1972).
97. N. Fuson and M. L. Josien, *J. Opt. Soc. Am.* **43**, 1102 (1953).
98. S. Bratoz, D. Hadzi, and N. Sheppard, *Spectrochim. Acta* **8**, 249 (1956).

99. D. Clague and A. Novak, *J. Mol. Struct.* **5**, 149 (1970).
100. L. W. Reeves, *Can. J. Chem.* **39**, 1711 (1961).
101. M. Kirszenbaum, J. Corset, and M. L. Josien, *J. Phys. Chem.* **75**, 1327 (1971).
102. I. S. Perelygin and A. M. Afanas'eva, *J. Struct. Chem.* (*Engl. Transl.*) **14**, 971 (1973).
103. T. S. S. R. Murty, *J. Phys. Chem.* **75**, 1330 (1971).
104. M. Kirszenbaum, J. Corset, and M. L. Josien, *C. R. Hebd. Seances Acad. Sci., Ser. B* **271**, 630 (1970).
105. J. W. O. Tam and W. O. Klotz, *Spectrochim. Acta, Part A* **29**, 633 (1973).
106. J. Bellanto and J. R. Barcelo, *Spectrochim. Acta* **16**, 1333 (1960).
107. G. Statz and E. Lippert, *Ber. Bunsenges. Phys. Chem.* **71**, 673 (1967).
108. P. M. Borodin, K. Dzhumabaev, M. K. Nikitin, A. S. Simonov, and G. Sherfeze, *Yad. Magn. Rezon.* **3**, 77 (1969); *C. A.* **72**, 30799w (1970).
109. P. M. Borodin and K. Dzhumabaev, *Yad. Magn. Reson.* **4**, 99 (1971); *C. A.* **76**, 92767a (1972).
110. F. Thyrion and D. Decroocq, *C. R. Hebd. Seances Acad. Sci.*, Ser. B, **260**, 2797 (1965).
111. W. S. Higazy and A. A. Taha, *J. Phys. Chem.* **74**, 1982 (1970).
112. A. A. Taha and S. D. Christian, *J. Phys. Chem.* **73**, 3430 (1969).
113. R. E. Jones and D. H. Templeton, *Acta Crystallogr.* **11**, 484 (1958).
114. F. Holtzberg, B. Post, and I. Fankuchen, *Acta Crystallogr.* **6**, 128 (1953).
115. M. Goldman, *J. Phys. Chem. Solids* **7**, 165 (1958).
116. B. A. Dunnel, L. W. Reeves, and K. O. Stromme, *Trans. Faraday Soc.* **57**, 381 (1961).
117. M. F. Collins and B. C. Haywood, *J. Chem. Phys.* **52**, 5740 (1970).
118. R. M. Fuoss and F. Accascina, "Electrolytic Conductance." Wiley (Interscience), New York, 1959.
119. R. M. Fuoss, *J. Am. Chem. Soc.* **57**, 488 (1935).
120. J. Bessière, *Bull. Soc. Chim. Fr.* p. 3353 (1969).
121. T. Shedlovsky, *J. Franklin Inst.* **225**, 739 (1938).
122. R. M. Fuoss and T. Shedlovsky, *J. Am. Chem. Soc.* **71**, 1496 (1949).
123. C. R. Witschonke and C. A. Kraus, *J. Am. Chem. Soc.* **69**, 2472 (1947).
124. D. F. Evans, J. A. Nadas, and M. A. Matesich, *J. Phys. Chem.* **75**, 1708 (1971).
125. H. C. Brookes, M. C. B. Hotz, and A. H. Spong, *J. Chem. Soc.* p. 2410 (1971).
126. G. R. Lester, T. A. Gover, and P. G. Sears, *J. Phys. Chem.* **60**, 1076 (1956).
127. E. Price, *in* "The Chemistry of Nonaqueous Solvents" (J. J. Lagowski, ed.), Vol. 1, p. 82. Academic Press, New York, 1966.
128. J. Barr, R. J. Gillespie, and E. A. Robinson, *Can. J. Chem.* **39**, 1266 (1961).
129. R. J. Gillespie and K. C. Malhotra, *J. Chem. Soc. A* p. 1994 (1967).
130. Yu. Ya. Fialkov and S. N. Kholodnikova, *J. Gen. Chem. USSR* (*Engl. Transl.*) **38**, 663 (1968).
131. G. Kresze and V. Schmidt, *Chem. Ber.* **90**, 1687 (1957).
132. M. A. Stake and I. M. Klotz, *Biochemistry* **5**, 1726 (1966).
133. J. E. B. Randles, J. C. Tatlow, and J. M. Tedder, *J. Chem. Soc.* p. 436 (1954).
134. Yu. Ya. Fialkov and V. S. Zhikharev, *J. Gen. Chem. USSR* (*Engl. Transl.*) **33**, 3728 (1963).
135. R. J. Gillespie, E. D. Hughes, and C. K. Ingold, *J. Chem. Soc.* p. 2552 (1950).
136. J. Long and B. Munson, *J. Am. Chem. Soc.* **95**, 2427 (1973).
137. Yu. Ya. Fialkov and Yu. Ya. Borovikov, *Vestn. Kiev. Politekhn. Inst., Ser. Khim. Mashinostr. Teknol.* **6**, 66 (1965); *C. A.* **65**, 1464e (1966).
138. Yu. Ya. Fialkov and V. S. Zhikharev, *J. Gen. Chem. USSR* (*Engl. Transl.*) **33**, 3402 (1963).
139. L. Quarterman, H. H. Hyman, and J. J. Katz, *J. Phys. Chem.* **65**, 90 (1961).
140. Yu. Ya. Fialkov and Yu. Ya. Borovikov, *Tr. Kiev. Politekh. Inst., Ser. Khim.* **1**, 66 (1965).
141. V. I. Ligus, Yu. I. Malov, V. Ya. Rosolovskii, and Yu. Ya. Fialkov, *J. Gen. Chem. USSR* (*Engl. Transl.*) **39**, 2569 (1969).

142. Yu. Ya. Fialkov and V. I. Ligus, *J. Gen. Chem. USSR* (*Engl. Transl.*) **42**, 256 (1972).
143. Yu. Ya. Fialkov, G. I. Yanchuk, and H. D. Krysenko, *Russ. J. Inorg. Chem.* (*Engl. Transl.*) **18**, 1080 (1973).
144. C. A. Kraus and W. C. Bray, *J. Am. Chem. Soc.* **35**, 1315 (1913).
145. M. G. Harriss and J. B. Milne, *Can. J. Chem.* **49**, 2937 (1971).
146. F. E. Harris and B. J. Alder, *J. Chem. Phys.* **21**, 1306 (1953).
147. I. M. Klotz, S. F. Russo, S. Hanlon, and M. A. Stake, *J. Am. Chem. Soc.* **86**, 4774 (1964).
148. R. A. Robinson and R. H. Stokes, "Electrolyte Solutions." Butterworth, London, 1965.
149. G. Petit and J. Bessière, *J. Electroanal. Chem.* **25**, 317 (1970).
150. G. Petit and J. Bessière, *J. Electroanal. Chem.* **31**, 393 (1971).
151. G. Petit and J. Bessière, *J. Electroanal. Chem.* **34**, 489 (1972).
152. R. Hara and G. H. Cady, *J. Am. Chem. Soc.* **76**, 4285 (1954).
153. J. M. Cleveland, *J. Inorg. Nucl. Chem.* **26**, 461 (1964).
154. A. L. Koch, W. A. Lamont, and J. J. Katz, *Arch. Biochem. Biophys.* **63**, 106 (1956).
155. D. Chapman, V. B. Kamat, J. de Grier, and S. A. Penkett, *J. Mol. Biol.* **31**, 101 (1968).
156. F. A. Bovey and G. V. D. Tiers, *J. Am. Chem. Soc.* **81**, 2870 (1959).
157. F. Weygand and R. Geiger, *Chem. Ber.* **89**, 647 (1956).
158. M. Markowitz, W. N. Hawley, D. A. Boryca, and R. F. Harris, *J. Chem. Eng. Data* **6**, 325 (1961).
159. R. G. Jones and J. R. Dyer, *J. Am. Chem. Soc.* **95**, 2465 (1973).
160. J. de Villepin, A. Lautie, and M. L. Josien, *Ann. Chim.* (*Paris*) [14] **1**, 365 (1966).
161. H. H. Cady and G. H. Cady, *J. Am. Chem. Soc.* **76**, 915 (1954).
162. C. C. Costain and G. P. Srivastava, *J. Chem. Phys.* **41**, 1620 (1964).
163. C. C. Costain and G. P. Srivastava, *J. Chem. Phys.* **35**, 1903 (1961).
164. O. I. Kachurin and L. M. Kapkan, *Ukr. Khim Zh.* **38**, 649 (1972); *C. A.* **77**, 131399z (1972).
165. W. Klemperer and G. C. Pimentel, *J. Chem. Phys.* **22**, 1399 (1954).
166. L. Golic and J. C. Speakman, *J. Chem. Soc.* p. 2530 (1965).
167. D. J. Millen and E. Vaal, *J. Chem. Soc.* p. 2913 (1956).
168. G. S. Denisov, A. L. Smolyanskii, A. A. Trusov, and M. I. Shiekh, *Zade*, *Izv. Vyssh. Uchebn., Zaved., Fiz.* **17**, 142 (1974); *C. A.* **81**, 62959a (1974).
169. N. G. Zarakhani and N. P. Vorob'eva, *Russ. J. Phys. Chem.* (*Engl. Transl.*) **45**, 369 (1971).
170. N. G. Zarakhani and N. P. Vorob'eva, *Russ. J. Phys. Chem.* (*Engl. Transl.*) **46**, 1392 (1972).
171. A. P. Genich, L. T. Eremenko, and L. A. Nikitina, *Bull. Acad. Sci. USSR, Div. Chem. Sci.* p. 733 (1967).
172. I. M. Kolthoff and S. Bruckenstein, *J. Am. Chem. Soc.* **78**, 1 (1956).
173. O. I. Kachurin, *Ukr. Khim. Zh.* **38**, 23 (1972); *C. A.* **76**, 117942u (1972).
174. J. Bessière, *Bull. Soc. Chim. Fr.* p. 3356 (1969).
175. G. A. Forcier and J. W. Olver, *Electrochim. Acta* **15**, 1609 (1970).
176. M. Hauptschein and A. V. Grosse, *J. Am. Chem. Soc.* **73**, 5139 (1951).
177. A. L. MacDonald, J. C. Speakman, and D. Hadzi, *J. Chem. Soc., Perkin Trans. 2* p. 825 (1972).
178. D. Hadzi, B. Orel, and A. Novak, *Spectrochim. Acta, Part A* **29**, 1745 (1973).
179. M. N. Tsarevskaya, *Sov. Prog. Chem.* (*Engl. Transl.*) **34**, 58 (1968).
180. A. Loewenstein and S. Meiboom, *J. Am. Chem. Soc.* **81**, 62 (1959).
181. T. Birchall and R. J. Gillespie, *Can. J. Chem.* **41**, 148 (1963).
182. P. Combelas and C. Garrigou-LaGrange, *Spectrochim. Acta, Part A* **30**, 550 (1974).
183. S. Hanlon, S. F. Russo, and I. M. Klotz, *J. Am. Chem. Soc.* **85**, 2024 (1963).
184. A. V. Purkina, A. I. Kol'tsov, and B. Z. Volchek, *J. Appl. Spectrosc.* **15**, 1051 (1971).
185. J. Bessière, *Anal. Chim. Acta* **52**, 55 (1970).
186. R. J. Gillespie, J. B. Milne, and J. B. Senior, *Inorg. Chem.* **5**, 1233 (1966).

187. G. Kresze and V. Schmidt, *Z. Anal. Chem.* **181**, 527 (1961).
188. J. E. de Vries, S. Schiff, and E. S. C. Gantz, *Anal. Chem.* **27**, 1814 (1955).
189. B. V. Lokshin, V. I. Zdanovich, N. K. Baranetskaya, V. N. Setkina, and D. N. Kursanov, *J. Organomet. Chem.* **37**, 331 (1972).
190. B. V. Lokshin, A. G. Ginzburg, V. N. Setkina, D. N. Kursanov, and I. B. Nemirovskaya, *J. Organomet. Chem.* **37**, 342 (1972).
191. D. N. Kursanov, V. N. Setkina, P. V. Petrovskii, V. I. Zdanovich, N. K. Baranetskaya, and I. D. Rubin, *J. Organomet. Chem.* **37**, 339 (1972).
191a. C. D. Garner and B. Hughes, *Adv. Inorg. Chem. Radiochem.* **17**, 1 (1975).
192. F. Swarts, *Bull. Soc. Chim. Belg.* **48**, 176 (1939).
193. K. W. Rillings and J. E. Roberts, *Thermochim. Acta* **10**, 285 (1974).
194. P. Sartori, J. Fazekas, and J. Schnackes, *J. Fluorine Chem.* **1**, 463 (1972).
195. R. O. C. Norman, C. B. Thomas, and P. J. Ward, *J. Chem. Soc., Perkin Trans. 1* p. 2914 (1973).
196. T. A. Stephenson, S. M. Morehouse, A. R. Powell, J. P. Heffer, and G. Wilkinson, *J. Chem. Soc.* p. 3632 (1965).
197. P. Sartori and M. Weidenbruch, *Angew. Chem., Int. Ed. Engl.* **4**, 1079 (1965).
198. W. Gerrard, M. F. Lappert, and R. Shafferman, *J. Chem. Soc.* p. 3648 (1958).
199. P. C. Wailes and H. Weigold, *J. Organomet. Chem.* **24**, 413 (1970).
200. M. J. Baillie, D. H. Brown, K. C. Moss, and D. W. A. Sharp, *J. Chem. Soc. A* p. 3110 (1968).
201. N. K. Hota and C. J. Willis, *Can. J. Chem.* **46**, 3921 (1968).
202. N. W. Alcock and T. C. Waddington, *J. Chem. Soc.* p. 4103 (1963).
203. R. N. Haszeldine and A. G. Sharpe, *J. Chem. Soc.* p. 993 (1952).
204. G. H. Crawford and J. H. Simons, *J. Am. Chem. Soc.* **77**, 2605 (1955).
205. M. Schmeisser, K. Dahmen, and P. Sartori, *Chem. Ber.* **100**, 1633 (1967).
206. P. V. Radeshwar, R. Dev, and G. H. Cady, *U.S. N.T.I.S., AD Rep.* **AD-736682** (1972).
207. P. V. Radeshwar, R. Dev, and G. H. Cady, *J. Inorg. Nucl. Chem.* **34**, 3913 (1972).
208. D. Naumann, H. Dolhaine, and W. Stopschinski, *Z. Anorg. Allg. Chem.* **394**, 133 (1972).
209. M. Schmeisser, P. Sartori, and D. Naumann, *Chem. Ber.* **103**, 312 (1970).
210. D. Naumann, M. Schmeisser, and R. Scheele, *J. Fluorine Chem.* **1**, 321 (1972).
211. M. Schmeisser, D. Naumann, and R. Scheele, *J. Fluorine Chem.* **1**, 369 (1972).
212. M. Schmeisser, P. Sartori, and K. Dahmen, *Chem. Ber.* **103**, 307 (1970).
213. V. V. Lyalin, V. V. Orda, L. A. Alekseeva, and L. M. Lagulop'skii, *Russ. J. Org. Chem. (Engl. Transl.)* **8**, 1027 (1972).
214. D. G. Lee and D. T. Johnson, *Can. J. Chem.* **43**, 1952 (1965).
215. J. I. Musher, *J. Am. Chem. Soc.* **90**, 7371 (1968).
216. M. Eisenberg and D. D. DesMarteau, *Inorg. Nucl. Chem. Lett.* **6**, 29 (1970).
217. F. O. Sladky, *Monatsh. Chem.* **101**, 1571 (1970).
218. R. E. Buckles and J. F. Mills, *J. Am. Chem. Soc.* **75**, 552 (1953).
219. L. J. Andrews and R. M. Keefer, *J. Am. Chem. Soc.* **79**, 1412 (1957).
220. R. E. Buckles and J. F. Mills, *J. Am. Chem. Soc.* **76**, 4845 (1954).
221. R. E. Buckles and J. F. Mills, *J. Am. Chem. Soc.* **76**, 6021 (1954).
222. P. Alcais, F. Rothenberg, and J. E. Dubois, *J. Chim. Phys.* **64**, 1818 (1967).
223. J. F. Dowdal, U.S. Patent 2,628,253 (1953); *C. A.* **48**, p. 1425i (1954).
224. W. Dmowski and R. Kolinski, *Roczniki. Chem.* **47**, 1211 (1973).
225. P. Sartori and M. Weidenbruch, *Chem. Ber.* **100**, 2049 (1967).
226. A. N. Sara and K. Taugbøl, *J. Inorg. Nucl. Chem.* **35**, 1827 (1973).
227. W. Rudorf and W. F. Sicke, *Chem. Ber.* **91**, 1348 (1958).
228. E. L. Muetterties, U.S. Patent 2,782,233 (1957); *C. A.* **51**, 9672i (1957).
229. H. U. Kibbel, *Z. Anorg. Allg. Chem.* **359**, 203 (1968).

230. M. Castagnola, B. Floris, G. Illuminate, and G. Ortaggi, *J. Organomet. Chem.* **60**, C17 (1973).
231. P. G. Blake and H. Pritchard, *J. Chem. Soc. B* p. 282 (1967).
232. A. T. Sadikova, N. S. Nikolaev, and T. A. Rasskazova, *Russ. J. Inorg. Chem.* (*Engl. Transl.*) **15**, 1039 (1970).
233. R. O. C. Norman, C. B. Thomas, and J. S. Willson, *J. Chem. Soc., Perkin Trans. 1* p. 325 (1973).
234. R. O. C. Norman, C. B. Thomas, and J. S. Willson, *J. Chem. Soc. B* p. 518 (1971).
235. N. C. Deno and D. G. Pohl, *J. Am. Chem. Soc.* **96**, 6681 (1974).
236. L. J. Andrews and R. M. Keefer, *J. Am. Chem. Soc.* **79**, 5169 (1957).
237. H. C. Brown and R. A. Wirkkala, *J. Am. Chem. Soc.* **88**, 1453 (1966).
238. H. C. Brown and R. A. Wirkkala, *J. Am. Chem. Soc.* **88**, 1456 (1966).
239. A. Himoe and L. M. Stock, *J. Am. Chem. Soc.* **91**, 1452 (1969).
240. V. N. Setkina and D. N. Kursanov, *Izv. Akad. Nauk SSSR, Ser. Khim.* p. 378 (1961); *C. A.* **55**, 18571f (1961).
241. S. Winstein, E. Grunwald, and H. W. Jones, *J. Am. Chem. Soc.* **73**, 2700 (1951).
242. T. W. Bentley, F. L. Schadt, and P. v. R. Schleyer, *J. Am. Chem. Soc.* **94**, 992 (1972).
243. P. E. Peterson, *J. Am. Chem. Soc.* **82**, 5834 (1960).
244. C. C. Lee and W. K. Chwang, *Can. J. Chem.* **48**, 1025 (1970).
245. J. E. Nordlander and W. J. Kelly, *J. Am. Chem. Soc.* **91**, 996 (1969).
246. J. E. Nordlander and W. G. Deadman, *J. Am. Chem. Soc.* **90**, 1590 (1968).
247. P. E. Peterson and R. J. Kamat, *J. Am. Chem. Soc.* **91**, 4521 (1969).
248. A. F. Diaz and S. Winstein, *J. Am. Chem. Soc.* **91**, 4300 (1969).
249. I. L. Reich, A. Diaz, and S. Winstein, *J. Am. Chem. Soc.* **91**, 5635 (1969).
250. P. E. Peterson and J. F. Coffey, *J. Am. Chem. Soc.* **93**, 5208 (1971).
251. T. J. Mason and R. O. C. Norman, *J. Chem. Soc., Perkin Trans. 2* p. 1840 (1973).
252. V. J. Shiner and W. Dowd, *J. Am. Chem. Soc.* **91**, 6528 (1969).
253. P. E. Peterson and F. J. Slama, *J. Am. Chem. Soc.* **90**, 6516 (1968).
254. J. D. Catt and W. L. Matier, *J. Org. Chem.* **39**, 566 (1974).
255. E. Lindner and H. Kranz, *Z. Naturforsch., Teil B* **20**, 1305 (1965).
256. E. Lindner and H. Kranz, *Chem. Ber.* **99**, 3800 (1966).
257. W. C. Firth, *J. Org. Chem.* **33**, 441 (1968).
258. J. D. Park, R. W. Rosser, and J. R. Lacher, *J. Org. Chem.* **27**, 1462 (1962).
259. R. Boschan, *J. Org. Chem.* **25**, 1450 (1960).
260. E. M. Bradbury, C. Crane-Robinson, H. Goldman, and H. W. E. Rattle, *Nature* (*London*) **217**, 812 (1968).
261. J. W. O. Tam and I. M. Klotz, *J. Am. Chem. Soc.* **93**, 1313 (1971).
262. M. G. Harriss and J. B. Milne, *Can. J. Chem.* **54**, 3031 (1976).
263. R. H. Fowler and E. A. Guggenheim, "Statistical Thermodynamics," Chapter IX. Cambridge Univ. Press, London and New York, 1949.
264. M. Ardon and G. Yahav, *Inorg. Chem.* **15**, 12 (1976).

2

Halosulfuric Acids

SARASWATHI NATARAJAN AND ALBERT W. JACHE

Marquette University
Milwaukee, Wisconsin 53233

I. FLUOROSULFURIC ACID

A. Introduction

Fluorosulfuric acid is one of the strongest of the simple protonic acids. It was first prepared in 1892 by Thorpe and Kirman[1] by reacting SO_3 and anhydrous HF

$$SO_3 + HF \rightarrow HSO_3F$$

In many respects fluorosulfuric acid is a convenient and useful solvent for systematic study. The acidity of fluorosulfuric acid is enhanced considerably when SbF_5 and SO_3 are added. These still more acidic media, known as "superacids," are extremely useful in the study of protonation reactions and for the formation of carbonium ions and related species.

There have been several reviews on fluorosulfuric acid and its derivatives by Lange,[2] Cady,[3] Williamson,[4] Thompson,[5] Gillespie,[6] and very recently by Jache.[7] The purpose of this review is to emphasize the solvent properties of both fluorosulfuric and chlorosulfuric acids. Bromosulfuric acid is also known but its solvent properties have not been investigated so far. Reactions where halosulfuric acids are used either as catalysts or as reagents for halogenation, sulfonation, and polymerization are beyond the scope of this chapter.

B. Physical Properties

Fluorosulfuric acid is a colorless, fuming liquid. It is easily purified by distillation. When free from HF, fluorosulfuric acid does not attack glass even at its boiling point so it can be easily and conveniently handled in glass apparatus. Fluorosulfuric acid is proved to be a more useful solvent than sulfuric acid as can be seen from several of its properties (see Table I).[1,8–12] The long liquid range permits the study of reactions over a wider range of temperatures. The freezing point (−88.98°C) is also considerably lower than that of sulfuric acid (10.371°C). This has enabled the NMR spectra to be obtained at low temperatures.

The viscosity of fluorosulfuric acid is much lower than that of H_2SO_4 (24.544 cP) and it is due to the fact that HSO_3F has only one hydroxyl group to form hydrogen bonds. It has a high dielectric constant and is therefore a more suitable, strongly acidic medium than H_2SO_4.

TABLE I

PHYSICAL PROPERTIES

Property	Value	Reference
Freezing point (°C)	−88.98	8
Boiling point (°C)	162.7	1
Viscosity (cp)	1.56	9
Density (g/cm³)	1.726	9
Dielectric constant (25°C)	120	9
Specific conductivity (ohm^{-1} cm^{-1})	1.085×10^{-4}	9
Heat of formation (kcal/mole)	189.4 ± 0.06	10, 11
Cryoscopic constant (deg $mole^{-1}$ kg)	3.93 ± 0.05	8
Hammett acidity function	15.07	12

C. Solvent System

In contrast to H_2SO_4, which undergoes extensive self-dissociation reactions, HSO_3F shows very limited self-dissociation which affects its properties to only a minor extent. Autoprotolysis is the only mode of self-dissociation for which definite evidence has been obtained,[10] although other modes have been suggested.[13] The autoprotolysis of the solvent is represented by:

$$2HSO_3F \rightleftharpoons H_2SO_3F^+ + SO_3F^-$$

The autoprotolysis constant is quite small (3.8×10^{-18} mole^2/kg^2) and does not give rise to any problem in the interpretation of conductometric and cryoscopic measurements.

Acids in fluorosulfuric acid solvent give the fluorosulfuric acidium ion $H_2SO_3F^+$:

$$HA + HSO_3F \rightarrow H_2SO_3F^+ + A^-$$

and bases give the fluorosulfate ion SO_3F^-:

$$B + HSO_3F \rightarrow BH^+ + SO_3F^-$$

Only a few acids are known in the solvent because of its high acidity and bases are much the larger class of electrolytes.

Hammett and Deyrup[14] showed that the Hammett acidity function, H_0, is a convenient quantitative measure of acidity. It is based on the relative degree of protonation of suitable neutral bases in very dilute solutions in an acid medium. The acidity function is given by

$$H_0 = pK_{BH^+} - \log \frac{[BH^+]}{[B]}$$

where $[BH^+]/[B]$ is the indicator ratio (ratio of the concentrations of protonated to unprotonated base). They made measurements on the H_2SO_4–H_2O system up to 100% H_2SO_4. Since the Hammett indicator approach has failed to produce an acidity function that is applicable to all classes of weak bases in aqueous sulfuric acid, several other workers[15–19] have sought other solvent–acid combinations. Following the work of Jorgenson and Harter,[20] Gillespie *et al.*[12,21] measured the Hammett acidity function H_0 for the systems H_2SO_4–SO_3, H_2SO_4–HSO_3F, H_2SO_4–HSO_3Cl, and H_2SO_4–$HB(HSO_4)_4$, using a set of aromatic and nitro compound indicators. The initial increase in H_0 values (see Fig. 1) for the superacid system is greatest for $HB(HSO_4)_4$–H_2SO_4 and successively less for the other systems in the order $HB(HSO_4)_4 > SO_3(H_2S_2O_7) > HSO_3F > HSO_3Cl$. This is reflected in the decreasing ionization constants of the acids $HB(HSO_4)_4$, 2×10^{-1}; $H_2S_2O_7$, 1.4×10^{-3}; HSO_3F, 2.3×10^{-3}; HSO_3Cl, 9×10^{-4}.

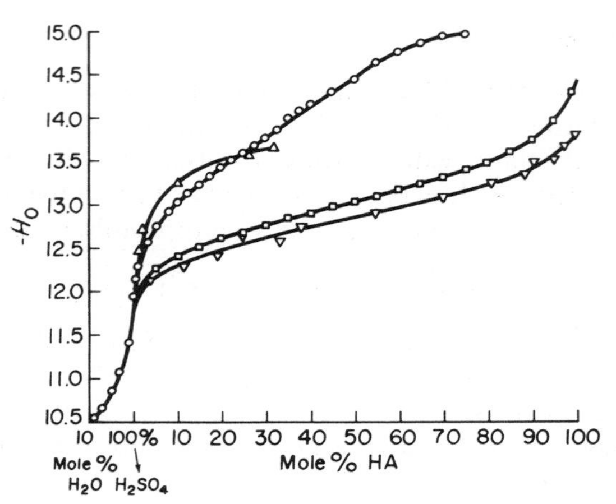

FIG. 1. H_0 values for the systems: ○, H_2O–H_2SO_4–$H_2S_2O_7$; △, H_2SO_4–$HB(HSO_4)_4$; □, H_2SO_4–HSO_3F; ▽, H_2SO_4–HSO_3Cl.

The initial behavior in the H_0 composition plots of H_2SO_4–HSO_3X (X = F, Cl) is explained on the basis that HSO_3X behaves as the weak acid of the H_2SO_4 solvent:

$$HSO_3X + H_2SO_4 \rightleftharpoons H_3SO_4^+ + SO_3X^-$$

$$X = F, Cl$$

Above 80 mole %, HSO_3F, H_2SO_4 may be considered as a solute in the acid solvent. In solutions in HSO_3F, H_2SO_4 behaves as a weak base and is partially protonated according to the equation:

$$HSO_3F + H_2SO_4 \rightleftharpoons SO_3F^- + H_3SO_4^+$$

The fluorosulfate ions produced shifts the equilibrium of the autoprotolysis reaction ($K_{ap} = 3.8 \times 10^{-8}$ mole^2/kg^2), decreasing the $H_2SO_3F^+$ concentration and thus decreasing the acidity. Gillespie and Peel[21] were not able to directly measure the acidity of 100% HSO_3F because the concentration of the basic indicator required for acidity measurements was of the same magnitude as that of the ions resulting from autoprotolysis. The SO_3F^- ion arising from the ionization of the indicator represses the autoprotolysis,

$$2H_2SO_3F \rightleftharpoons H_2SO_3F^+ + SO_3F^-$$

reducing the $H_2SO_3F^+$ concentration and lowering the acidity below that of pure 100% HSO_3F. It was, however, possible to extrapolate the H_0 curve for the H_2SO_4–HSO_3F system to 100% HSO_3F. This gave an H_0 value of −15.07. The same value was obtained for the HSO_3F–SO_3 and KSO_3F–HSO_3F systems.

To overcome some inherent difficulties in determining the pK_a's of weak organic bases by acidity function method, Arnett, Quirk, and Burke[22] established a new basicity scale. This was based on enthalpies of protonation of 35 amines in pure fluorosulfuric acid. They found a good linear correlation between these enthalpies and the pK_a value of the corresponding conjugate acids in water. Arnett *et al.*[23] also reported the application of the enthalpy basicity scale to the determination of the base strengths of 52 carbonyl compounds. They found that cyclopropyl and α,β-unsaturated ketones are very basic compared to aliphatic or aromatic ketones. Aliphatic ketones were more basic on their enthalpimetric scale than the previous pK_a's had suggested. The pK_a's of aromatic ketones fitted well into the correlation of pK_a's vs. ΔH established for the amines. Within the alicyclic ketone series, the basicity increased with increasing ring strain.

Conductometric and potentiometric studies have been made on solutions of a number of inorganic acids in the solvents formamide,[24] dimethyl × formamide,[25] ethyl acetate,[26] ethyl formate,[27] acetic acid,[28] nitromethane,[29] *n*-butanol,[30] molten acetamide,[31] methane sulfuric acid,[32] and nitrobenzene.[33] In each solvent, fluorosulfuric acid was found to be the strongest acid studied, stronger than even perchloric, chlorosulfuric, and sulfuric acids. Paul and his co-workers also showed by thermochemical studies in acetone,[34] alcohol,[34a] and *n*-hexane,[34b] that fluorosulfuric acid is a strong acid. The acid strength order is $H_2S_2O_7 > HSO_3F > HSO_3Cl > H_2SO_4$. Paul demonstrated the usefulness of HSO_3F as an acid titrant in acetic acid[28] and alcohol solutions.

Benoit *et al.*[35] have studied sulfolane solutions of HSO_3F, $HClO_4$, $HSbCl_6$, and $H_2S_2O_7$ and have concluded that the order of acid strength is $HClO_4$ ($K = 10^{-2.7}$), HSO_3F ($K = 10^{-3.3}$), $H_2S_2O_7$ ($K = 10^{-5}$), while $HSbCl_6$ is a strong acid. This order differs from that shown in H_2SO_4.

Engelbrecht and Rode[36] determined the dissociation constant of HI, HSO_3F, and CF_3SO_3H in glacial acetic acid: $HI = 10^{-5.8}$, $HSO_3F = 10^{-6.1}$, and $CF_3SO_3H = 10^{-4.7}$. These values showed that CF_3SO_3H is an even stronger acid than HSO_3F, and HSO_3F is intermediate between HI and H_2SO_4.

D. Inorganic Solutes

1. Alkali and Alkaline Earth Metals

Electrical conductivity and cryoscopic techniques have been extensively used for the investigation of solutions in fluorosulfuric acid in addition to various spectroscopic techniques such as UV, visible, and NMR spectroscopy.

Electrical conductivities of solutions of acids and bases in HSO_3F occurs very largely by proton-transfer conduction involving the ions $H_2SO_3F^+$ and SO_3F^-, respectively.[9] Mobilities involving this mechanism are much greater than would be predicted for ordinary diffusion mechanisms. The conductivities of the alkali metal fluorosulfates are all very similar, which can be explained on the basis of the high mobility of SO_3F^-.

Although the conductivities of the alkali metal fluorosulfates are very similar at any given concentration, they decrease slightly in this order: $NH_4 > Nb \sim K > Na \sim Li$. The conductivities of Ba and Sr fluorosulfates are also similar, that of Sr salt being somewhat less than that of Ba. The small differences were attributed to an increase in the extent of solvation of the cations in the series $NH_4 < Rb \sim K < Na \sim Li < Ba < Sr$.

Sulfuric acid was found to be weakly conductive, insignificant when compared to the total conductivity of K_2SO_4, with a basic dissociation constant of 10^{-4}. This is inconsistent with Woolf's[37] earlier report that sulfuric acid forms solutions comparable in conductivity with KSO_3F. Fluorosulfuric acid is a weak acid in H_2SO_4 system.

Hydrogen fluoride is a base in this solvent system.[9,37]

$$HF + HSO_3F \rightarrow H_2F^+ + SO_3F^-$$

The behavior of $HClO_4$ in this solvent system is not yet clear. Woolf[37] found a minimum conductivity in a titration with $HClO_4$, suggesting a basic behavior. Rather than concluding that $HClO_4$ is a proton acceptor, the following was suggested:

$$2HSO_3F + HClO_4 \rightarrow ClO_3^+ + H_3O^+ + 2SO_3F^-$$

This reaction is inconsistent with the findings of Barr *et al.*[9] that the conductivities of $KClO_4$ solution were only slightly greater than that of KSO_3F:

$$KClO_4 + HSO_3F \rightarrow K^+ + SO_3F^- + HClO_4$$

Small increases in conductivity were attributed to differences in viscosity, caused by the presence of $HClO_4$, rather than to protonation of $HClO_4$.

Paul *et al.*[38] found that HCl gas is only slightly soluble in HSO_3F and behaves as a nonelectrolyte in HSO_3F. However, HCl gas behaves as a weak base in the superacid system HSO_3F–SbF_5–SO_3.

Since fluorosulfuric acid is such a strong acid, salts of other inorganic acids undergo complete solvolysis:

$$KF + HSO_3F \rightarrow K^+ + SO_3F^- + HF$$

$$KCl + HSO_3F \rightarrow K^+ + SO_3F^- + HCl$$

$$K_2SO_4 + 2HSO_3F \rightarrow 2K^+ + 2SO_3F^- + H_2SO_4$$

Some compounds undergo more complex reactions resulting in base behavior. Water undergoes both protonation and solvolysis reaction on HSO_3F:

$$H_2O + HSO_3F \rightarrow H_3O^+ + SO_3F^-$$

Paul *et al.*[39] investigated the H_2O–HSO_3F system and reported the existence of three hydrates, $HSO_3F \cdot H_2O$ (mp −34°C), $HSO_3F \cdot 2H_2O$ (mp −6°C), and $HSO_3F \cdot 4H_2O$ (mp 14°C). The infrared spectrum of $HSO_3F \cdot 4H_2O$ shows the presence of H_3O^+ as well as the ionic nature of these hydrates. Conductance measurements, however, indicate the existence of monohydrate only.

Potassium nitrate[40] ionizes according to the equation:

$$KNO_3 + 3HSO_3F \rightarrow K^+ + NO_2^+ + H_3O^+ + SO_3F^-$$

The nitronium ion has been identified by its Raman spectrum.

Seely and Jache[41] have investigated the solubilities of the bases group I and II fluorosulfates in HSO_3F. The solubilities (Table II) have been compared with those of the corresponding fluorides in HF and rationalized on the basis of lattice energy and solvation energy considerations.

The smooth trend of increasing solubility with increasing atomic number is spoiled by Na (or K) fluorosulfate if one considers the solubility on a gram basis or by Na (or K or Rb fluorosulfate) on a mole basis. These solubilities

TABLE II

SOLUBILITIES OF FLUOROSULFATES IN FLUOROSULFURIC ACID[a]

	Solubility	
Fluorosulfate	HSO_3F (g/100 g)	HSO_3F (moles/100 g)
Li	33.78 ± 0.64	0.319
Na	80.21 ± 0.99	0.658
K	63.83 ± 0.51	0.461
Rb	89.48 ± 0.99	0.486
Cs	132.4 ± 1.5	0.992
Mg	0.12 ± 0.04	5.4×10^{-4}
Ca	16.39 ± 0.46	6.86×10^{-2}
Sr	14.52 ± 0.33	5.10×10^{-2}
Ba	4.67 ± 0.35	1.39×10^{-2}

[a] Reproduced from *The Journal Fluorine Chemistry* **2**, 225. Copyright 1972 by Elsevier Scientific Publishing Company. Reprinted with permission of the copyright owner.

suggest that a crossover of the predominance of lattice energy effect on the solvation energy considerations occurs close to the middle of the group. The solubilities of Co, Sr, and Ba fluorosulfates with the exception of Mg fluorosulfate decreases with increasing atomic number. This is consistent with a decrease in solvation energy of metal ions. It may also be that within the group lattice energies may be larger with larger cations than with very small cations, reflecting a more desirable radius ratio (anion–anion repulsion).

Brazier and Woolf[42] investigated qualitatively the reaction of several metals in HSO_3F. They found that several metals like Au, Mg, Zn, Cd, B, Al, Ce, Sm, Lu, Ga, V, Cr, W, Mn, Re, Fe, Co, Ni, Ru, Os, and Pt were inert to boiling acid. Copper and Bi gave white insoluble fluorosulfates whereas Ag, As, and Sb gave soluble fluorosulfates. Sodium, K, Ca, Ir, Tl, and Sn, in addition to giving green solutions like Nb, Ta, U, and Pb, also gave a white precipitate. The white precipitates were fluorosulfates or decomposition products, while the green solutions were paramagnetic. The ESR behavior and UV spectrum of the green solutions are like those of sulfur in oleum. The metals which gave the green solutions are those which are good reducing agents in highly acidic solutions and reduce the sulfur to elemental form, whereas the less potent reducing metals go only as far as SO_2.

2. Tin and Lead

Carter, Milne, and Aubke[43] studied the conductivity of $Pb(SO_3F)_4$ in HSO_3F as well as the vibrational spectra of $Pb(SO_3F)_4$. They found that both $Pb(SO_3F)_4$ and $Sn(SO_3F)_4$ are likely to have a polymeric structure with a distorted octahedral configuration of the fluorosulfate groups about the central metal atoms. The recent synthesis of the compound containing the hexakisfluorosulfate stannate(IV) anion suggested the feasibility of forming the $Pb(SO_3F)_6^{2-}$ anion in HSO_3F. However, a conductometric titration of a solution of $Pb(SO_3F)_4$ in HSO_3F with KSO_3F showed that $Pb(SO_3F)_4$ undergoes basic solvolysis producing more than 1 mole of fluorosulfate anion per mole of solute. Their results were not consistent with the cryoscopic and conductometric measurements of Gillespie and co-workers[44,45] on lead(IV) acetate in 100% H_2SO_4 where $H_2Pb(HSO_4)_6$ and its anions are formed. They attributed the observed contrast in solvolysis behavior of two lead(IV) compounds possibly to the use of different solvent systems.

Lead(IV) trifluoroacetate, $Pb(CF_3COO)_4$, also exhibits basic behavior in HSO_3F. ^{19}F-NMR of both $Pb(SO_3F)_4$ and $Pb(CF_3COO)_4$ shows a single peak in the fluorosulfate region at -41.0 ppm relative to $CFCl_3$, indicating a rapid fluorosulfate exchange between $Pb(SO_3F)_4$ and HSO_3F.

Mailer[46] has shown by conductivity measurements that both $Sn(SO_3F)_2$ and $Pb(SO_3F)_2$ are weaker bases than $Ca(SO_3F)_2$. Mössbauer parameters of $Sn(SO_3F)_2$ indicated the most ionic nature of the tin(II) compound yet reported.

3. Nitrogen, Phosphorus, Arsenic, Antimony, and Bismuth

The behavior of a number of acid anhydrides of N, P, As, Sb, and Bi in HSO_3F has been studied by cryoscopy and conductometry by Paul and his co-workers.[47] Both nitrous and nitric oxide behave as nonelectrolytes in HSO_3F. The mode of ionization of dinitrogen trioxide, tetraoxide, and pentoxide in HSO_3F is represented as:

$$N_2O_3 + 3HSO_3F \rightarrow 2NO^+ + H_3O^+ + 3SO_3F^-$$

$$N_2O_4 + 3HSO_3F \rightarrow NO^+ + NO_2^+ + H_3O^+ + 3SO_3F^-$$

$$N_2O_5 + 3HSO_3F \rightarrow 2NO_2^+ + H_3O^+ + 3SO_3F^-$$

Both nitronium and nitrosyl fluorosulfate have been isolated and characterized.

Hayek and Engelbrecht[48] reported that P_2O_5 reacts with HSO_3F to form phosphoryl fluoride:

$$P_2O_5 + 3HSO_3F \rightarrow POF_3 + HPO_3 + 2SO_3 + H_2SO_4$$

This reflects the strong solvolyzing power of the acid. However, Paul *et al.*[47] showed that the same products were not obtained when SO_3-free HSO_3F is treated with P_2O_5. Instead, disulfuryl difluoride is obtained. They confirmed this by analysis and molecular weight measurements.

$$P_2O_5 + 6HSO_3F \rightarrow 3S_2O_5F_2 + 2H_3PO_4$$

H_3PO_4[49] is fully protonated as is evident from the exchange of its K salt which is completely solvolyzed:

$$KH_2PO_4 + 2HSO_3F \rightarrow K^+ + P(OH)_4^+ + 2SO_3F^-$$

Paul *et al.*[47] also suggested the possibility that the H_3PO_4 formed may, in the presence of excess of P_2O_5, be dehydrated to form condensed phosphates. The behavior of H_3AsO_4 is similar to that of H_3PO_4. It also gives $As(OH)_4^+$ which is evident from the solvolysis of $(NH_3)_4AsO_4$:

$$(NH_3)_4AsO_4 + 4HSO_3F \rightarrow 3NH_4^+ + As(OH)_4^+ + 4SO_3F^-$$

The behavior of both As_2O_3 and Sb_2O_3 are similar in HSO_3F:

$$As_2O_3 + 9HSO_3F \rightarrow 2As(SO_3F)_3 + 3H_2SO_4 + 3HF$$

$$Sb_2O_3 + 9HSO_3F \rightarrow 2SbF_3 \cdot 3SO_3 + 3H_2SO_4 + 3HF$$

Since hydrogen fluoride behaves almost like a nonelectrolyte and H_2SO_4 is a very weak base in HSO_3F, there is no substantial increase in the conductance of the solution. Perhaps arsenic trifluorosulfate behaves as a weak acid in the system and further decreases the conductance of the solution by removing the SO_3F^- ions:

$$SO_3F^- + As(SO_3F)_3 \rightarrow As(SO_3F)_4^-$$

The compound $2As(SO_3F)_3$, or $(2AsF_3 \cdot 3SO_3)$ at higher concentrations, separates from the solution possibly due to polymerization through fluorosulfate bridges.

Metallic Sb dissolves slowly in HSO_3F at room temperature[50] to give a colorless solution which, contrary to an earlier report,[51] has a conductivity which is considerably greater than that of the pure solvent. Antimony reacts with HSO_3F at room temperature according to the equation:

$$2Sb + 4HSO_3F \rightarrow 2Sb(SO_3F) + H_3O^+ + SO_3F^- + HF + SO_2$$

The colorless monovalent antimony salt $Sb \cdot SO_3F$ formed from the metal and HSO_3F slowly reduces the solvent to S cations (blue and yellow). Eventually, elemental sulfur is formed.

Phosphorus oxyhalides and thiohalides ($POCl_3$, $POBr_3$, $PSCl_3$, $PSBr_3$) all behave as weak bases[49] in HSO_3F. The fact that the degree of ionization of the thiohalides was greater than that of oxyhalides suggested the less basic behavior of P=S group compared to the P=O group.

Woolf[37] and Barr *et al.*[9] have reported that AsF_3 and SbF_3 behave as weak bases in HSO_3F. Barr *et al.*[9] reported that the conductivity of these compounds increases on standing, and that the conductivity and rate of increase in conductivity with time is greater for the Sb compound. This is explained by the following reactions:

Initially,

$$AsF_3 + HSO_3F \rightarrow HAsF_3^+ + SO_3F^-$$

On standing

$$AsF_3 + HSO_3F \rightarrow AsF_2SO_3F + HF$$

Conductometric, cryoscopic, and NMR studies on solutions of SbF_5, $SbF_4(SO_3F)$, and SbF_5–SO_3 mixtures in HSO_3F by Thompson *et al.*[52] showed that there exists a series of acids with the general formula $H[SbF_{(5-n)}(SO_3F)_{1+n}]$ where $n = 0$, 1, 2, and 3, which increase in strength with increasing value of n. The acid $H[SbF_2(SO_3F)_4]$ is a strong acid of the fluorosulfuric acid solvent system. NMR studies show the possibility of higher polymer forms of these acids and polymerization occurs via fluorosulfate bridges. Addition of SO_3 to this system still increases the acidity as SbF_3 is replaced by the fluorosulfates $SbF_4(SO_3F)$, $SbF_3(SO_3F)_2$, and $SbF_2(SO_3F)_3$. The reaction of $SbF_2(SO_3F)_4$ in HSO_3F solvent is represented as:

$$SbF_2(SO_3F)_3 + HSO_3F \rightarrow H[SbF_2(SO_3F)_4]$$

$$H[SbF_2(SO_3F)_4] + HSO_3F \rightarrow H_2SO_3F^+ + SbF_2(SO_3F)_4^-$$

The formation of the acids $H[SbF_4(SO_3F)_2]$, $H[SbF_3(SO_3F)_3]$, and $H[SbF_2(SO_3F)_4]$ has also been confirmed by ^{19}F-NMR. The structure of SbF_5–HSO_3F solutions has been represented by six equilibria by Commeyras and Olah[53]:

$$HSO_3F + SbF_5 \rightleftharpoons H[SbF_5SO_3F]$$

$$H[SbF_5SO_3F] + HSO_3F \rightleftharpoons H_2SO_3F^+ + [SbF_5SO_3F]^-$$

$$2H[SbF_5SO_3F] \rightleftharpoons H_2SO_3F^+ + [Sb_2F_{10}SO_3F]^-$$

$$HSO_3F \rightleftharpoons SO_3 + HF$$

$$2HF + 3SbF_5 \rightleftharpoons HSbF_6 + HSb_2F_{11}$$

$$3SO_3 + 2HSO_3F \rightleftharpoons HS_2O_6F + HS_3O_9F$$

It has been possible to differentiate these species at low temperatures. When SO_2 is used as a solvent, the acidity of the system decreases.

The reactions of SO_3 and SbF_5 with fluorosulfuric acid has been investigated, using electrochemical methods, by Badoz-Lambling *et al.*[54] Their conclusions agreed more with that of Commeyras and Olah[53] than with those of Gillespie *et al.*,[52] where Gillespie *et al.* showed that SO_3 is present in SO_3–SbF_5–HSO_3F mixtures only in concentrated solutions.

$$HSbF_2(SO_3F)_4 \rightleftharpoons HSbF_3(SO_3F)_3 + SO_3$$

However, Gillespie's study was limited to solutions of less than 20 mole % SbF_5. Olah *et al.* used higher concentrations for their investigations. Badoz-Lambling *et al.* obtained results that agreed with the work of Olah and Commeyras, who found evidence for the formation of SbF_6^- and $Sb_2F_{11}^-$ species on the basis of ^{19}F-NMR spectra of SbF_5–HSO_3F solutions.

$$SbF_5 + HSO_3F \rightleftharpoons HSbF_6 + SO_3$$

They[54] considered both the autodissociation and autoprotolysis of HSO_3F:

Autodissociation: $HSO_3F \rightleftharpoons HF + SO_3$

Autoprotolysis: $2HSO_3F \rightleftharpoons H_2SO_3F^+ + SO_3F^-$

Conductivity measurements on solutions of PF_5, AsF_5, BiF_5, NbF_5, PF_5–SO_3, NbF_5–SO_3, and AsF_3–SO_3 were made by Gillespie *et al.*[55] All the other fluorides gave much smaller conductivities than SbF_5. NbF_5 and PF_5 were essentially considered to be nonelectrolytes. The conductivity of BiF_5 was slightly greater than that of AsF_5 solutions. But both are considered weaker acids than SbF_5. The behavior of AsF_5 and BiF_5 is explained according to the equations:

$$AsF_5 + 2HSO_3F \rightleftharpoons H_2SO_3F^+ + AsF_5(SO_3F)^-$$

$$BiF_5 + 2HSO_3F \rightleftharpoons H_2SO_3F^+ + BiF_5(SO_3F)^-$$

A possible mode of ionization of TiF_4 is

$$TiF_4 + 2HSO_3F \rightleftharpoons H_2SO_3F^+ + TiF_4(SO_3F)^-$$

The proposed acid behavior of BiF_5 and AsF_5 was confirmed by conductometric titration with the base KSO_3F.

Sulfur trioxide was found to have no effect on the conductivities of NbF_5 and PF_5, but there was a considerable effect on AsF_5 solutions. Solutions of SO_3 in HSO_3F–BiF_5 are unstable. On the basis of these investigations it was concluded that the acid strength increases in the order $PF_5 \sim NbF_5 < TiF_4 \sim AsF_5$ $BiF_5 < AsF_4(SO_3F)$ $SbF_5 < AsF_2(SO_3F) < SbF_2(SO_3F)$. It is interesting to note that in these series of pentafluorides SbF_5 is the strongest Lewis acid. Replacement of fluorine by the fluorosulfate group appears to cause an increase in the acceptor strength. However, the maximum acidity cannot be obtained as the pentafluorosulfates are not stable. Olah and McFarland[56] studied the behavior of fluoro- and oxyphosphorus compounds in fluorosulfuric and HSO_3F–SbF_5 solutions by NMR methods. They found that with these phosphorus compounds reactions other than proton transfer can also take place. H_2PO_3F is essentially completely protonated and HPO_2F_2 is also substantially protonated, whereas both POF_3 and PF_3 are not protonated. In excess HSO_3F, however, POF_3 is protonated in HSO_3F–SbF_5. These results are in agreement with the expected decreasing order of basicity ($H_2PO_3F > H_2PO_2F_2 > POF_3$) which is a consequence of the effect of successive replacement of the hydroxyl groups of orthophosphoric acid by more electronegative fluorine atoms. Fluorination reactions in addition to protonation take place in $H_2PO_2F_2$, and in phosphonic and phosphinic fluorides. Pyrophosphates give protonated phosphoric acid and protonated fluorophosphoric acid. They suggested the following protolysis mechanism:

$$H_4P_2O_7 + HSO_3F \rightleftharpoons (HO)_3P^+OP(O)(OH)_2 + FSO_3^-$$

$$(HO)_3P^+OP(O)(OH)_2 \rightleftharpoons (HO)_3PO + (O)P^+(OH)_2$$

$$(O)P^+(OH)_2 + FSO_3^- \rightarrow FSO_3P(O)(OH)_2 \rightarrow FP(O)(OH)_2 + SO_3$$

Since SO_3 is a strong dehydrating agent, it is likely that some polyphosphate condensation occurs. Polyphosphate species are protolytically cleaved by HSO_3F to fluorophosphates and H_3PO_4.

Arnett and Wolf[57] have shown the effect of alkyl substitution on the heats of protonation of H_2S and PH_3 in fluorosulfuric acid. The difference of as much as 14 kcal/mole between the heats of protonation of H_2S and CH_3SH or between PH_3 and $C_6H_{11}PH_2$ has been attributed to electrostatic and alkyl group polarization stabilization in solution.

4. Sulfur, Selenium, and Tellurium

Direct evidence for the formation of the polyfluorosulfuric acids $H(SO_3)_nF$ in solutions of SO_3 in HSO_3F or HSO_3F–SO_2ClF has been obtained by Dean and Gillespie[58] and Gillespie and Robinson[59] by low-temperature ^{19}F-NMR. Unlike SO_3, SeO_3 has a limited solubility in HSO_3F,[47] but when the solution is slightly heated the following reactions take place:

$$HSO_3F + SeO_3 \rightarrow HSeO_3F + SO_3$$

$$HSeO_3F + HSO_3F \rightarrow H_2{}^+SeO_3F + SO_3F^- \rightarrow SeO_2F_2 + H_2SO_4$$

Selenium oxyfluorosulfate, $SeO(SO_3F)_2$, is miscible with HSO_3F in which it behaves as a medium strong base.[60]

Fluorosulfuric acid has been used as a solvent for several redox reactions.[61] Some new cations of selenium have been discovered when Se is oxidized with $S_2O_6F_2$ in solution in HSO_3F. These species are identified as $Se_8{}^{2+}$ (green) and $Se_4{}^{2+}$ (yellow) by conductometric and spectrophotometric methods. $Se_4(SO_3F)_2$ has been isolated from a solution of $Se_4{}^{2+}$ in HSO_3F. $Se_8{}^{2+}$ and $Se_4{}^{2+}$ in HSO_3F are formed according to the equations:

$$8Se + S_2O_6F_2 \rightarrow Se_8{}^{2+} + 2SO_3F^-$$

$$4Se + S_2O_6F_2 \rightarrow Se_4{}^{2+} + 2SO_3F^-$$

A red species, $Te_4{}^{2+}$, is formed by the oxidation of Te with $S_2O_6F_2$ in solution in HSO_3F. Herlem, Thiebault, and Adhami[62] observed four different species of S ($S_2{}^+$, S^+, S^{2+}, and S^{4+}), three different species of Se ($Se_2{}^+$, Se^{2+}, and Se^{4+}), and three different species of Te ($Te_2{}^{2+}$, Te^{2+}, and Te^{4+}) in reactions of S, Se, and Te with HSO_3F. These species were identified using spectrophotometry and electrochemical methods by varying the composition of the solvent from basic (nonoxidizing) medium (0.1 *M* $NaSO_3F$) to oxidizing medium (0.5 *M* SO_3) (see Table III). Their results differ from those of Gillespie *et al.*[50,61,63–67] Gillespie *et al.* have not identified the species $S_2{}^+$, S^+, and $Se_2{}^{2+}$, but they have found evidence from their investigations in acidic media[63,65] for the ability of S, Se, and Te to form species of low positive oxidation states such as $S_{16}{}^{2+}$, $S_8{}^{2+}$, $S_4{}^{2+}$, $Se_8{}^{2+}$, $Se_4{}^{2+}$, $Te_6{}^{2+}$, $Te_4{}^{2+}$, and $Te_4{}^+$. Herlem *et al.*[62] have explained these differences as being caused by the differences in the experimental conditions.

Both isomers of disulfur difluoride, SSF_2 and FSSF, dissolve in HSO_3F and in SO_3 containing H_2SO_4 (30% oleum) at low temperature.[68] The UV and ESR spectra of these solutions indicated that these solutions contain species of S cations as found in solutions of elemental S in oleum. It was also

TABLE III

ELECTROCHEMICAL OXIDATION[a]

Compound	Basic media containing $NaSO_3F$	Oxidizing media containing SO_3
Sulfur	S is oxidized by the solvent S_2^+ radical cation	S is oxidized by the solvent to S^+ radical
Selenium	Selenium is oxidized by the	solvent to Se_2^{2+}
Tellurium	Tellurium is oxidized by the	solvent to Te_2^{2+}

[a] The electrochemical oxidation of every cation to the final (IV+) oxidation takes place through the (II+) oxidation state which is not stable.

found that reaction of S_2F_2 with HSO_3F gives disulfur difluoride and thionylfluoride which are also formed by reactions of SF_4 with HSO_3F.

Both $SeCl_4$ and $TeCl_4$ show basic behavior[49]:

$$TeCl_4 + HSO_3F \rightarrow TeCl_3^+ + HCl + SO_3F^-$$

$$SeCl_4 + HSO_3F \rightarrow SeCl_3^+ + HCl + SO_3F^-$$

Conductometric titrations of $SeCl_4$ and $TeCl_4$ in a superacid medium (SbF_5–SO_3–HSO_3F) show that the HCl formed also behaves as a base:

$$TeCl_4 + 2HSbF_2(SO_3F)_4 \rightarrow TeCl_3^+ + 2[SbF_2(SO_3F)_4]^- + H_2^+Cl$$

5. HALOGENS

Oxidation of I_2 with $S_2O_6F_2$ in fluorosulfuric acid has given several new cations, I_2^+, I_3^+, I_5^+, and I_4^{2+}.[52] Gillespie and Milne[69,70] reported the results of NMR, freezing point, and conductivity measurements on 1 : 7 and 1 : 3 I_2–$S_2O_6F_2$ solutions in HSO_3F. They show that $I(SO_3F)_3$ is the highest trisfluorosulfate formed in solution in HSO_3F. The conductance measurement of 1 : 3 I_2–$S_2O_6F_2$ solutions using KSO_3F as base shows that $I(SO_3F)_3$ behaves as an ampholyte. Initially, the added fluorosulfate represses a basic ionization which is represented as

$$I(SO_3F)_3 \rightleftharpoons I(SO_3F)_2^+ + SO_3F^-$$

But as the concentration of SO_3F^- increases, the acidic behavior of $I(SO_3F)_3$ gradually becomes more important.

$$I(SO_3F)_3 + SO_3F^- \rightarrow I(SO_3F)_4^-$$

The acid and base ionization constants are $K_a = 10$ mole/kg and $K_b = 2.9 \times 10^{-5}$ mole/kg at 25°C and 2.1×10^{-6} mole/kg at 78°C. $I(SO_3F)_3$ reacts with water at low temperature to give iodosyl fluorosulfate:

$$I(SO_3F)_3 + H_2O \rightarrow IOSO_3F + 2HSO_3F$$

Iodine in the 1+ oxidation state is completely disproportionated in solution in HSO_3F to the I_2^+ ion and $I(SO_3F)_3$.[70] Bromine(I) fluorosulfate is disproportionated to an appreciable extent to the Br_2^+ ion and $Br(SO_3F)_3$ in the superacid SbF_5–$3SO_3$–HSO_3F.[71]

No evidence was obtained for Cl_2^+ in the reaction of chlorine in the 1+ oxidation state in the superacid medium HSO_3F–SbF_5–$3SO_3$.[72] The reaction of ClF or Cl_2–ClF was, however, found to produce $ClOSO_2F$. These results do not agree with the results of Olah and Comisarow,[73] who studied the ESR spectrum of ClF in HSO_3F–SbF_5. The results of Olah and Comisarow[73] have also been criticized by Symons *et al.*[74] Gillespie and Morton[72] showed the possibility of other species like ClO_2F^+ other than the one $ClOF^+$ suggested by Symons *et al.*[74] Furthermore, Gillespie and Morton think that the claim of Cl_2^+ by Olah and Comisarow[73] in the reaction of Cl_2 and IrF_6 to give the salt $Cl_2^+IrF_6^-$ may be incorrect.

The solvolysis of dichloryl trisulfate $(ClO_2)_2S_3O_{10}$ and chloryl fluorosulfate $(ClO_2)OSO_2F$ was studied by Aubke *et al.*[75,76] by NMR, UV, and visible-light spectrophotometry and by conductometry. These results reveal the formation of the chloronium cation ClO_2^+ in solution. They compare their conductivity results to those for nitryl and nitrosyl fluorosulfates. All solutes behave as strong bases in HSO_3F and show an increase in conductivity similar to that of KSO_3F. The slight decrease in conductivity and lower γ values of $NOOSO_2F$, NO_2OSO_2F, and $ClO_2(OSO_2F)$ were explained by differences in cation mobilities. The observed order $K^+ > NO^+ > NO_2^+ > ClO_2^+$ at molalities $> 1.5 \times 10^{-2}$ moles/kg is in agreement with the expected trend based on cation size. The conductivity increase for $(ClO_2)_2S_3O_{10}$ was twice as large as that found for chloryl fluorosulfate. The following solvolysis reactions are suggested:

$$(ClO_2)_2S_3O_{10}^- + 2HSO_3F \rightarrow 2ClO_2^+ \text{ (solv)} + 2SO_3F^- \text{ (solv)} + H_2S_3O_{10} \text{ (solv)}$$

$$(ClO_2)OSO_2F \rightarrow ClO_2^+ \text{ (solv)} + SO_3F^-$$

In this equation the acid $H_2S_3O_{10}$ is regarded as a nonelectrolyte in HSO_3F. This has been confirmed by studies using the K salt $K_2S_3O_{10}$ in this solvent.

The redox reactions between Br_2, ICl, Cl_2 or NOCl, and PBr_3, PCl_3, PPh_3, and $AsCl_3$ or $AsPh_3$ in HSO_3F were followed both conductometrically and visually by Paul *et al.*[77] HSO_3F stabilizes the oxidized products as cations. Phosphorus(V) compounds were fairly stable even in the solid state. The arsenic(V) compounds are formed only in solution.

Gillespie and Pez[78] investigated the behavior of N_2, O_2, Ne, Xe, H_2, NF_3, CO, CO_2, SO_2, and 1,3,5-trichlorobenzene in the system HSO_3F–SbF_5–SO_3. They made solubility and conductivity measurements and reported on observations of the infrared spectrum of CO_2 and SO_2 (also the UV and NMR spectra of CO_2). There was no significant base behavior.

Although the lack of basicity of NF_3 to common acids is well known, it is worthwhile to note that the substitution of fluorine for hydrogen in NH_3 reduces the basicity of the lone pair so much that they are not basic even to this extremely acid medium.

Small changes in the conductivities of SO_2 in HSO_3F–SbF_5 and HSO_3F–SbF_5–SO_2 were attributed to changes in viscosity rather than to protonation. IR and Raman data were consistent with this view.

In the same paper the authors demonstrated the acidity of HSO_3F–SbF_5–SO_3 vs. HSO_3F by showing conductometrically that 1,3,5-trinitrobenzene, a weak base in HSO_3F, is completely ionized in the mixed solvent system.

6. Xenon

The ionization of XeF_6 in HSO_3F, SbF_5, HF, BrF_5, and ClF_5 solvents was studied by Gillespie and Schrobilgen[79,79a] ny ^{19}F-NMR. They pointed out that an earlier report by Des Marteau and Eisenberg[79b] on XeF_6 reactions with HSO_3F to give $Xe(SO_3F)_2F_4$ was in error, since the XeF_6 used by them was contaminated with XeF_4. The subsequent work of Des Marteau and Eisenberg[79c] on the SeF_6–HSO_3F system is in agreement with Gillespie's earlier results.[79a] The ionization of a 1 : 3 mixture of XeF_6 and SbF_5 in the solvent HSO_3F is approximately represented by the equations:

$$XeF_6 + 3SbF_5 \rightarrow XeF_5^+ + Sb_3F_{16}$$

$$Sb_3F_{16}^- + 2HSO_3F \rightleftharpoons Sb_2F_{11}^- + SbF_5(SO_3F)^- + H_2SO_3F^+$$

$$Sb_2F_{11}^- + 2HSO_3F \rightleftharpoons SbF_6^-(SO_3F)^- + H_2SO_3F^+$$

A solution of XeF_6 in HSO_3F at $-50°C$ gave the XeF_5^+ spectrum and a line due to HF, as well as a line due to the solvent undergoing rapid exchange with SO_3F^-. The ionization is represented by:

$$HSO_3F + XeF_6 \rightarrow XeF_5^+ + SO_3F^- + HF$$

The compound $Xe(SO_3F)F_5$ has been reported independently by Des Marteau and Eisenberg[79c] and Gillespie *et al.*[79a] A solution of $Xe(SO_3F)F_5$ in HSO_3F gave a low temperature spectrum identical with that obtained for XeF_6 in HSO_3F except that no HF peak was observed.

E. Organic Solutes: Experimental Methods

Cryoscopic, conductometric, NMR, IR, Raman, and electronic spectroscopy have been the most important physical methods that have been used to study protonation. Cryoscopic and conductometric methods have been valuable in the detection of protonated species, but have given little or no information as to the site of protonation. In IR and Raman spectroscopy difficulties both in obtaining and interpreting such spectra have limited the application of these techniques. Electronic spectroscopy (UV and visible) has been of considerable usefulness in the study of equilibria, particularly between the organic base and its conjugate acid, but again, detailed interpretation is difficult and has been restricted to certain classes of conjugated compounds. NMR spectroscopy offers a unique possibility for examining in detail the structure of protonated molecules in solution.

Sulfuric acid and oleum as solvents for the study of protonation have the serious disadvantage that they are quite viscous and possess relatively high freezing points. Investigations in this solvent are generally limited to 10°C where no stable protonated species or its cleavage products, carbonium ions, and related products are observed.

Olah and his associates[80–82] and Gillespie[6] found it possible to stabilize the carbonium ions by using fluorosulfuric acid HSO_3F and HSO_3F containing SbF_5 as coacid. HSO_3F–SbF_5 and HF–SbF_5 are probably the most acidic solvent systems yet found, although their acidities have not been accurately established.[21,83] The estimated H_0 for (HSO_3F–10% SbF_5) is −18.94. Certain solvents like SO_2, SO_2ClF, or SO_2F_2 used as dilutents along with HSO_3F–SbF_5 considerably decrease the viscosity of the medium at low temperatures without appreciably diminishing the acidity of the system. These acidic systems are called superacids. The term "superacid" is currently used in an indiscriminate manner for a number of extremely strong acids that have come into use in the studies of stable, long-lived carbonium ions. These consist generally of mixtures of HF or HSO_3F with SbF_5 in varying mole ratios.

This chapter is limited only to the use of HSO_3F and its superacid system as solvent for the study of organic systems.

Cryoscopic and conductometric studies of the ionization of a number of weak bases, e.g., nitro compounds in HSO_3F, have shown that they are more extensively ionized than in H_2SO_4[8,9] (Table IV).

These results suggest that fluorosulfuric acid is more acidic than sulfuric acid. This same conclusion has been reached from spectroscopic study of the ionization of the same weak bases. This study has shown that the Hammett acidity function, H_0, has a value of −13.96, (recent value = −15.07)[21] for fluorosulfuric acid as compared with −12.1 for H_2SO_4.

The electronic spectra of a number of organic systems such as alkyl (aryl) carbonium ions and alkenyl cations with cyclic π systems in various solvents, including that in HSO_3F, have been discussed by Olah, Pittman, and Symons.[84]

TABLE IV

IONIZATION OF WEAK BASES

Base	$10^2 K_b$ in HSO_3F	$10^2 K_b$ in H_2SO_4
Nitrobenzene	Fully ionized	1.0
m-Nitrotoluene	Fully ionized	2.3
p-Nitrochlorobenzene	76	0.4
m-Nitrochlorobenzene	7.9	Too weak to be measured
Nitromethane	2.7	0.25
2,4-Dinitrotoluene	1.4	Too weak to be measured
2,4-Dinitrochlorobenzene	0.16	Too weak to be measured
2,4-Dinitrofluorobenzene	0.16	Too weak to be measured
1,3,5-Trinitrobenzene	0.004	Too weak to be measured

1. ALCOHOLS AND THIOLS

A series of aliphatic alcohols have been investigated by NMR spectroscopy in the superacid system HSO_3F–SbF_5–SO_2 at $-60°$.[85,86] The following alcohols were protonated: methyl, ethyl, *n*-propyl, isopropyl, *n*-butyl, isobutyl, *sec*-butyl, *n*-amyl, isoamyl, neopentyl, *n*-hexyl, and neohexyl alcohols. All normal and secondary aliphatic alcohols were observed as stable O-protonated species with negligible exchange rates at temperatures ranging from $-60°$ to $+60°$C. Protonated *tert*-butyl and *tert*-amyl alcohols were not observed due to very fast cleavage to the corresponding tertiary carbonium ions:

$$(CH_3)_3C{-}OH \xrightarrow{SbF_5\ HSO_3} (CH_3)_3C^+\ SbF_5FSO_3^-\ +\ H_3O^+$$

The kinetics of cleavage of these protonated (normal) alcohols to carbonium ions are first order. The reaction rate is enhanced by chain branching. The stability of the protonated primary alcohols decreases as the chain length is increased. This is shown by the comparison of the rate constants of the cleavage at 15° (Table V). The activation energy determined from the Arrhenius equation $K = Ae^{Ea/RT}$ between 0° and +25°C were shown in Table III. The activation energy of the cleavage seems to be identical for the primary alcohols studied (Table VI).

Aliphatic diols are diprotonated in HSO_3F–SbF_5–SO_2 solutions at $-60°$C.[87]

$$\underset{\displaystyle OH\quad OH}{R{-}CH{-}CH{-}R'} \xrightarrow[-60°C]{FSO_3H{-}SbF_5{-}SO_2} \underset{\displaystyle {}^+OH_2\ OH_5}{R{-}CH{-}CH{-}R'}$$

In diprotonated diols the protons on the oxygen are found at lower fields than in protonated alcohols, reflecting the presence of two positive charges.

TABLE V

RATE OF CLEAVAGE OF PROTONATED NORMAL ALIPHATIC ALCOHOLS AT 15°C

Protonated alcohol	Rate constant $\times 10^{-3}$/min
Methyl	Stable up to +50°
Ethyl	Stable up to +30°
Propyl	20.5
Butyl	48.1
Pentyl	68.4
Hexyl	91.4

When the solutions of diols are allowed to warm up, different rearrangement reactions occur, depending on the diol studied.

α-glycols and β-glycols undergo specific pinacolic rearrangement with hydride shifts:

$$R-CH(\overset{+}{O}H_2)-CH(\overset{+}{O}H_2)-R' \xrightarrow[-H_3O^+]{\text{slow}} R-CH(\overset{+}{O}H_2)-\overset{+}{C}H-R' \xrightarrow{\text{fast}} R-C(=\overset{+}{O}H)-CH_2-R'$$

Diprimary γ-, δ-, and ε-glycols are very stable and no rearrangement was observed.

Disecondary γ-glycols were observed to undergo cyclization to tetrahydro-furan (THF) derivatives.

$$CH_3-CH(\overset{+}{O}H_2)-CH_2-CH_2-CH(\overset{+}{O}H_2)-CH_3 \xrightarrow{\text{slow}} \text{2,5-dimethyl-THF-}\overset{+}{O}H + H_3O^+$$

Ditertiary alcohols could not be observed as protonated species. The cleavage reaction is too fast at −70°C and the rearrangement depends on the carbonium ion formed, for example, pinacol $(CH_3)_2C(OH)C(OH) \times (CH_3)_2$ at −70°C in HSO_3F–SbF_5–SO_2 solutions rearranges immediately and the only species observed in the solution is protonated pinacolone.

TABLE VI

ACTIVATION ENERGY OF CLEAVAGE OF PROTONATED ALCOHOLS

Protonated alcohol	Cleavage of activation energy (kcal/mole)
n-Propyl	19.8 ± 1.5
n-Butyl	20.0 ± 0.8
n-Pentyl	18.8 ± 0.9
n-Hexyl	18.6 ± 0.9

a. *Thiols.* S-protonation of aliphatic thiols was observed at $-60°C$ in the superacid medium[88]

$$RSH \xrightarrow{HSO_3F-SbF_5} RSH_2^+SbF_5\text{-}FSO_3^-$$

They found that the protonated thiols were surprisingly more stable to increases in temperature relative to the corresponding protonated alcohols. The stability of these protonated thiols followed the order primary thiol > *sec*-thiol > *tert*-thiol.

For example: protonated *n*-butyl thiol slowly cleaves to carbonium ion only at $+25°C$:

$$CH_3CH_2CH_2CH_2SH_2^+ \xrightarrow{+25°C} [CH_3CH_2CH_2CH_2^+] \rightarrow (CH_3)_3C^+$$

Protonated *sec*-butyl thiol cleaves to $(CH_3)_3C^+$ at 0°C:

$$\underset{\displaystyle CH_3}{CH_3CH_2\overset{|}{C}HSH_2^+} \xrightarrow{0°C} [CH_3CH_2\overset{+}{C}HCH_3] \longrightarrow (CH_3)_3C^+$$

Whereas *tert*-butyl thiol cleaves even at $-30°C$ ($t_{\frac{1}{2}} \sim 15$ min)

$$(CH_3)_3CSH_2^+ \xrightarrow{-30°C} (CH_3)_3C^+ + H_3S^+$$

2. ETHERS AND SULFIDES

a. *Aliphatic Ethers.* These ethers are protonated in the acid[89]

$$R_2O \xrightarrow[-60°C]{HSO_3F-SbF_5-SO_2} R_2OH^+SbF_5FSO_3^-$$

Protonated *n*-alkyl ethers are stable up to $+40°C$ in (1 : 1) $HSO_3F\text{-}SbF_5$. Above 40°C *n*-butyl methyl ether cleaves to form trimethylcarbonium ion. Protonated *sec*-butyl methyl ether cleaves cleanly at $-30°C$ to protonated methanol and $(CH_3)_3C^+$.

Ethers in which one of the alkyl groups is tertiary cleave rapidly even at

$-70°C$. The kinetics of cleavage of protonated *sec*-butyl methyl ether was measured by following the disappearance of the methoxy doublet in the NMR spectrum with simultaneous formation of protonated methanol and trimethyl carbonium ion. The cleavage showed pseudo-first-order kinetics (Table VII) indicating that the rate-determining step is the formation of methyl ethyl carbonium ion followed by rapid rearrangement to the more stable trimethyl carbonium ion ($k_1 \ll k_2$).

$$CH_3CH_2\underset{\underset{CH_3}{|}}{CH}\overset{H}{\underset{+}{O}}CH_3 \xrightarrow{k_1} CH_3OH_2^+ + CH_3CH_2\underset{+}{C}HCH_3 \xrightarrow{k_2} (CH_3)_3C^+$$

b. *Sulfides*. Alkyl sulfides are quantitatively protonated in $HSO_3F–SbF_5–SO_2$ at $-60°C$[88]

$$R_2S \xrightarrow[-60°C]{HSO_3F\ SbF_5\ SO_2} R\overset{+}{\underset{H}{S}}R\ SbF_5SO_3F^-$$

Protons on S in protonated thiols and sulfides are at considerably higher field (-6.5 to 5.8 ppm) than those on the corresponding oxygen in protonated alcohols and ethers (-9.2 to -7.9 ppm). Protonated sulfides are more stable to cleavage than the corresponding protonated ethers and also more stable than the protonated thiols (for example, protonated methyl *tert*-butyl sulfide cleaves slowly at $-15°C$).

$$(CH_3)_3C\overset{H}{\underset{+}{S}}CH_3 \xrightarrow[H^+]{-15°C} (CH_3)_3C^+ + CH_3SH_2^+$$

The corresponding methyl butyl ether cleaves even at $-70°C$. Protonated *sec*-isopropyl sulfide is appreciably stable even up to $+70°C$ in $HSO_3F–SbF_5$ (1 : 1).

c. *Alicyclic Ethers and Sulfides*. Olah and Szilagyi[90] found that four-, five-, and six-membered alicyclic ethers (oxiranes) and sulfides (thiaranes) form most stable protonated intermediates in $HSO_3F–SbF_5–SO_2$ at low temperatures. The fate of the three-membered ring ethers and sulfides was

TABLE VII

KINETICS OF CLEAVAGE OF PROTONATED *sec*-BUTYL METHYL ETHER

Temp. (°C)	k (sec^{-1} × 10^4)	E_a
−29.0	4.14 ± 0.64	8.5 ± 3 kcal
−17.5	9.08 ± (1.2)	8.5 ± 3 kcal

uncertain. The more highly substituted oxiranes showed no evidence of forming any long-lived intermediate in the temperature range studied. Ethylene oxide, propylene oxide, and dimethyl oxiranes show, at low temperatures in superacid, the presence of oxiranium intermediates contaminated that are, however, with polyones and undergo rapid exchange process, for example, propylene oxide (2-methyl oxirane).

$$\text{(I)} \xrightarrow{HSO_3F\text{-}SbF_5} \text{(II)} \longrightarrow \text{(III)}$$

The exclusive production of propionaldehyde **(III)** from the reaction of propylene oxide with "magic acid" (HSO_3F–SbF_5) in SO_2 solution at low temperature indicates, however, that a protonated intermediate oxiranium ion **(II)** is undergoing a ring-opening reaction followed by a 1,2-hydride shift. This shift is further assisted by the contribution of the lone electron pair of oxygen placing the positive charge on the thermodynamically most favorable position.

Reactions of ethylene sulfide and propylene sulfide with HSO_3F–SbF_5 yielded only polymeric products insoluble in SO_2 at $-60°C$ in SO_2. When the reaction mixture was prepared by extracting the thiaranes from *n*-pentane into the magic acid–SO_2 solution at $-78°C$ or lower, NMR indicated the presence of protonated thiarane.

$$\xrightarrow[-78°C]{HSO_3F\text{-}SbF_5} \quad + \quad \text{polymeric products}$$

NMR indicated the presence of isomers of protonated propyl sulfide.

$$\xrightarrow[-78°C]{HSO_3F\text{-}SbF_5}$$

As a model compound for thiaranium ion the protonation of ethylene sulfoxide was also carried out. Protonation takes place on sulfur.

$$\xrightarrow[-65°C]{HSO_3F\text{-}SbF_5}$$

The other sulfides studied include compounds of the type:

S → S^+–H

$n = 3, 4, 5$

NMR shifts of alicyclic protonated ethers and sulfides have been compared with ethylene and tetramethylene halonium ions. Regular changes in the series of halonium, oxonium, and sulfonium ions have been shown.

d. *Alkyl Phenyl Ethers: Hydroxybenzenes.* Phenols and aromatic ethers have been studied by several investigators using both NMR and UV methods in strong acids.[91-94] The discrepancy of O vs. C protonation seemed to depend on several factors. The general trend showed that oxygen protonation is observed for unsubstituted phenols and unsubstituted alkyl phenyl ethers at relatively low temperatures. Substitution of electron-releasing groups on the aromatic[95,96] ring causes carbon protonation to predominate. Olah and Mo, using both ^{1}H- and ^{13}C-NMR, found that the sites of protonation (*O*- vs. *C*-) in a number of hydroxybenzenes and their methyl ethers were dependent upon the acid media used. Protonation of isomeric hydroxybenzenes and their methyl ethers was studied in four different superacid media: [SbF_5–HF (1 : 1) *M/M*]–SO_2ClF **(IV)**, SbF_5–HSO_3F (1 : 1 *M/M*)–SO_2ClF **(V)**, [SbF_5–HSO_3F (1 : 4) *M/M*]–SO_2ClF **(VI)**, and HSO_3F–SO_2ClF **(VII)**. Generally O-protonation was favored in weaker acid media, while C-protonation was achieved in stronger acid media. In some cases a mixture of O- and C-protonated species were observed in the same solvent. The ratio was related to the acid media used. Based on their protonating ability, the acidity of the four superacid systems used decreases in the order **(IV)** > **(V)** > **(VI)** > **(VII)**. The ratio of the stereoisomers (*cis* and *trans*) was also dependent on the acid media used. The stability of hydroxy(alkoxy)benzenium ions, including isomeric ion forms derived from the same precursors and the relative ease of protonation was discussed in terms of steric, resonance, and inductive effects.

The course of protonation of trihydroxybenzenes and their methyl ethers in super acids was found to be more complicated than those of mono- and dihydroxy benzenes. Two different types of diprotonation were found.

Diprotonation (*i*)

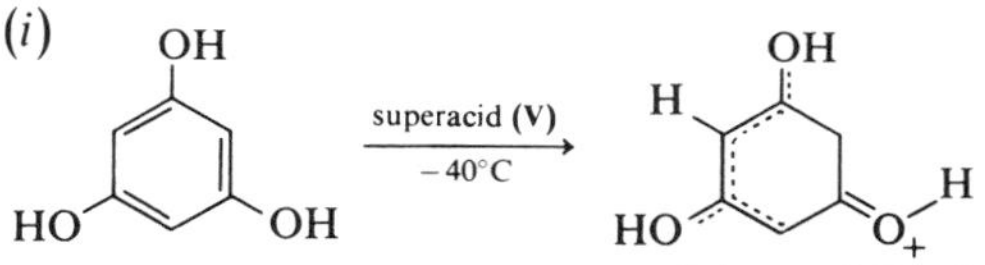

OH, HO, OH → superacid (V), −40°C → OH, H, HO, O^+–H

oxocyclohexenyl dication

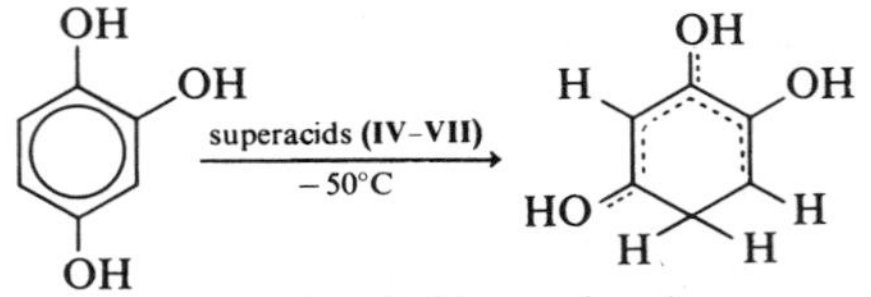

superacids (**IV–VII**)
$-50°C$

monoprotonated benzenium ions

Diprotonation (*ii*)

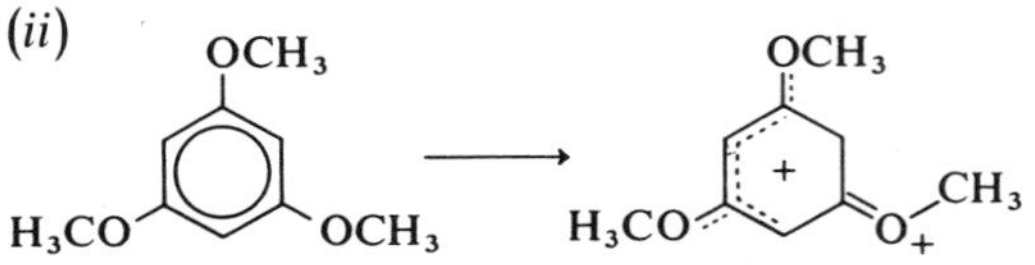

a benzenium ion with a $CH_3O(H)^+$ or OH_2^+ at the *meta* position.

The same authors[97] have studied halophenols and haloanisoles in the same four acid systems. The protonation of haloarenes has also been investigated independently by Brouwer *et al.*[98–100] The site of protonation of halophenols and haloanisoles was found to be dependent upon the four superacids (**IV–VII**) used as well as the nature of halogen atom. In general, C-protonation of halophenols and haloanisoles to give the corresponding halogenated hydroxy(methoxy)benzenium ions was achieved in the strongest superacid (**IV**), while O-protonated oxonium ion were formed in the weakest acid (**VII**). Since OH and OCH_3 are stronger activating groups than halogen atoms, protonation of ring always takes place at the position *para* to the hydroxy or methoxy group. In the case when the *para* position is substituted by a halogen atom (example: *p*-halophenols), no C-protonation was observed. Instead the corresponding O-protonated oxonium ions were formed even in superacid (**IV**).

Protonation of O-halophenols (**VIII–X**) in superacid (**VI**) is particularly interesting.

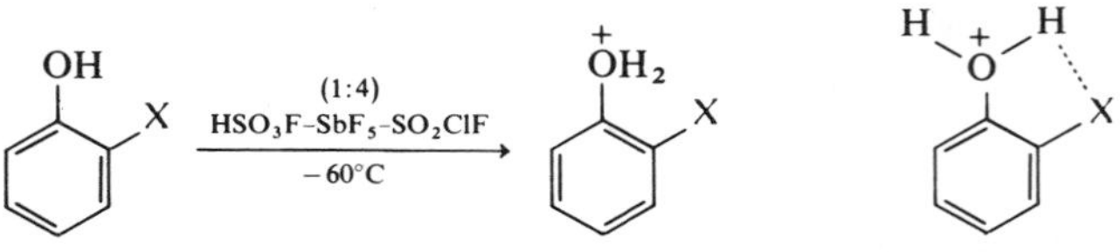

(1:4) HSO_3F-SbF_5-SO_2ClF
$-60°C$

VIII-*X* **VIII**-X_a **VIII**-X_b

VIII-Br and **VIII**-I were completely C-protonated in superacid (**VI**) at $-60°C$ while **VIII**-F and **VIII**-Cl were both C- and O-protonated. These results reflect the strong negative inductive effects of fluorine and chlorine atoms respectively. O-protonation to give oxonium ions **VIII**-F_a and **VIII**-Cl_a is favored through involvement of hydrogen bonding (**VIII**-X_b). The study of the protonation of halophenols and haloanisoles in various superacid media also gives useful information relating to the electrophilic aromatic substitution of these compounds. They suggested that kinetic vs. thermodynamic control can be responsible for O- or C-substitution.

The importance of steric and resonance effects on the stability of arenium ions was indicated by the study of protonation of 2,4,6-trimethoxytoluene and *m*-xylene in superacid solution.[101] The protonation of these compounds led to the formation of benzenium ions and methylated oxocyclohexenyl dications. 2,4,6-trimethoxytoluene in $HSO_3F–SO_2ClF$ gives benzenium ions.

$$\xrightarrow{HSO_3F-SO_2ClF} \text{(IX)} + \text{(X)}$$

The ratio of these two products **(IX)** and **(X)** was temperature dependent. The same compound in the stronger acid media $HSO_3F–SbF_5–SO_2ClF$ and $HF–SbF_5–SO_2ClF$ solution was diprotonated to give methylated oxocyclohexenyl dications **(XI)** and **(XII)** in the ratio 4 : 1. This reaction is temperature dependent.

$$\xrightarrow[\text{at } -60°C]{HF-SbF_5-SO_2ClF \text{ or } HSO_3F-SbF_5-SO_2ClF} \text{(XI) (20\%)} \quad \text{(XII) (80\%)}$$

3. Carboxylic Acids, Esters, and Anhydrides

One of the first applications of the $HSO_3F–SbF_5$ medium was the observation for the first time of the PMR spectra of the conjugate acids of acetic, propionic, and benzoic acids at temperatures of $-70°C$ and below.[102,103] Conductivity measurements showed that acetic acid and benzoic acid behave as strong bases in fluorosulfuric acid solvent. Olah and White[104] showed that all the aliphatic carboxylic acids were completely protonated and they were able to study their cleavage to the corresponding alkyloxocarbonium ions under nonequilibrium conditions.

$$RCOOH \xrightarrow[-60°C]{HSO_3F\ SbF_5\ SO_2} RCO_2H_2^+ \xrightarrow[\Delta,\ -H_2O]{HSO_3F\ SbF_5\ SO_2} RCO^+$$

The protons on oxygen resonances occur at lower fields than in protonated alcohols and ethers, but are more shielded than those in protonated aliphatic ketones and aldehydes. This has been attributed to the partial double bond character of the protonated acids.

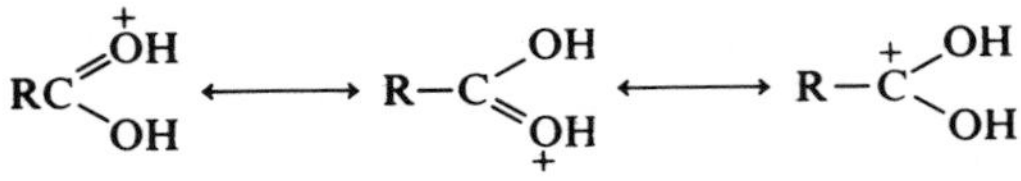

$$\mathrm{RC}\begin{matrix}\overset{+}{\mathrm{OH}}\\ \mathrm{OH}\end{matrix} \longleftrightarrow \mathrm{R{-}C}\begin{matrix}\mathrm{OH}\\ \underset{+}{\mathrm{OH}}\end{matrix} \longleftrightarrow \mathrm{R{-}\overset{+}{C}}\begin{matrix}\mathrm{OH}\\ \mathrm{OH}\end{matrix}$$

The cleavage of protonated acid to give an oxocarbonium ion in HSO_3F–SbF_5 shows simple first-order kinetics. The rate of disappearance of the protonated acid and the appearance of oxocarbonium ion was followed by integration of the appropriate peaks in the NMR species (Table VIII).

$$RCO_2H_2^+ \xrightarrow[k_1]{HSO_3F\ SbF_5} RC{=}O + H_3O^+$$

The differences in the rate constants for primary, secondary, and tertiary carboxylic acid cleavage are small and contrast both in magnitude and reactivity sequence with the cleavage of ethers and alcohols in the same acid system (Table IX).

The entropies of activation found for the cleavage of protonated acids were consistent with the strongly acidic solvent behavior. The negative entropy changes have been interpreted as indicating a greater difference in solvation than in the weaker acid system.

The presence of isomers in protonated aliphatic acids has been shown by Brookhart *et al.*[103] and Olah *et al.*[105,106]

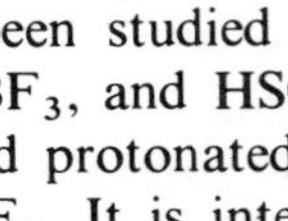

$$\mathrm{R{-}C}\overset{+}{\begin{matrix}\mathrm{O{-}H}\\ \mathrm{O{-}H}\end{matrix}}\qquad \mathrm{R{-}C}\overset{+}{\begin{matrix}\mathrm{O{-}H}\\ \mathrm{O{-}H}\end{matrix}}$$

(XIII) 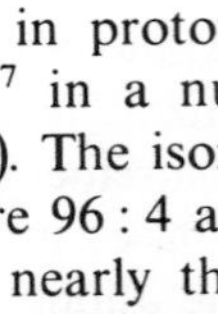 **(XIV)**

The configurational equilibria (**XIII** ⇌ **XIV**) in protonated formic and acetic acid has been studied by Hogeveen[107] in a number of solvents (HF–SbF_5, HF–BF_3, and HSO_3F–SbF_5–SO_2). The isomer ratios of protonated acetic and protonated formic acids are 96 : 4 and 77 : 23, respectively, in HF–SbF_5. It is interesting to note nearly the same activation energy is required for the equilibration of protonated acetic acid in different solvent systems (Table X). This implies that differences in solvation effects are nearly the same in initial and transitional states, and that the energy barrier measured is to be associated with intrinsic property of the ion.

TABLE VIII

RELATIVE RATES OF CONVERSION OF PROTONATED ACID TO OXOCARBONIUM ION

Protonated acid	Rate of conversion (sec^{-1}) (Rate constant)
$CH_3CO_2H_2^+$	1.00
$CH_3CH_2CO_2H_2^+$	1.46 ± 0.05
$(CH_3)_2CHCO_2H_2^+$	1.23 ± 0.06
$(CH_3)_3CCO_2H_2^+$	0.56 ± 0.02
$(CH_3)_2CHCH_2CO_2H_2^+$	1.15 ± 0.06

Brouwer and Kiffen[108] studied the conformational equilibration of protonated pivalic acid.

$$(CH_3)_3C{-}C(O{-}H_S)(O{-}H_A)^+ \overset{(i)}{\rightleftharpoons} (CH_3)_3C{-}C(O{-}H_A)(O{-}H_S)^+$$

A = Anti S = Syn

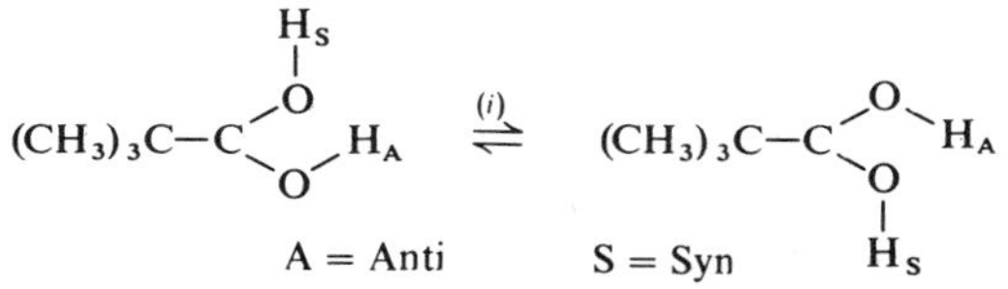

They found that different solvents can have markedly different effects on the rate of reaction (i), which is not surprising. Some disagreement exists in the literature as to the predominant isomer present in protonated pivalic acid. In an $HF{-}BF_3$ solution at $-75°C$, only a single OH absorption has been reported.[109] This was attributed to the large bulk of the tertiary butyl group causing the hydroxyl protons to adopt an all-*trans* configuration

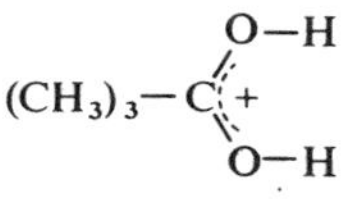

$$(CH_3)_3{-}C^+(O{-}H)(O{-}H)$$

Since no other carboxylic acid has been found to exist in this conformation and since in an $HSO_3F{-}SbF_5$ solutions at $-60°C$ two OH absorption are

TABLE IX

ACTIVATION PARAMETERS CALCULATED AT 0°C

	$\Delta H \pm$	$\Delta S \pm$
$CH_3CO_2H_2^+$	16.3 ± 1.5	-13.1 ± 5
$CH_3CH_2CO_2H_2^+$	14.8 ± 0.5	-18.6 ± 3
$(CH_3)_2CHCO_2H_2^+$	14.7 ± 0.3	-19.0 ± 3
$(CH_3)_3CCO_2H_2^+$	15.9 ± 0.5	-15.5 ± 3

TABLE X

ACTIVATION ENERGY

Protonated acid	E_a	Solvent
CH_3COOH	13.5 ± 0.6	HF–BF_3
CH_3COOH	12.6 ± 0.6	HSO_3F–SbF_5SO_2
CH_3COOH	11.2 ± 0.2	HF–SbF_5
$HCOOH$	15.3 ± 0.9	HF–SbF_5

found,[105] there is no reason to believe that protonated pivalic acid exists in other than the conformer

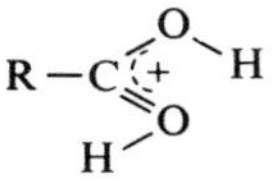

$$R-C\overset{\diagup O \diagdown H}{\underset{\diagdown O \diagup H}{\cdot\!+}}$$

and that the result obtained in HF–BF_3 is a result of the rotation about the $C{=}O^+H$ bands not being "frozen out" on the NMR time scale.

The secondary and tertiary oxocarbonium ions, $C_5H_{11}CO^{\oplus}$ (obtained by initial dehydration of any secondary or tertiary carboxylic acid $C_5H_{11}COOH$), interconvert quite readily at room temperature in an HSO_3F–SbF_5 (1 : 1 m/m) solution by a process of decarbonylation, rearrangement, and carbonylation.[110] The free enthalphy of activation (ΔG) for the conversion of 2-methylbutyl-2-oxocarbonium ion to 2-methybutyl-3-oxocarbonium ion is 21.2 kcal/mole at 20°C. The rate of carbonylation is 3×10^7 liter mole^{-1} sec^{-1} at 20°C.

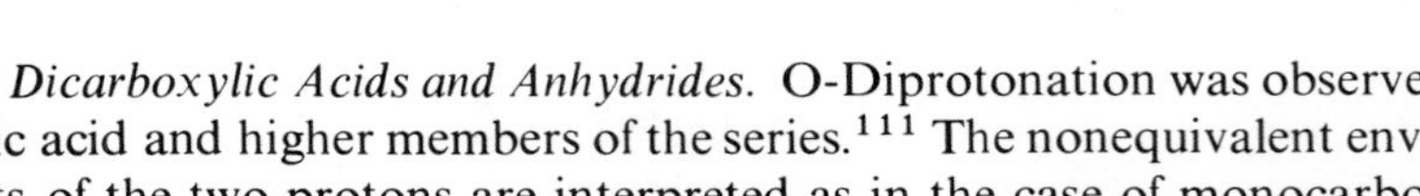

$$\underset{\textbf{(XV)}}{C-\underset{CO^+}{\overset{C}{C}}-C-C} \underset{CO}{\overset{-CO}{\rightleftharpoons}} C-\overset{C}{\underset{+}{C}}-C-C \rightleftharpoons C-\overset{C}{C}-\underset{+}{C}-C \underset{-CO}{\overset{CO}{\rightleftharpoons}} \underset{\textbf{(XVI)}}{C-\overset{C}{C}-\underset{CO^+}{C}-C}$$

The enthalpy diagram of the interconversion **(XV)** ⇌ **(XVI)** in HOS_3F–SbF_5 at 20°C is shown in Fig. 2.

a. *Dicarboxylic Acids and Anhydrides.* O-Diprotonation was observed for oxalic acid and higher members of the series.[111] The nonequivalent environments of the two protons are interpreted as in the case of monocarboxylic acids due to the structure **(XVII)** being the predominant species.

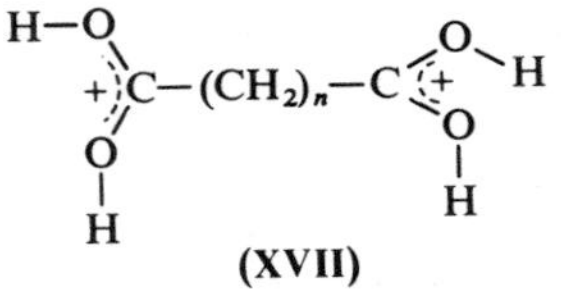

$$\underset{\textbf{(XVII)}}{\overset{H-O}{\underset{O-H}{\cdot\!+}}C-(CH_2)_n-C\overset{O-H}{\underset{O-H}{\cdot\!+}}}$$

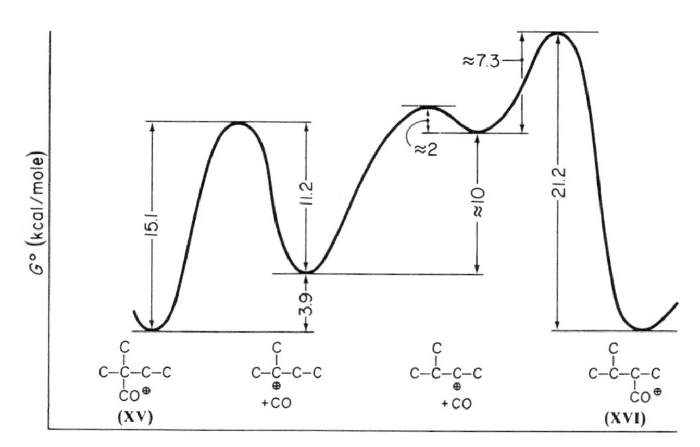

FIG. 2. Free enthalpy diagram of the interconversion (XV)⇌(XVI) in $FHSO_3/SbF_5$ at 20°C. (Concentrations expressed in mole/liter.) H. Hogeveen and C. F. Roobeek, *Recl. Trav. Chim. Pays-Bas* **89**, 1121 (1970).

All the aliphatic dicarboxylic acids (oxalic, malonic, succinic, glutaric, adipic, and pimelic) were completely diprotonated and they were able to observe, in certain cases, cleavage at higher temperatures to the corresponding monocarboxonium-monooxocarbonium ion and the dioxocarbonium ions.

$$HOOC(CH_2)_nCOOH \xrightarrow[-60°C]{HSO_3F-SbF_5-SO_2} H_2O_2C(CH_2)_nCO_2H_2 \xrightarrow[\Delta(-H_2O)]{HSO_3F-SbF_5-SO_2}$$

$$H_2O_2{}^+(CH_2)_nCO^+ \xrightarrow[\Delta(-H_2O)]{HSO_3F-SbF_5-SO_2} OC(CH_2)_nCO$$

(monocarboxonium-monooxocarbonium ion) dioxocarbonium ion

The ease with which alkylene dicarboxonium (diprotonated dicarboxylic acids) can be dehydrated diminishes with decreasing separation between the acid functions.

Squaric acid (1,2-dihydroxycyclobutenedione) is considered to be a dicarboxylic acid. The NMR of this squaric acid showed a diprotonated species which is represented in the following resonance forms:

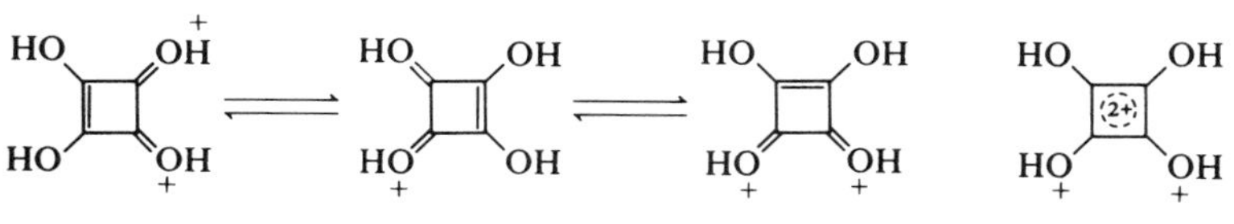

b. *Hydroxycarboxylic Acids.* NMR spectra of a number of aliphatic hydroxycarboxylic acids showed complete diprotonation[112]:

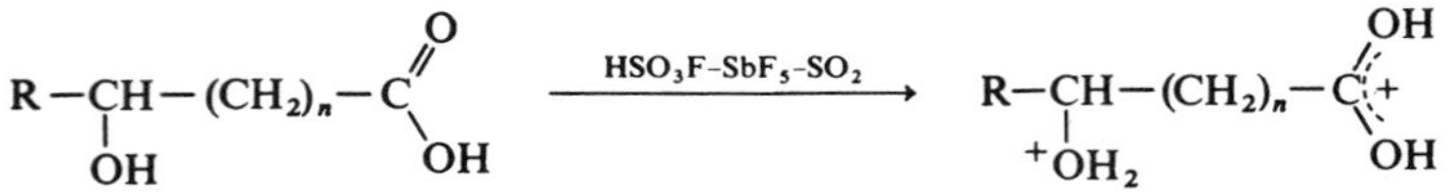

$$R-CH(OH)-(CH_2)_n-C(=O)OH \xrightarrow{HSO_3F-SbF_5-SO_2} R-CH(^+OH_2)-(CH_2)_n-C^+(OH)_2$$

Depending on the relative position of the OH and COOH group, the dehydration of protonated α-, β-, and γ-hydroxycarboxylic acid takes place to form the corresponding lactide, α,β-unsaturated carboxylic acid and lactone, respectively.

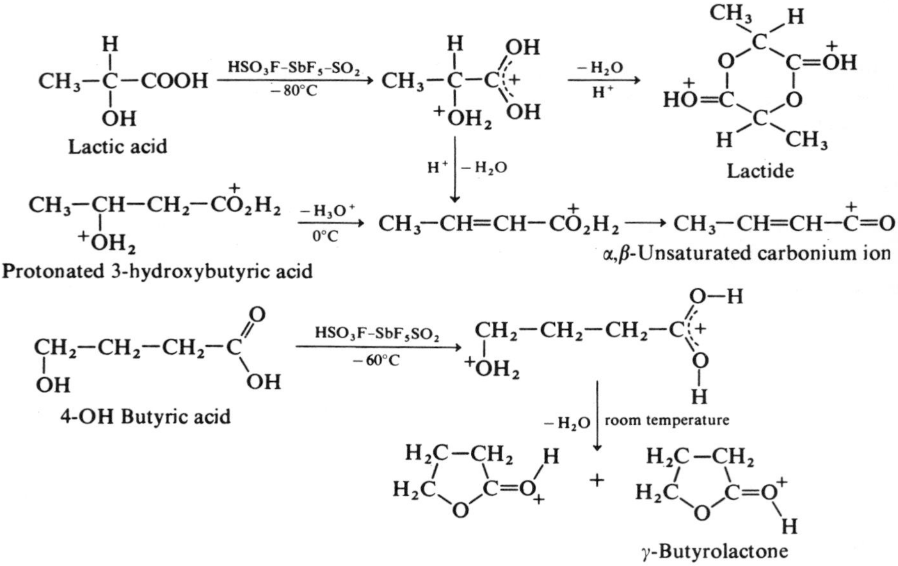

$$CH_3-CH(OH)-COOH \xrightarrow[-80°C]{HSO_3F-SbF_5-SO_2} CH_3-CH(^+OH_2)-C^+(OH)_2 \xrightarrow[H^+]{-H_2O} \text{Lactide}$$

Lactic acid

Lactide

$$CH_3-CH(^+OH_2)-CH_2-C\overset{+}{O}_2H_2 \xrightarrow[0°C]{-H_3O^+} CH_3-CH=CH-C\overset{+}{O}_2H_2 \longrightarrow CH_3-CH=CH-\overset{+}{C}=O$$

Protonated 3-hydroxybutyric acid

(from the protonated lactic acid: H^+, $-H_2O$)

α,β-Unsaturated carbonium ion

$$CH_2(OH)-CH_2-CH_2-C(=O)OH \xrightarrow[-60°C]{HSO_3F-SbF_5SO_2} CH_2(^+OH_2)-CH_2-CH_2-C^+(OH)_2$$

4-OH Butyric acid

$$\xrightarrow[\text{room temperature}]{-H_2O} \text{protonated } \gamma\text{-butyrolactone (two isomers)}$$

γ-Butyrolactone

c. *Aromatic Hydroxy Acids.* Salicylic acid and salicylic aldehyde are only monoprotonated while the other isomers of hydroxybenzoic acids are diprotonated in magic acid solution.[113] The protonation occurs on the carboxyl and aldehyde groups, respectively. The nonbonded electron pairs of the phenol oxygen atom undergo hydrogen bonding with the protonated acid and aldehyde group, preventing their own protonation. This results, in the case of salicylic acid, in the nonequivalence of the two acidic OH protons. The diprotonated *m*- and *p*-hydroxy benzoic acids on warming cleave to give a dication

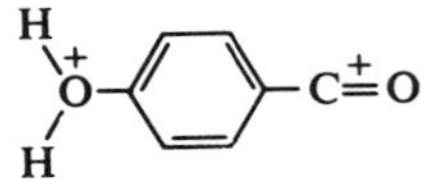

$$H_2\overset{+}{O}-C_6H_4-\overset{+}{C}\equiv O$$

Protonated salicylic acid gives benzyne and its polymeric product.

d. *α,β-Unsaturated Acids.* Olah and Calin[114] observed the O-protonation of acrylic, α-methylacrylic, β-methylacrylic and β,β-dimethylacrylic acids in $HSO_3F–SbF_5–SO_2$ solutions by NMR methods at −20° and −95°C. The change in the OH shift (which occurs at a higher field than in protonated saturated carboxylic acids) as a function of the temperature proved to be a valuable method for investigating the stereoisomerism of the protonated compounds. Furthermore, the kinetics of cleavage to alkenyloxocarbonium ions were measured and compared with those of protonated saturated carboxylic acids (Tables XI and XII).

$$R_2C{=}CRCOOH \xrightarrow[-20^\circ \text{ to } -95^\circ C]{HSO_3F-SbF_5-SO_2} R_2C{=}CRCO_2H_2^+ \xrightarrow[0^\circ-40^\circ C]{HSO_3F-SbF_5} R_2C{=}CRCO^+ + H_3O^+$$

Stable crystalline salts of these ions have been observed by ionizing alkenoyl fluorides in SbF_5.[115]

e. *Keto Acids.* In the fluorosulfuric superacid system keto acids are completely diprotonated.[116] Like hydroxy acids, the mode of cleavage of diprotonated aliphatic keto acids to protonated ketooxocarbonium ions is dependent on the distance between the keto and the carboxyl group. Acetylbutyric and acetylvaleric acids undergo complete dehydration at 0°C to give the corresponding protonated ketooxocarbonium ions and equimolar amounts of H_3O^+

$$CH_3{-}\overset{\overset{+}{O}H}{\overset{\|}{C}}{-}(CH_2)_n{-}COOH_2^+ \xrightarrow[0^\circ C]{HSO_3F-SbF_5-SO_2} CH_3{-}\overset{\overset{+}{O}H}{\overset{\|}{C}}{-}(CH_2)_n{-}\overset{+}{C}{=}O + H_3O^+$$

Diprotonated levulinic acid was stable up to +60°C. Acetoacetic acid gives a complex and yet unidentified products at 0°C. Protonated 2-benzoyl benzoic acid undergoes dehydration, and acylation at room temperature to give diprotonated anthraquinone.

f. *Acid Anhydrides.* (i) Acyclic carboxylic acid anhydrides. Protonation of acyclic carboxylic anhydrides in $HSO_3F–SbF_5–SO_2$ solutions, which

TABLE XI

RELATIVE RATES OF DEHYDRATION OF PROTONATED α,β-UNSATURATED CARBOXYLIC ACIDS TO OXOCARBONIUM IONS

Carboxylic acid	Rate of dehydration (sec^{-1})
$CH_3CO_2H_2^+$	1.00
$(CH_3)_2C{=}CHCO_2H_2^+$	0.48
$CH_2{=}C(CH_3)CO_2H_2^+$	0.12
$CH_3CH{=}CHCO_2H_2^+$	0.09

TABLE XII

ACTIVATION PARAMETERS CALCULATED AT 0°C

	$\Delta H \pm$	$\Delta S \pm$
$CH_3CO_2H_2^+$	16.3	−13.1
$(CH_3)_2C{=}CHCO_2H_2$	16.7	−8.8
$CH_2{=}C(CH_3)CO_2H_2^+$	19.0	+1.7
$CH_3CH{=}CHCO_2H_2^+$	19.8	+3.7

leads to the formation of intra- and intermolecularly rapid proton exchanging protonated anhydrides, was studied by PMR spectroscopy.[104,117] The anhydrides undergo mono protonation in less than 1 mole $HSO_3F{-}SbF_5{-}SO_2$ solution relative to acid anhydrides at −78°C.

A typical one of the intermolecular exchange with precursor **(XVIII)**

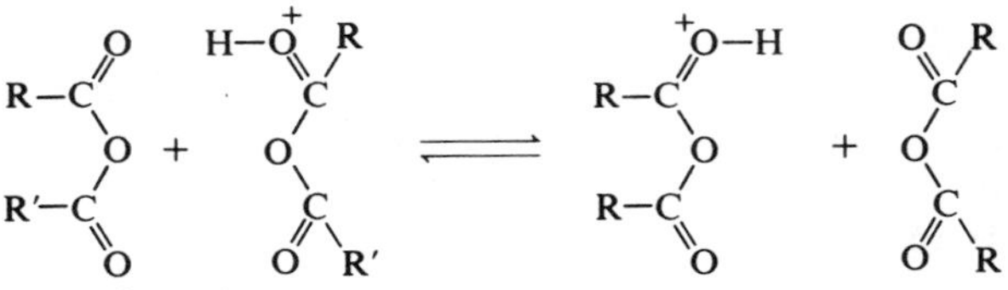

$$\underset{\textbf{(XVIII)}}{R{-}C({=}O){-}O{-}C({=}O){-}R'} + R{-}C({=}\overset{+}{O}{-}H){-}O{-}C({=}O){-}R' \rightleftharpoons R{-}C({=}\overset{+}{O}{-}H){-}O{-}C({=}O){-}R + R{-}C({=}O){-}O{-}C({=}O){-}R$$

PMR shows more than one proton and additional intermolecular exchange process, when these anhydrides were protonated in 1 : 1–1 : 5 mole equivalents of acid solution. When acyclic anhydrides were protonated in a large excess of superacid, cleavage reactions took place to produce oxocarbonium ions and protonated carboxylic acid:

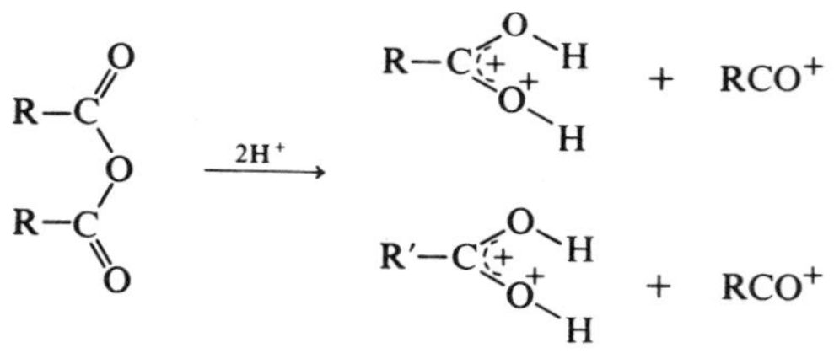

$$R{-}C({=}O){-}O{-}C({=}O){-}R \xrightarrow{2H^+} R{-}C(OH)_2^+ + RCO^+ \qquad R'{-}C(OH)_2^+ + RCO^+$$

Symmetrical anhydrides give a single protonated acid and the corresponding oxocarbonium ion. Mixed anhydrides give both protonated acids and oxocarbonium ions.

Formic anhydride and trifluoroacetic anhydride behave differently from the other acyclic anhydrides. The formyl cation, HCO^+, was not observed as a long-lived species in superacid solution because of the ease of proteolytic cleavage to CO. Attempts to obtain the formyl cation by several routes including one under CO pressure were unsuccessful. Protonation of trifluoroacetic anhydride was not achieved at low temperatures, indicating that the CF_3 methyl group is highly electron withdrawing.

(ii) Cyclic anhydrides. Cyclic carboxylic anhydrides behave differently from their acyclic analogs in super-acid solutions. Aromatic and unsaturated cyclic anhydrides, e.g., maleic anhydride, dimethylmaleic, and 3,4,5,6-tetrahydrophthalic anhydride show remarkable stability in super-acid media. They explained this on the basis that they contain four adjacent sp^2 carbon atoms and thus favor the formation of a five-membered ring. Even if this could be cleaved in superacid media, the recycling process to the corresponding anhydrides was very rapid. In contrast, acyclic saturated dicarboxylic acids (such as glutaric acid) are dehydrated to diacyl cations. The reactions of succinic and glutaric anhydrides[111] have been reexamined by Olah, Mo, and Grant[118] in $HSO_3F–SbF_5–SO_2$ solutions. Protonated anhydrides could not be observed even when the solutions were prepared and examined at $-80°C$, as they were cleaved in magic acid, giving the corresponding alkylene–oxocarbonium ions $OC(CH_2)_nCOOH_2^+$ ($n = 2$ and 3, respectively). The reactions of glutaric, succinic, and diglycolic anhydrides were carried out in SO_2 solutions containing different amounts of HSO_3F and SbF_5.

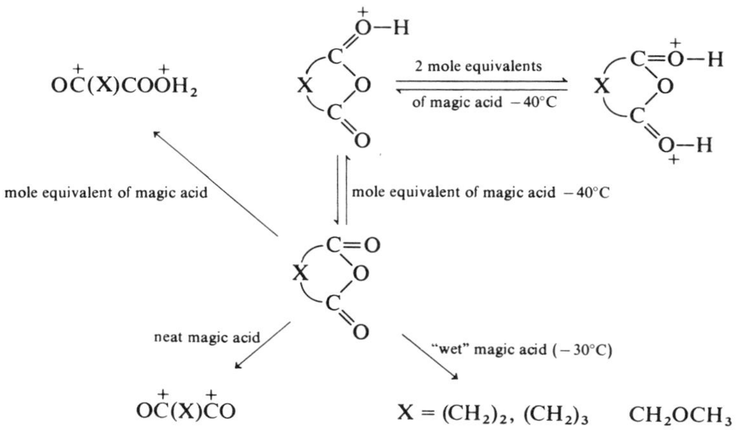

$\overset{+}{O}C(X)CO\overset{+}{O}H_2$

2 mole equivalents of magic acid −40°C

mole equivalent of magic acid

mole equivalent of magic acid −40°C

neat magic acid

"wet" magic acid (−30°C)

$\overset{+}{O}C(X)\overset{+}{C}O$

$X = (CH_2)_2, (CH_2)_3 \quad CH_2OCH_3$

In phthalic anhydride intermolecular exchange process involving mono-, or diprotonated pthalic anhydride, such as **(XIX)** + **(XIXa)** ⇌ **(XIXa)** +

(XIX), (XIXa) + (XIXb) ⇌ (XIXb) + (XIXa); and **(XIXa) + H$^+$ ⇌ (XIXc) ⇌ (XIXb) + H$^+$** takes place

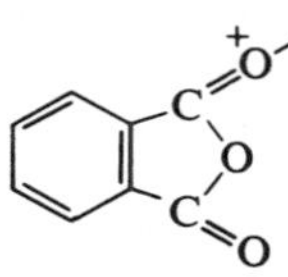

(XIXa) 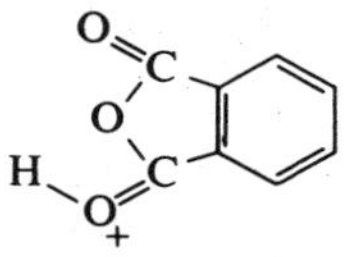 **(XIXb)** 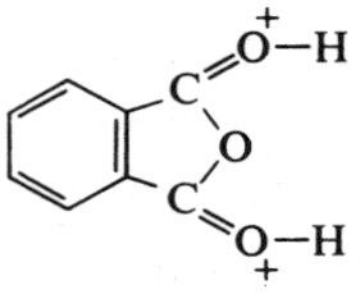 **(XIXc)**

g. *Esters.* The protonation of esters was studied by Olah, O'Brien, and White.[119] The chemical shifts of the proton on oxygen in protonated esters (−12.2 to −13.5 ppm) was intermediate between that of protonated aldehydes and ketones (~ −15 to −16 ppm) and that of protonated ethers (~ −9 to 0 ppm). Spectral evidence was found for the existence of isomers of protonated formates.

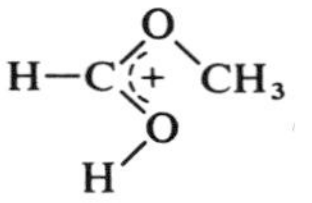

$$H-C(\overset{+}{\cdots})(OCH_3)(OH) \quad (90\%) \qquad\qquad H-C(\overset{+}{\cdots})(OCH_3)(O-H)$$

90%

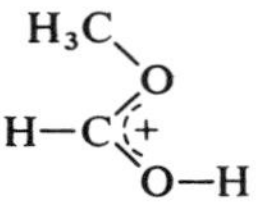

Both acyl oxygen and alkyl oxygen cleavage was observed for esters of primary aliphatic alcohols.

Methyl acetate cleaves via acyl oxygen fission

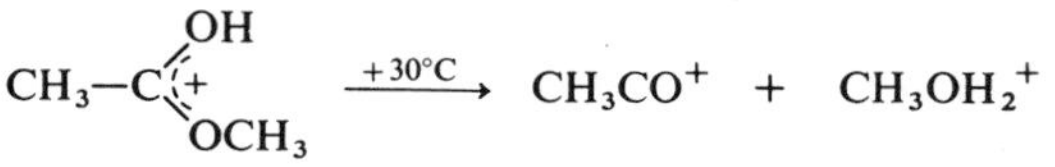

$$CH_3-C^{+}(OH)(OCH_3) \xrightarrow{+30°C} CH_3CO^+ + CH_3OH_2^+$$

Where as *sec*-butyl acetate undergoes alkyl oxygen cleavage at −30°C.

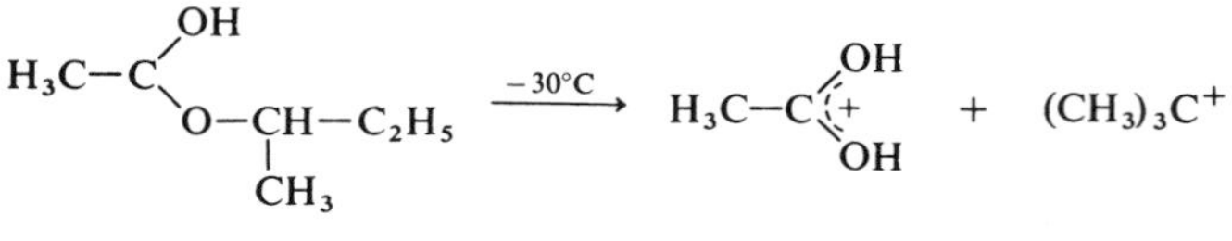

$$H_3C-C(OH)-O-CH(CH_3)-C_2H_5 \xrightarrow{-30°C} H_3C-C^{+}(OH)(OH) + (CH_3)_3C^+$$

They found that the enthalpy of activation for cleavage of the investigated series of methyl esters were about 6 kcal/mole higher than for the corresponding carboxylic acids. This is due to the electron donor properties of the methyl group.

Esters of tertiary alcohols such as anhydrides of carboxylic acids cleave so rapidly that only the protonated acid and tertiary carbonium ion could be observed, even at −80°C.

Ethyl esters (e.g., ethyl acetate) undergo alkyl oxygen cleavage in a 1 : 1 acid system. The reaction sequence is:

$$CH_3C(OH)(OCH_2CH_3)^+ \xrightarrow{1:1\ \text{acid}} CH_3C(OH)_2^+ + [CH_3CH_2]^+$$

$$CH_3C(OH)_2^+ \longrightarrow CH_3\overset{+}{C}{=}O + H_3O^+$$

$$[CH_3CH_2]^+ \longrightarrow CH_3CH_2F \cdot SbF_5 \longrightarrow (CH_3)_3C^+$$

$$[CH_3CH_2]^+ \longrightarrow (CH_3)_3C^+$$

Under conditions of lower acidity [2 : 1, 3 : 1, and 4 : 1, etc.] acyl oxygen cleavage was also observed.

4. Thiocarboxylic Acids and Their Esters (*S*-Alkyl)

Thiocarboxylic acids were protonated in HSO_3F–SbF_5 solution.[120] Three isomers of protonated thio acids were found and identified by NMR spectroscopy.

The protonation of a series of thio acid esters and the rates and mechanism of their cleavage was also investigated.[121] Primary and secondary *S*-alkylthioacetates cleave via acyl–sulfur fission, whereas alkyl–sulfur fission was found for *S*-*tert*-alkylthioacetates.

$$R{-}C(OH)(SR')^+ \longrightarrow R{-}C(OH)(SH)^+ + R'^+ \quad + \quad R'{=}C(CH_3)_3$$

$$R{-}C(OH)(SR')^+ \longrightarrow R{-}C{\equiv}O^+ + R'OH_2^+$$

$$[R'{=}CH_3 \text{ and } (CH_3)_2CH]$$

In the case of *O*-alkyl thioesters, only esters of tertiary alcohol were found to cleave yielding *tert*-alkylcarbonium ions and protonated thioformic acid:

$$R{-}C(OC(CH_3)_3)(SH)^+ \longrightarrow R{-}C(OH)(SH)^+ + C(CH_3)_3^+$$

Two isomeric species in the ratio (95 : 5) were found for protonated methyl and ethylthion acetates.

$$\underset{\text{(XX)}}{R{-}C(\cdots S{-}H)(\cdots O{-}R)} \qquad \underset{\text{(XXI)}}{R{-}C(\cdots S{-}H)(\cdots O{-}R)}$$

A number of dithioesters methyl, ethyl, *n*-propyl, isobutyl, and *tert*-butyldithioacetates were protonated on the thiocarbonyl S atom in a 4 : 1 HSO_3F–SbF_5–SO_2 solution at $-60°C$.

$$R{-}C({=}S){-}SR' \longrightarrow RC^{+}(\cdots SH)(\cdots SR')$$

Protonated primary and secondary dithioacetates were stable up to 100°C. Protonated *tert*-butyldithioacetate and dithiopropionate, however, undergo alkyl sulfur cleavage at 30° to $-20°C$ to the corresponding protonated dithiocarboxylic acids.

5. Carbamic Acids and Alkylcarbamates

Olah and Calin[122] observed carbonyl protonation of alkyl carbamates and carbamic acids in HSO_3F–SbF_5 independently of the nature of the N and O substituents. The cleavage products (alkyl oxygen cleavage) were the corresponding carbonium ion and protonated carbamic acid.

$$RR'N{-}C({=}O){-}OR'' \xrightarrow[-60°C]{HSO_3F\text{–}SbF_5\text{–}SO_2} RR'N{\cdots}C^{+}(OH){-}OR'' \xrightarrow{\Delta} RR'N{\cdots}C^{+}(OH)(OH) + R^{+}$$

The rate of cleavage followed the order *tert*-alkylcarbamate > secondary > primary > methyl. Both protonated alkyl carbamates and carbamic acids show *cis–trans* isomerism due to the partial double bond character of the C–N bond.

In contrast to the usual O-protonation of carbamate esters, *N,N*-diisopropyl-carbamate esters (methyl and ethyl) have been shown to be N-protonated in HSO_3F and 98% H_2SO_4 solution.[123] Olah, White, and Ku[124] suggested that this observation by Moodie[123] was caused by $=\overset{+}{O}H \rightarrow \overset{+}{N}H$ rearrangement following usual O-protonation of the carbamates. The same authors[123] reexamined alkylcarbamates with bulky substituents on N (such as isopropyl, *tert*-butyl) in HSO_3F-SO_2 or HSO_3F-SO_2ClF. In these strong acid solutions at low temperatures, the carboxyl oxygen atoms of carbamic acid esters are kinetically protonated first. However, carbamates with two isopropyl group substituents on N, such as methyl *N,N*-(diisopropyl)carbamate and ethyl *N,N*-(diisopropyl)-carbamate rearrange slowly to the N- protonated species upon raising the temperature. They have also studied acidity function dependence and the kinetic vs. thermodynamic aspects of these alkyl carbamates.

A series of dialkylhydrazo diformates $[RO_2CNH]_2$ which are bisalkylcarbamates were protonated on carbonyl oxygen as are alkyl carbamates in $HSO_3F-SbF_5-SO_2$ and/or $HF-SbF_5-SO_2$ solution regardless of substituents.[125] At higher temperatures these diprotonated dialkylhydrazo diformates cleave to give stable alkyl carbonium ions and diprotonated hydrazo diformic acid $[(NHCO_2H_2^+)_2]$. A representative example is di-*tert*-butylhydrazodiformate which when treated in magic acid solution cleaves at −80°C to give *tert*-butyl cation and diportonated hydrazodiformic acid.

$$t\text{-}C_4H_9OOCNHNHCOO\text{-}t\text{-}C_4H_9 \xrightarrow[-80^\circ C]{HSO_3F-SbF_5-SO_2} (HO)_2\overset{+}{C}{=}NH{-}NH{=}\overset{+}{C}(OH)_2 + 2(CH_3)_3C^+$$

6. Amides and Thioamides

For a long time there has been uncertainty as to the site of protonation in amides. Oxygen protonation of amides and ureas and sulfur protonation of thioamides and thioureas has been challenged by several investigators.[126–129] The principal evidence favoring O-protonation has been based on NMR spectroscopy while the evidence for N-protonation was provided by IR spectroscopy and some basicity measurements of substituted aromatic amides using UV measurements.[129,130] These conflicting arguments have been reviewed and it is concluded that the weight of the experimental evidence favors O-protonation.[131] It was not possible to observe separate signals in NMR for amides in H_2SO_4, owing to rapid exchange with the solvent. The use of stronger acid systems, like fluorosulfuric acid, and of lower temperatures has confirmed the oxygen protonation of amides and thioamides.[132–134] Acetamide, *N*-methylacetamide, *N,N*-dimethylacetamide, formamide, and *N,N*-dimethylformamide in HSO_3F were all O-protonated. The protons on the oxygen in amides occurred at a more shielded position ($\sim\delta 10$) compared to that in protonated carbonyls ($\sim\delta 14$–16) and protonated esters ($\sim\delta 12$). This has been explained on the basis of charge delocalization and the considerable contribution from resonance form

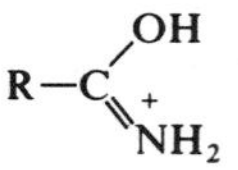

$$R-C(OH)=\overset{+}{N}H_2$$

where the positive charge resides on nitrogen.

The S-protonation of thioacetamide and thioacetanilide in HSO_3F at low temperatures has been observed by Birchall and Gillespie.[133] The proton on S ($\delta 5.72$) is slightly shielded compared to the protons on protonated thiols and sulfides.[88] As with the protonated amides considerable contribution to the resonance form comes from the one where the positive charge is on nitrogen atom.

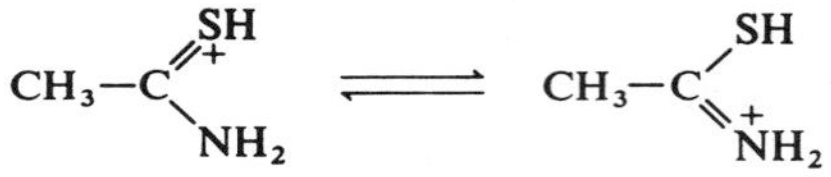

$$CH_3-C(=\overset{+}{S}H)-NH_2 \rightleftharpoons CH_3-C(SH)=\overset{+}{N}H_2$$

(XXII) (XXIII)

7. Urea and Guanidine

The behavior of urea and guanidine in strong acids has not, however, been clearly established. In H_2SO_4, an earlier investigation,[133] using cryoscopic methods, found tetramethyl urea to be diprotonated and guanidine to be tetraprotonated. Subsequent investigations in the same acid system showed only monoprotonation for both bases. Birchall and Gillespie,[133] using a 20 : 1 mixture of HSO_3F–SbF_5, studied urea and dimethyl urea. The absence of NH and OH protons was attributed to diprotonation of the bases. Olah and White[135] using a much higher acid system (1 : 1 mixture of HSO_3F–SbF_5) reexamined urea and substituted ureas. In addition, they also examined the behavior of some guanidine bases, biologically important urea base "biotin" in the superacid system.

Urea was diprotonated in HSO_3F–SbF_5 and no restricted rotation around the C–N or C–O bonds was observed.

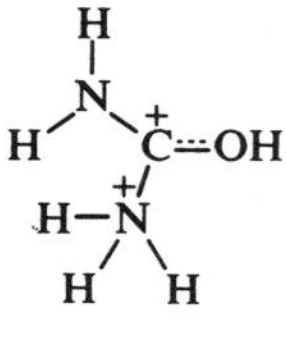

$$H_2N-\overset{+}{C}(\cdots OH)-\overset{+}{N}H_3$$

In asymmetrically substituted ureas the site of protonation was at the least substituted nitrogen. In 1,1-dimethylurea and tetramethylurea nonequivalence of the methyl groups suggested the barrier to rotation about the C–N bond. The ΔG for rotation (about 20 kcal/mole) suggested the predominant contribution of the immonium form rather than the resonance structure, which places the positive charge on either the central C atom or on the O atom.

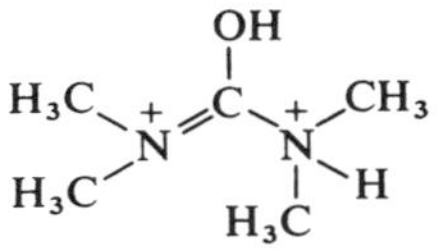

$$(H_3C)_2\overset{+}{N}{=}C(OH){-}\overset{+}{N}(CH_3)(H)(CH_3)$$

Guanidines were also diprotonated in the magic acid system:

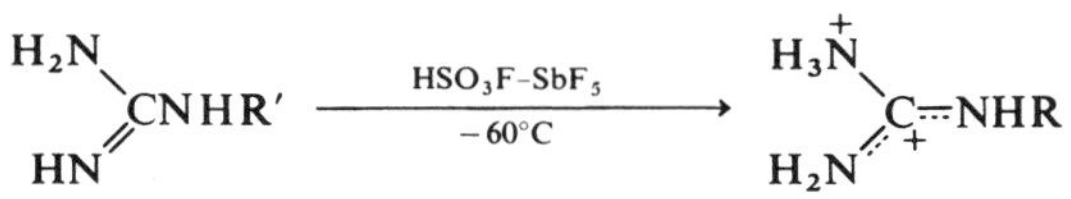

$$\begin{matrix} H_2N \\ HN \end{matrix}\!\!>\!CNHR' \xrightarrow[-60^{\circ}C]{HSO_3F\text{-}SbF_5} \begin{matrix} H_3\overset{+}{N} \\ H_2N \end{matrix}\!\!>\!\overset{+}{C}{=}NHR$$

The lower ΔG value (15 kcal/mole) for hindered rotation around C–N bond for diprotonated guanidine compared to diprotonated ureas demonstrated the greater charge delocalization and lower C–N bond order.

Protonation at all three sites (sulfide, carbonyl, and urea carbonyl oxygen) of biotin, a biologically important compound related to urea, takes place in the superacid medium. It is interesting to note that only monoprotonation is found for the urea base in the molecule.

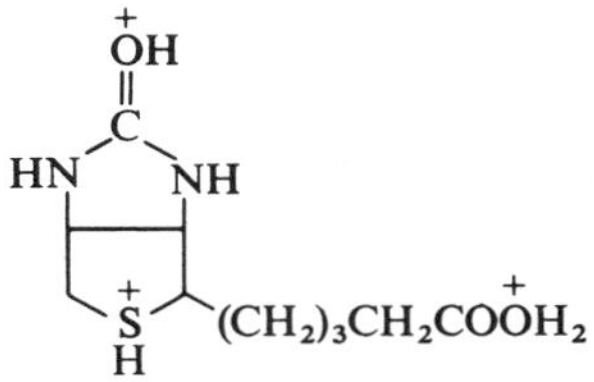

$$\text{HN}{-}C({=}\overset{+}{O}H){-}\text{NH (fused ring)};\quad \overset{+}{S}H{-}(CH_2)_3CH_2CO\overset{+}{O}H_2$$

S–H coupling constants indicated *trans* S–H proton conformation to the valeric acid chain. At higher temperatures, the protonated acid group is dehydrated to the oxocarbonium ion, without any other changes in the molecule. This study demonstrates the versatility of the magic acid solvent system for structural studies in relatively complex molecules in addition to other simple molecules.

NMR evidence shows diprotonation of thiourea both in 100% H_2SO_4 and in HSO_3F.[133] In magic acid both seleno and thioureas are diprotonated.

8. Amino Acids and Peptides

Protonation of amino and carboxyl groups as well as other available basic sites were found for amino acids.[136] No dehydration of the protonated α- and β-amino acids to oxocarbonium ions as with the aliphatic carboxylic acids[104] was observed by Olah *et al.*[136] However, they observed some cleavage of γ-protonated and complete cleavage of δ-protonated amino acids.

$$H_3N^+(CH_2)_nCO_2H_2^+ \xrightarrow{HSO_3F-SbF_5} H_3N^+(CH_2)_nCO^+ + H_3O^+$$

$$n = 3 \text{ or } 4$$

Glycine, L-alanine, L-valine, L-leucine, L-phenylalanine, L-proline, L-lysine, and L-glutaric acids are representative acids studied. The ease with which diprotonated aminocarboxylic acids dehydrated in the acid system increased with the separation of the protonated amino group from the protonated carboxyl group. This is very similar to the dehydration of diprotonated dicarboxylic acids.[111]

No protonation of the disulfide linkage was found in L-cystine, but S-protonation was observed at δ6.52 in L-methionine. Diaminomonocarboxylic and monoaminodicarboxylic acids such as L-lysine, aspartic acid, and glutaric acids are all triprotonated. Amino acids with guanidine groups (arginine and homoarginine) were tetraprotonated. In asparagine and glutamine the amide groups on side chain were O-protonated. Amino acids containing aromatic groups reacted with the acid system.

Protonation on carbonyl oxygen of peptide linkages was observed in addition to protonation of other basic sites. Simple peptides like di-, tri-, and tetraglycines were found to be tri-, tetra-, and pentaprotonated, respectively.

$$H_2NCH_2C(=O)NHCH_2COOH \xrightarrow[-60^\circ C]{HSO_3F-SbF_5-SO_2} H_3\overset{+}{N}CH_2C(=\overset{+}{O}H)NHCH_2CO_2\overset{+}{H}_2$$

The imidazole ring in histidylhistidine were observed to be protonated as in histidine itself, but over the temperature range of −20° to −90°C, protonated carboyl group, protonated peptide linkage (OH and NH), and imidazole ring NH protons could not be detected. This is probably because of their rapid chemical exchange.

$$\begin{array}{c} \overset{+}{N}H_3 \quad O \quad\quad CO_2H_2^+ \\ HC{=}C{-}CH_2{-}CH{-}\overset{\|}{C}NH\overset{|}{C}CHCH_2C{=}CH \\ HN\overset{+}{\cdots}C(H)\cdots NH \quad\quad\quad\quad HN\overset{+}{\cdots}C(H)\cdots NH \end{array}$$

Lactams like 2-pyrrolidinone, δ-valerolactam,. and ε-caprolactam in HSO_3F–SbF_5–SO_2 were all O-protonated and were quite stable even when maintained at 40° for 4 hr.

Protonation of a number of *N*-alkoxycarbonyl-substituted amino acids (of glycine, L-alanine, L-valine, L-leucine, L-aspartic acid, etc.) in HSO_3F–SbF_5–SO_2 solution have been examined.[137] Alkyl–oxygen cleavage was observed for both benzyloxycarbonyl and *tert*-butyloxycarbonyl derivatives at −70°C in HSO_3F–SbF_5–SO_2 solutions (dication having both protonated carboxylic and protonated carbamic acid function).

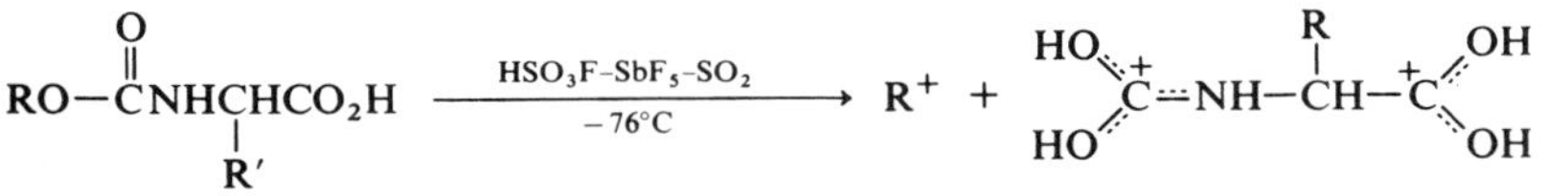

$$RO-\overset{O}{\overset{\|}{C}}NH\underset{R'}{\underset{|}{C}}HCO_2H \xrightarrow[-76°C]{HSO_3F-SbF_5-SO_2} R^+ + (HO)_2\overset{+}{C}{-}NH{-}\overset{R}{\overset{|}{C}}H{-}\overset{+}{C}(OH)_2$$

(dication having both protonated carboxylic and protonated carbamic acid functions)

9. IMINES

Imines were protonated in HSO_3F–SbF_5, HSO_3F, and D_2SO_4–SbF_5 solution.[138] Protonated ketimines like ketones and aldehydes showed hindered rotation about the C=N bond on the NMR time scale. In *N*-propylidenemethylimine, the C-methyl groups were nonequivalent.

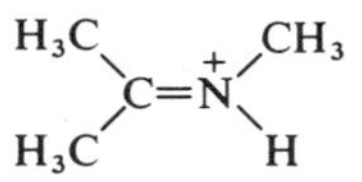

$$(H_3C)_2C{=}\overset{+}{N}(CH_3)H$$

The NMR studies of these protonated imines indicated the predominance of immonium structure ($R_1R_2C{=}N^+HR_3$), with only limited contribution of aminocarbonium forms ($R_1R_2C^+{-}NHR_3$).

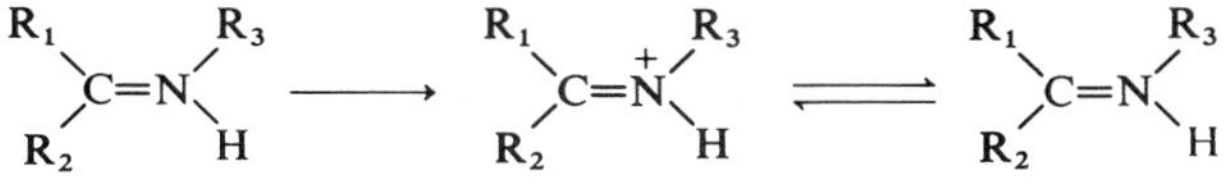

$$R_1R_2C{=}NR_3 \longrightarrow R_1R_2C{=}\overset{+}{N}R_3H \rightleftharpoons R_1R_2C{=}NR_3H$$

In aryl conjugated systems, despite the fact that the amino carbonium ion should be stabilized by the neighboring phenyl group,

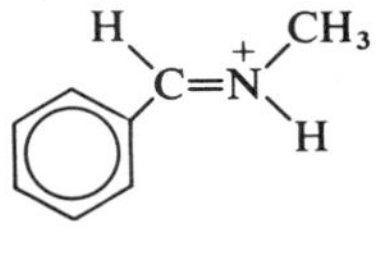

$$C_6H_5(H)C{=}\overset{+}{N}(CH_3)H$$

the NMR chemical shift and coupling data indicate the predominance of immonium form.

10. Aldehydes and Ketones

Protonation reactions of aldehydes and ketones in HSO_3F and $HSO_3F–SbF_5(SO_2)$ have been studied by a number of workers.[102,103,139] Protonated formaldehyde, the simplest of the aldehydes, was generated by passing formaldehyde from the pyrolysis of paraformaldehyde over a stirred solution of $HSO_3F–SbF_5–SO_2$ at −76°C or by the reaction of methylene chloride in the same acid system at −10°C.[140]

$$(CH_2O)_x \xrightarrow[-76°C]{HSO_3F-SbF_5-SO_2} CH_2{=}\overset{+}{O}H$$

$$CH_2Cl_2 \xrightarrow[-10°C]{HSO_3F-SbF_5-SO_2} CH_2{=}\overset{+}{O}H$$

The presence of isomers syn (*cis*) and anti (*trans*) in the aldehydes (in superacids) has been shown by a number of investigators.[102,139,141–145]

$$R{-}C(H){=}\overset{+}{O}{-}H \ \text{(Syn)} \qquad R{-}C(H){=}\overset{+}{O}{-}H \ \text{(Anti)}$$

Syn Anti

Protonated pivaldehyde was found to undergo rapid rearrangement even at −70°C to give methylisopropylketone probably through a tertiary carbonium ion followed by a hydride shift.[139]

$$(CH_3)_3C{-}CH{=}\overset{+}{O}H \longrightarrow (CH_3)_2\overset{+}{C}{-}CH(OH){-}CH_3 \longrightarrow (CH_3)_2CH{-}C({=}\overset{+}{O}{-}H){-}CH_3$$

a. *Aromatic Aldehydes.* The charge distribution in protonated aldehydes and ketones have been evaluated using ^{13}C-NMR spectroscopy.[146] In connection with the evaluation of the charge distribution in protonated aldehydes and ketones the NMR spectra of protonated benzaldehyde and acetophenone are quite revealing. The differences in the behavior of benzaldehyde and acetophenone has been explained in a different manner from ^{19}F chemical shifts in *meta*- and *para*-substituted fluorobenzenes in which the substituent can be protonated. These shifts are related to electronic effects (inductive and resonance) of the substituents.[147] This approach has been applied to *m*- and *p*-fluorobenzaldehyde and acetophenone.

b. *Ketones.* Protonation of ketones in acidic solutions has been indicated by IR and UV measurements.[148] Birchall and Gillespie[102] studied the NMR spectra of protonated acetone and aromatic ketones. A number of cyclic ketones, aromatic ketones, 7-norbornanone, and 7-norbornenone were studied by Brookhart *et al.*[149] Substituted benzophenone was investigated by Sekkur and Kranenburg.[150] A systematic NMR study of a series of aliphatic ketones has been made by Olah, O'Brien and Calin.[139]

For symmetrical ketones, only one proton on oxygen resonance appears. For unsymmetrical ketones, two proton on oxygen resonances appear.

$$R_1R_2C{=}\overset{+}{O}H \qquad R_1R_2C{=}\overset{+}{O}H \rightleftharpoons R_1R_2C{=}\overset{+}{O}H$$

$$R_1 = R_2$$

Oxygen diprotonation was observed for a number of hydroxyketones[151] in $HSO_3F–SbF_5$ solution. Hydroxyacetone also gives monoprotonated acetone.

$$CH_2(OH){-}CO{-}CH_3 \xrightarrow{HSO_3F-SbF_5-SO_2} CH_3{-}\overset{+OH}{\overset{\|}{C}}{-}CH_3 + CH_2(OH){-}\overset{+OH}{\overset{\|}{C}}{-}CH_3$$

or

$$H_2C{-}C(=O){-}CH_3 \text{ with } HO{\cdots}\overset{+}{H}{\cdots}O \text{ (chelated)}$$

c. *Alicyclic Ketones.* From the PMR data of protonated saturated alicyclic ketones[143] in $HSO_3F–SbF_5–SO_2$ and $HSO_3F–SO_2$ solutions at $-60°C$, it has been concluded that the structure of these protonated ketones are best represented by onium ion resonance form **(XXIV)** with little contribution from the hydroxy carbonium ion **(XXV)**

$$(CH_2)_n\,C{=}O \xrightarrow[-60°C]{HSO_3F-SbF_5-SO_2} (CH_2)_n\,\overset{+}{C}{=}OH$$

$$\underset{\textbf{(XXIV)}}{(CH_2)_n\,C{=}\overset{+}{O}H} \rightleftharpoons \underset{\textbf{(XXV)}}{(CH_2)_n\,\overset{+}{C}{-}OH}$$

Hogeveen[152] found a novel ring contraction of cyclohexenones to cyclopentenones in the solvents $HF–BF_3$, $HF–SbF_5$, and $HSO_3F–SbF_5$. The isomerization of cyclohex-2-enone to 3-methylcyclopent-2-enone took place at much higher temperature (130°C) compared to that (50°C) in $HF-SbF_5$. The proposed mechanism is via a dication:

$$\text{cyclohex-2-enone}(=\overset{+}{O}H) + H^+ \rightleftharpoons {}^{+}\text{cyclohexyl}(=\overset{+}{O}H) \longrightarrow H_3\overset{+}{C}\text{-cyclopentyl}(=\overset{+}{O}H) \rightleftharpoons H_3C\text{-cyclopentenyl}(=OH)$$

For a solution containing 46% M SbF_5 the activation parameters were found to be $E_a = 19.2$ kcal/mole. $A = 10^{6.8}$ sec^{-1}. In HSO_3F or H_2SO_4 solution cyclohex-2-enone did not isomerize but gradually decomposed at 150°C.

A number of protonated alkenones were found to cyclize to alkyl tetrahydrofuryl and pyryl cations. The cyclization and transformations of these five- and six-membered rings were carried out in different acids by Brouwer.[153] He found that the rates of all transformation reactions and the ratios of the products formed were found to depend strongly on the acid used (Table XIII). For example, 2-hepten-6-one forms both tetrahydrofuryl and pyryl cations

$$C-\overset{O}{\overset{\|}{C}}-C-C-C=C-C \longrightarrow \text{(XXVI)} \rightleftharpoons \text{(XXVII)}$$

(XXVI) (XXVII)

These ratios are kinetically and not thermodynamically controlled (Table XIV).

Brouwer and VanDoorn[154] estimated the relative acidities of 1 : 1 $HF-SbF_5$, 9 : 1 $HF-SbF_5$, 1 : 1 HSO_3F-SbF_5, and 5 : 1 HSO_3F-SbF_5 as $> 500 : 1 : 10^{-1} : 10^{-5}$ by studying the rates of interconversions of trimethyltetrahydrofuryl ions. Other interesting systems studied in HSO_3F

TABLE XIII

RATES OF INTERCONVERSIONS OF **(XXVI)** ⇌ **(XXVII)**

Solvent	% **(XXVII)**	°C	10^4k/sec[a]
HF	% **(XXVI)** is 80	−10	7
H_2SO_4	45	69	6.2
HSO_3F	55	71.5	19
		53	2.0
HSO_3F-SbF_5	80	41	5.8

[a] First-order or pseudo-first-order rate constant.

TABLE XIV

RATIO OF **(XXVI)** AND **(XXVII)** IN DIFFERENT SOLVENTS

	% **(XXVII)**
H_2SO_4 (at 0° and 35°C)	45
HSO_3F (−78° and 20°C)	55
HSO_3F–SbF_5 (−78° and 20°C)	80
HF (−78° and −50°C) % of **(XXVI)**	80

by Brouwer *et al.* involve the kinetics of rearrangements of protonated aldehydes and ketones,[155] dehydration of protonated ketones to allylic cations,[156] protonated methyl neopentyl ketones.[157]

The most salient facts observed by Brouwer and VanDoorn[158] concerning the dehydration of α-dibranched, α-monobranched, and α-unbranched ketones are

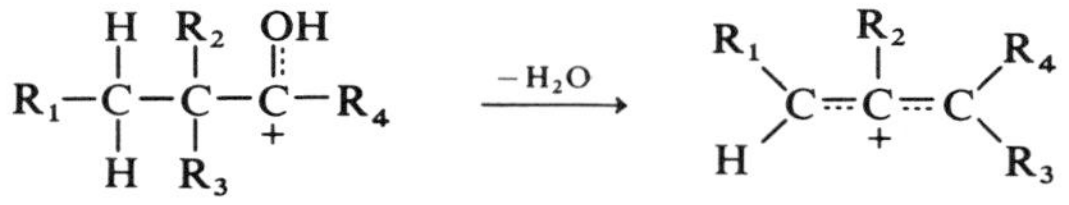

$$R_1-\overset{H}{\underset{H}{C}}-\overset{R_2}{\underset{R_3}{C}}-\overset{OH}{\underset{+}{C}}-R_4 \xrightarrow{-H_2O} (R_1)(H)C\text{=}\overset{R_2}{\underset{+}{C}}\text{=}C(R_4)(R_3)$$

R_1, R_2, R_3 = H or alkyl; R_4 = alkyl

(1) The rate decreases considerable from α-dibranched to α-monobranched to α-unbranched ketones. Assuming a normal activation energy (20–25 kcal/mole) the relative rates are of the order $10^4 : 10^2 : 1$.

(2) Generally, the rate increases upon substitution at the β-carbon atom

tert-amyl > *tert*-butyl; *sec*-butyl > isopropyl > isobutyl > *n*-propyl > ethyl)

an exception being the equal rates of the different α-unbranched ketones in 1 : 1 HF–SbF_5.

(3) The rate decreases with decreasing acidity of the solvent but the variations in rate (1 : 1 HF–SbF_5) ≫ (9 : 1 HF–SbF_5) > 1 : 1 HSO_3F–SbF_5 are rather small compared with the changes in acidity (as measured by the diprotonation of 1,3-diketones and unsaturated ketones). Brouwer and VanDoorn[158] found that protonated mesityl oxide (1,3,3-trimethyl-1-hydroxyalkyl cation,

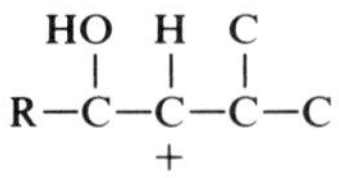

$$R-\overset{HO}{C}-\overset{H}{\underset{+}{C}}-\overset{C}{C}-C$$

$R=CH_3$

undergoes isomerization, cracking, and dehydrogenation depending on the acid used as the solvent.

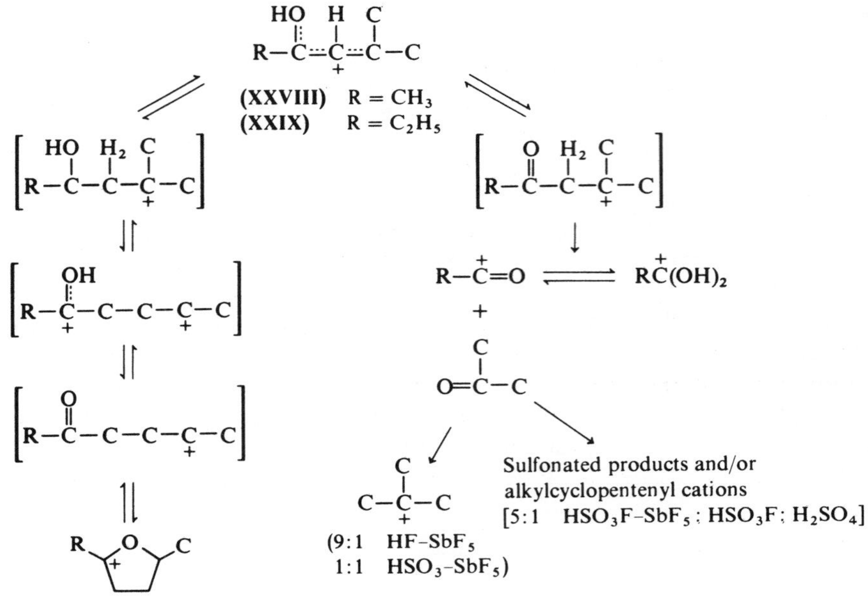

HO H C
R—C⋯C⋯C—C
+

(XXVIII) R = CH_3
(XXIX) R = C_2H_5

[HO H_2 C
R—C—C—C—C
+]

[O H_2 C
R—C—C—C—C
+]

[OH
R—C—C—C—C—C
+ +]

[O
R—C—C—C—C—C
+]

$R-\overset{+}{C}=O \rightleftharpoons R\overset{+}{C}(OH)_2$

+

C
O=C—C

C
C—C—C
+

Sulfonated products and/or alkylcyclopentenyl cations
[5:1 HSO_3F-SbF_5; HSO_3F; H_2SO_4]

(9:1 $HF-SbF_5$
1:1 HSO_3-SbF_5)

R O C
+

(XXXI) R = CH_3
(XXXI) R = C_2H_5
1:1 $HF-SbF_5$
9:1 $HF-SbF_5$
1:1 HSO_3F-SbF_5)

11. Lactones

The protons on the oxygen of the protonated lactones[159] occurred at lower field (NMR) than those in protonated alcohols and ethers, but are more shielded than those in protonated aliphatic ketones and aldehydes similar to those in protonated alkyl carboxylic acids and esters. This is consistent with the partial double bond character in the protonated lactones.

C=O (O) $\xrightarrow[-80°C]{HSO_3F-SbF_5-SO_2}$ $C=\overset{+}{O}H$ (O)

$C=\overset{+}{O}H$ (O) $\rightleftharpoons$ $\overset{+}{C}-OH$ (O) $\longleftrightarrow$ C—OH (O_+)

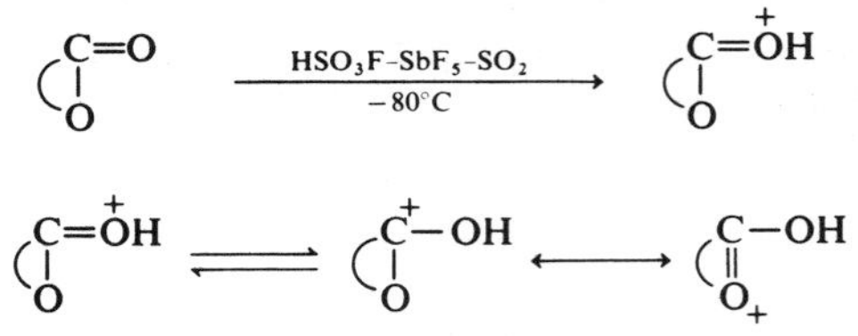

Two isomeric species were found in most of the protonated lactones studied.

Example: α-Angelicalactone

$$\text{H}_2\text{C}{-}\text{C}{=}\text{O},\ \text{HC}{=}\text{C}(\text{CH}_3){-}\text{O}\ (\text{ring}) \xrightarrow[-80^\circ\text{C}]{\text{HSO}_3\text{F-SbF}_5\text{-SO}_2} \text{H}_2\text{C}{-}\text{C}{=}\overset{+}{\text{O}}{-}\text{H}\ (\text{ring}) + \text{H}_2\text{C}{-}\text{C}{=}\text{O}^{+}\text{H}\ (\text{ring})$$

α-Angelicalactone

α-Angelicalactone at higher temperature cleaves to give ketooxocarbonium ion.

$$\text{H}_2\text{C}{-}\text{C}{=}\overset{+}{\text{O}}\text{H},\ \text{HC}{=}\text{C}(\text{CH}_3){-}\text{O}\ (\text{ring}) \xrightarrow[-60^\circ\text{C}]{\text{HSO}_3\text{F-SbF}_5\text{-SO}_2} \text{H}_2\text{C}{-}\text{C}{=}\overset{+}{\text{O}}\text{H},\ \text{H}_2\text{C}{-}\overset{+}{\text{C}}(\text{CH}_3){-}\text{O}\ (\text{ring}) \longrightarrow \text{CH}_3{-}\overset{\overset{+}{\text{O}}\text{H}}{\overset{\|}{\text{C}}}{-}\text{CH}_2{-}\text{CH}_2{-}\overset{+}{\text{C}}{=}\text{O}$$

Ketooxocarbonium ion

12. Sulfoxides and Sulfones

A series of sulfoxides (dimethyl, diethyl, dipropyl, dibutyl, and diphenyl) and some tetrahydrosulfones and sulfolenes were protonated in $HSO_3F–SbF_5$ diluted with SO_2ClF.[160] Sulfoxides were protonated on S, whereas in sulfones the site of protonation was on oxygen.

$$\text{R}{-}\ddot{\text{S}}(\text{=O}){-}\text{R} \xrightarrow[-80^\circ\text{C}]{\text{HSO}_3\text{F-SbF}_5\text{-SO}_2\text{ClF}} \text{R}{-}\overset{\text{H}}{\ddot{\text{S}}^{+}}(\text{=O}){-}\text{R}$$

Sulfoxides

$$\text{R}_2\text{S}(\text{=O})_2 \xrightarrow[-80^\circ\text{C}]{\text{HSO}_3\text{F-SbF}_5\text{-SO}_2\text{ClF}} \text{R}_2\text{S}(\text{=O})(\text{=}\overset{+}{\text{O}}\text{H})$$

Sulfones

Protonated sulfoxides and sulfones (except protonated benzyl and *tert*-butyl sulfones) were stable up to 65°C in $HSO_3F–SbF_5–SO_2ClF$. The exceptional case was cleaved to *tert*-butyl cation and phenylmethanesulfonic acid even at a temperature as low as −78°C.

13. Sulfonic Acid, Sulfonates, Sulfinic Acids, and Sulfinates

Alkyl and aryl sulfonic acids and sulfinic acids and alkyl sulfonates and sulfinates were protonated in an $HSO_3F–SbF_5$ solution[161]

$$\text{RSO}_3\text{H} \xrightarrow[-60^\circ\text{C}]{\text{HSO}_3\text{F-SbF}_5\text{-SO}_2\text{ClF}} \text{RSO}_3{}^{+}\text{H}_2$$

$$\text{RSO}_3\text{R} \xrightarrow[-60^\circ\text{C}]{\text{HSO}_3\text{F-SbF}_5\text{-SO}_2\text{ClF}} \text{R}{-}\overset{\overset{+}{\text{O}}\text{H}}{\overset{\|}{\text{S}}}{-}\text{OR}\ (\text{S=O})$$

At higher temperatures, protonated methane, benzene, and toluene sulfonic acids undergo dehydration to give the corresponding sulfonylium ion which is not observed as it quickly picks up a fluoride ion from the acid system to form the corresponding sulfonyl fluorides.

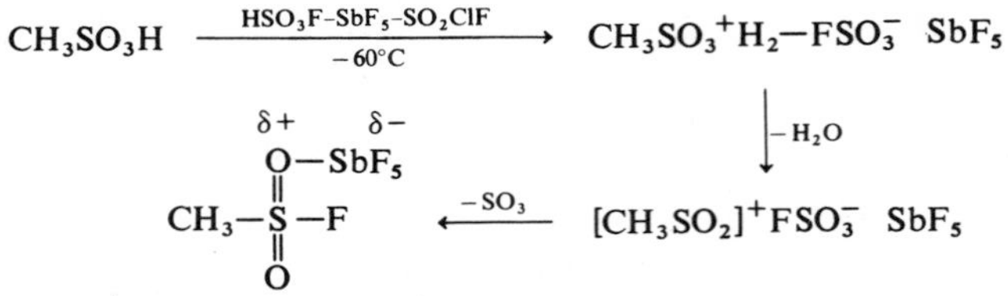

$$CH_3SO_3H \xrightarrow[-60°C]{HSO_3F\text{-}SbF_5\text{-}SO_2ClF} CH_3SO_3{}^{+}H_2\text{—}FSO_3^{-}\ SbF_5$$

$$\downarrow -H_2O$$

$$\underset{\text{donor–acceptor complex}}{CH_3\text{—}\overset{\overset{\delta+}{O}\text{—}\overset{\delta-}{SbF_5}}{\overset{\|}{\underset{\underset{O}{\|}}{S}}}\text{—}F} \xleftarrow{-SO_3} [CH_3SO_2]^{+}FSO_3^{-}\ SbF_5$$

The cleavage (alkyl–sulfur cleavage) products for the higher homologs were carbonium ions.

Two isomers were found for both protonated methane sulfonic acid and methyl methane sulfonates. As with the corresponding sulfonic acids, methyl methane sulfonate and methyl benzene sulfonate gives donor–acceptor complexes. Here again higher homologs gave carbonium ions.

Protonated sulfinic acids

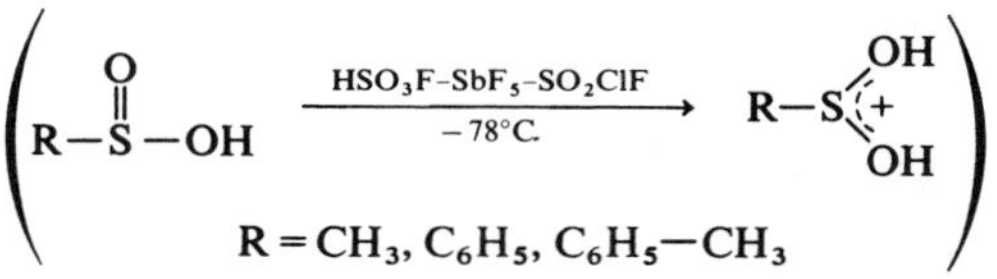

$$\left(R\text{—}\overset{\overset{O}{\|}}{S}\text{—}OH \xrightarrow[-78°C]{HSO_3F\text{-}SbF_5\text{-}SO_2ClF} R\text{—}\overset{+}{S}\begin{matrix}OH\\OH\end{matrix} \quad R = CH_3,\ C_6H_5,\ C_6H_5\text{—}CH_3 \right)$$

and sulfinates

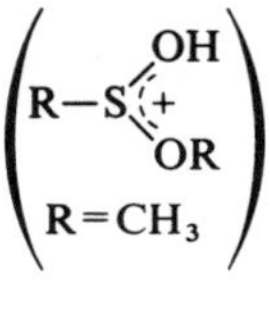

$$\left(R\text{—}\overset{+}{S}\begin{matrix}OH\\OR\end{matrix} \quad R = CH_3 \right)$$

were extremely stable up to 65°C.

14. Carbonic Acid, Carbonates, and Their Thio Analogs

In the strong acid system 1 : 1 HSO_3F–SbF_5 containing an equal volume of SO_2 at −78°C, Na, K, and Ba carbonates, as well as sodium hydrogen carbonates, were found to dissolve with no evolution of CO_2.[162] The peak at −12.05 ppm in all cases was attributed to trihydroxy carbonium ion (protonated carbonic acid):

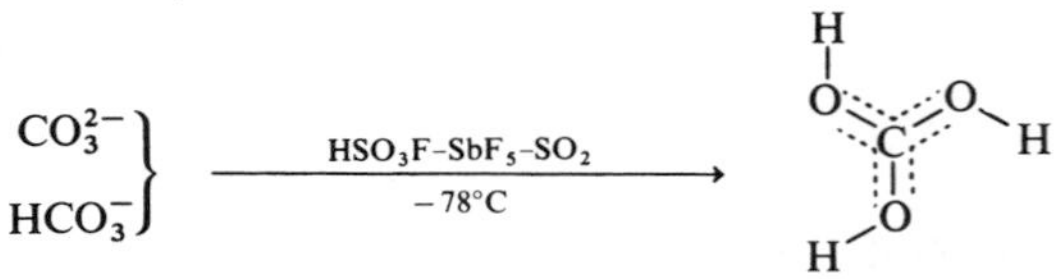

$$\left.\begin{matrix}CO_3^{2-}\\HCO_3^{-}\end{matrix}\right\} \xrightarrow[-78°C]{HSO_3F\text{-}SbF_5\text{-}SO_2} C(OH)_3^{+}$$

(XXXII)

This ion $C^{+}(OH)_3$ was stable up to 0°C in the absence of SO_2. Decomposition of **(XXXII)** in acid media gives $CO_2 + H_3O^+$:

$$H_3CO_3^+ \rightarrow H_3O^+ + CO_2$$

Protonation and dialkyl oxygen cleavage via alkyl–oxygen fission were also studied in the same acid system.[163] Cleavage products are carbonium ions and protonated carbonic acids:

$$(RO)_2C^{+}\text{—}OH \longrightarrow (HO)_2C^{+}\text{—}OH + R^+$$

These protonated carbonic acids were found to be effective carbonylating agents. It is suggested that they may play an important role in certain biological processes.

The formation of mono-, di-, and trithiol analogs of protonated carbonic acid has been observed in 1 : 1 HSO_3F–SbF_5 solution at −60°C.[120,121] Protonated trithiocarbonic acid was formed in solution of barium trithiocarbonate in 1 : 1 HSO_3F–SbF–SO_2 solution at −60°C. The reactions of potassium *tert*-butylxanthate and 0-*tert*-butyl-*S*-potassium thiocarbonate in the same acid system gave protonated dithiocarbonic and protonated thiocarbonic acid respectively.

$$BaCS_3 \xrightarrow[-60°C]{HSO_3F\text{-}SbF_5,\ SO_2} (HS)_3C^+$$

$$S{=}C(K)OC(CH_3)_3 \longrightarrow (HS)_2C^{+}(OH) + C(CH_3)_3^+$$

$$O{=}C(SK)OC(CH_3)_3 \longrightarrow (HO)_2C^{+}(SH) + C(CH_3)_3^+$$

$$BaCO_3 \longrightarrow (HO)_3C^+$$

The NMR shifts for the OH and SH protons in protonated thiocarbonic acids are summarized in Table XV. The increased deshielding of both the OH and SH protons as the number of thiol groups in the ion is increased is consistent with the lesser ability of sulfur compared with oxygen to delocalize the positive charge on the central atom.

Both CMR and PMR studies have been used for protonation and cleavage of reactions of dialkyl pyrocarbonates.[164] The cleavage products were identified by the addition of authentic products to the sample.

TABLE XV

CHEMICAL SHIFTS OF THE THIOL ANALOGS OF PROTONATED CARBONIC ACID

	OH	SH
$C(OH)_3^+$	11.55	—
$C(OH)_2SH^+$	11.99	6.73
$C(SH)_2OH^+$	12.56	7.19
$C(SH)_3$	—	7.66

Dialkylpyrocarbonates in magic acid give diprotonated dications of the type.

$$R{-}O{-}C({=}O){-}O{-}C({=}O){-}O{-}R \xrightleftharpoons{H^+} R{-}O^+{=}C(OH){-}O{-}C({=}O^+H){-}O{-}R \rightleftharpoons \ldots$$

$R = CH_3, C_2H_5, C_3H_7, i\text{-}C_3H_7$

The proposed mechanism for cleavage was confirmed by the addition of authentic products to the sample. The cleavage of dimethyl pyrocarbonate products were identified and confirmed by the results of a study involving methyl chloroformate.[165]

$$CH_3{-}O{-}C({=}O){-}Cl \xrightarrow{HSO_3F\text{-}SbF_5} CH_3{-}O{-}C({=}{}^+OH){-}Cl \xrightarrow{-HCl} [CH_3{-}O^+{=}C{=}O] \xrightarrow{-CO_2} CH_3Cl \longrightarrow SbF_5$$

$$[CH_3{-}O^+{=}C{=}O] \xrightarrow{HSO_3F} CH_3OSO_2F \xleftarrow{HSO_3F} CH_3Cl$$

15. ALLOPHANATES AND ALLOPHANIC ACIDS

The behavior of alkyl allophanates has been investigated in HSO_3F–SbF_5–SO_2 solution.[166] Olah *et al.* observed for the first time the elusive stable protonated allophanic acid as one of the cleavage products of diprotonated alkyl allophanates.

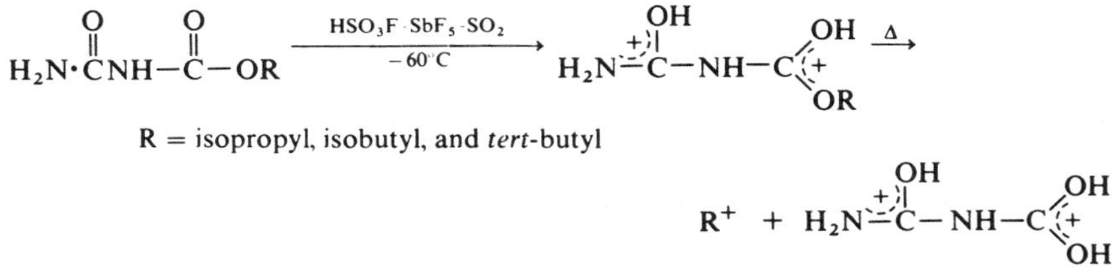

$$H_2N{\cdot}\overset{O}{\overset{\|}{C}}NH{-}\overset{O}{\overset{\|}{C}}{-}OR \xrightarrow[-60°C]{HSO_3F{\cdot}SbF_5{\cdot}SO_2} H_2N\overset{+}{=}\overset{OH}{\overset{|}{C}}{-}NH{-}C^{+}(OH)(OR) \xrightarrow{\Delta}$$

R = isopropyl, isobutyl, and *tert*-butyl

$$R^+ + H_2N\overset{+}{=}\overset{OH}{\overset{|}{C}}{-}NH{-}C^{+}(OH)(OH)$$

The proton on oxygen NMR signals occur at lower field than do those in protonated amides and carbamates. However, these protons are more shielded than are those in diprotonated ketocarboxylic acids. Two isomeric species were observed only for methyl and ethyl allophanates at $-90°C$ owing to restricted rotation around the C=N bond.

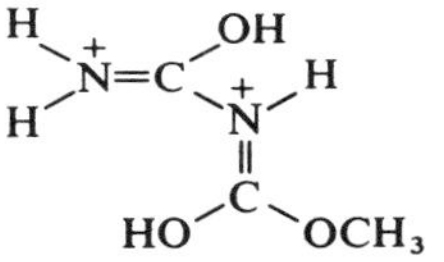

$$H_2\overset{+}{N}{=}C(OH){-}\overset{+}{N}H{=}C(OH)(OCH_3) \quad \textbf{(XXXIII)}$$

$$H_2\overset{+}{N}{=}C(OH){-}\overset{+}{N}H{=}C(OCH_3)(OH) \quad \textbf{(XXXIV)}$$

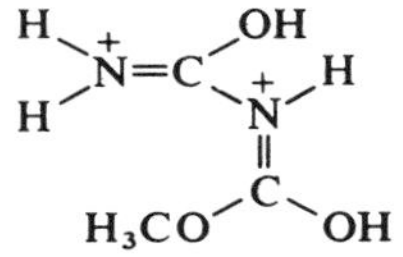

16. Carbamyl Halides, Alkyl (Aryl) Isocyanates, and Isothiocyanates

Complex formation of carbamyl chlorides and fluorides with Lewis acid halides and protonation of carbamyl halides both take place at the carbonyl oxygen atom.[167]

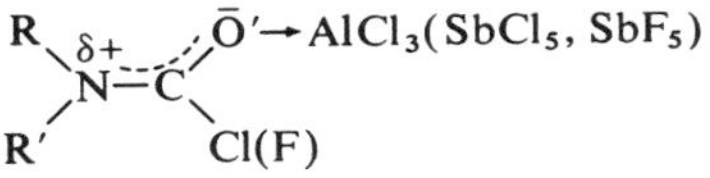

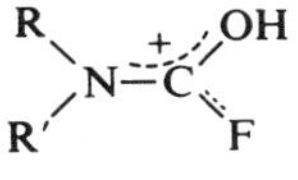

$$RR'\overset{\delta+}{N}{=}C(\bar{O}{\rightarrow}AlCl_3(SbCl_5, SbF_5))(Cl(F)) \qquad RR'N\overset{+}{=}C(OH)(F)$$

These observations were closely related to protonated carbamic acid, diprotonated urea, and guanidine.

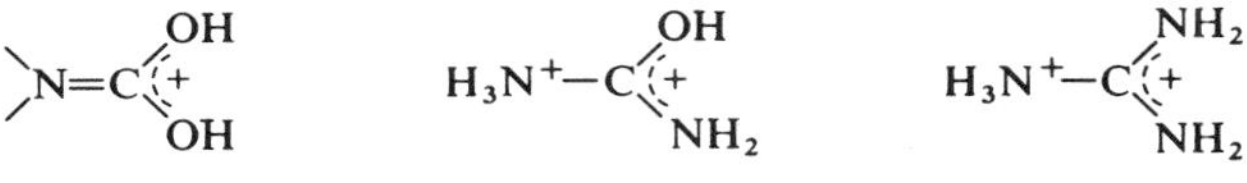

$${>}N{=}C^{+}(OH)(OH) \qquad H_3N^{+}{-}C^{+}(OH)(NH_2) \qquad H_3N^{+}{-}C^{+}(NH_2)(NH_2)$$

The remarkable stability of protonated carbamyl fluorides has been compared with that of protonated fluoromethyl alcohol[168] which contains a highly stable hydroxyl group and fluorine atom at the same carbon.

When formaldehyde is dissolved in 1 : 1 $HSO_3F{-}SbF_5/SO_2ClF$ at $-78°C$ a stable solution of protonated formaldehyde (hydroxycarbonium ion) is formed. Upon addition of fluoride ion (as NaF or HF) at $-78°C$ protonated fluoromethyl alcohol is obtained.

$$CH_2O \xrightarrow[SO_2ClF;\ -78^\circ C]{HSO_3F\text{-}SbF_5} CH_2{=}OH^+ \leftrightarrow \overset{+}{C}H_2OH \xrightarrow[2.\ H^+]{1.\ F^-} FCH_2OH_2^+$$

The protonation of carbamyl fluoride has also been correlated with protonation of formamides.

A significant difference between carbamyl fluorides and acyl fluorides is that no protonated acyl fluorides were observed. The acyl fluorides form only acyl cations with the elimination of HF.

$$CH_3{-}C(=O)F \xrightarrow[-100^\circ C]{HSO_3F\text{-}SbF_5\text{-}SO_2ClF} \left[CH_3{-}C(=\overset{+}{O}H)F\right] \xrightarrow{-HF} CH_3CO^+$$

The preparation of carbamyl cations was attempted by an alternate route, the protonation of isocyanic acid and alkyl (aryl) isocyanates in super acid. They found that instead of carbamyl cations they all formed allophanyl cations:

$$RNCO \xrightarrow{HSO_3F\text{-}SO_2} [RN\overset{+}{H}CO] \xrightarrow{RNCO} (R)(H)N{-}C(=O){-}N(R){=}\overset{+}{C}{=}O$$

Isothiocyanic and thiocyanic acids, on the other hand, gave the corresponding thiocarbamyl cation. This has been considered to be due to the stabilization of the thiocarbamyl cations by the more positive sulfur.

$$RNCS \xrightarrow{HSO_3F\text{-}SO_2} RNH\overset{+}{C}S$$

thiocarbamyl cation

17. Nitriles

Generally N-protonated nitriles were observed in HSO_3F–SbF_5–SO_2 solution.[169] Protonated nitrile $H\text{-}C{\equiv}N^+H$ was obtained by adding a mixture of HSO_3F–SbF_5 to an SO_2 solution of HCN or by dissolving AgCN in the acid–SO_2 solution. The Ag^+ in the HCN–Ag ion complex formed initially is displaced by HSO_3F–SbF_5 at $-90^\circ C$.

$$AgCN \xrightarrow{HSO_3F\text{-}SbF_5\text{-}SO_2} HCN \qquad Ag^+[\text{or}\,(HCN)_2Ag^+]$$

$$HCN \xrightarrow[-SO_2]{HSO_3F\text{-}SbF_5} HC\overset{+}{N}H$$

Hogeveen[170] has carried out some NMR measurements on propionitrile in a number of solvents of varying acidity. Protonation of ^{15}N-propionitrile leads to the formation of ^{15}N-iminoethylcarbonium ion which occurs in one configuration only $R\text{-}\overset{+}{C}{=}NH$. He also studied protonated ^{14}N-propionitrile since in protonated nitriles the N- protonation cannot be observed by NMR due to the electric quadrupole moment of nitrogen.

18. Ketoximes

Nitrogen protonation of acetone and acetophenone oximes has been observed in $HSO_3F\text{–}SbF_5\text{–}SO_2$ solutions.[169] On heating a solution of protonated acetone oxime at 100°C for 30 min, conversion to *N*-methylacetonitrilium ion was observed via a Beckmann rearrangement:

$$(CH_3)_2C{=}\overset{+}{N}(H)\text{–}O\text{–}H \longrightarrow CH_3\text{–}C{=}\overset{+}{N}\text{–}CH_3$$

Protonated cyclohexanone oxime has been observed on dissolution of nitrosocyclohexane in $HSO_3F\text{–}SbF_5\text{–}SO_2$.[171]

$$C_6H_{11}\text{–}N{=}O \longrightarrow C_6H_{10}{=}\overset{+}{N}(H)OH$$

19. Nitro Compounds

A number of nitro compounds are O-protonated in $HSO_3F\text{–}SbF_5$.[172] It is of interest that the spectra of protonated nitrobenzenes are temperature dependent. Thus, at low temperatures, the ortho protons in protonated 3,5-dichloro-4-methyl nitrobenzene are nonequivalent due to restricted rotation of the protonated nitro group, but the ortho protons are equivalent at higher temperatures. The barrier to rotation ($\Delta G = 7$ kcal/mole) shows the importance of charge delocalization by the aromatic nucleus in the protonated species.

Nitroalkanes have been shown to cleave in $HSO_3F\text{–}SbF_5$ solutions, leading to formation of carbonium,[172] nitrosonium, and hydronium ions. Thus 2-fluoro-2-nitropropane gives the fluorodimethylcarbonium ion, and 1-nitro-2-methylpropane gives the *tert*-butyl cation.

$$F\text{–}C(CH_3)_2\text{–}\overset{+}{N}({=}O)OH \xrightarrow[SO_2,\ -80°C]{HSO_3F\text{–}SbF_5} F\text{–}\overset{+}{C}(CH_3)_2 + NO^+ + H_3O^+$$

$$(H_3C)_2CH\text{–}CH_2\text{–}NO_2^+H \longrightarrow (CH_3)_3C^+ + NO^+ + H_3O^+$$

20. Aromatic Hydrocarbons

The heats of formation of a number of alkyl and polyalkyl benzenonium ions with methyl, ethyl, isopropyl, and *tert*-butyl substituents and for two polycyclic systems are reported by Arnett and Larsen.[173,174] The heats of formation of these ions followed the Baker–Nathan order (i.e., methyl > ethyl > isopropyl > *tert*-butyl). Brouwer and VanDoorn[175] measured competitive protonation of the 9- and 10-positions of 9-ethyl-10-methylanthracene and found them to be equally reactive, contrary to what might have been expected from Arnett and Larsen's results. Olah and his associates[176] found that aromatic hydrocarbons (benzene, toluene, ethylbenzene, the isomeric propyl and butylbenzenes, neopentylbenzene, 2,2-diphenylpropane, 1,1,1-triphenylethane, and tetraphenyl methane) were protonated either in HSO_3F–SbF_5 with HSO_3F or in 1 : 1 M HF–SbF_5 acid systems with a slightly larger than equal volume of SO_2ClF as diluent. Carbon-13 magnetic resonance spectroscopic studies were also carried out for these ions by indor double resonance methods (INDR). These compounds form stable, long-lived benzenium ions (cyclohexadienyl cations). Both the CMR and PMR spectra were indicative of the anticipated charge distribution in the benzenium and alkylbenzenium ions. The most stable form of the latter was the 4-alkylbenzenium ions. Neither Baker–Nathan nor an inductive order for the 4-methyl-, 4-ethyl-, and 4-isopropylbenzenium ions could be implied from the observed magnetic resonance parameters of these ions.

The protonation of mesitylene, durene, pentamethyl benzene, hexamethylbenzene, xylenes, and anisole takes place on the ring C atom.[177] Rates and activation energies for intermolecular proton exchange with the solvent have been measured.

Naphthalenium ions are formed by the protonation of naphthalene and substituted naphthalenes in HF-SbF_5-SO_2ClF or HSO_3F-SO_2ClF.[178] Temperature-dependent PMR studies of these naphthalenium ions indicated the presence of rapidly equilibrating ions via a simple 1,2-hydrogen shift.

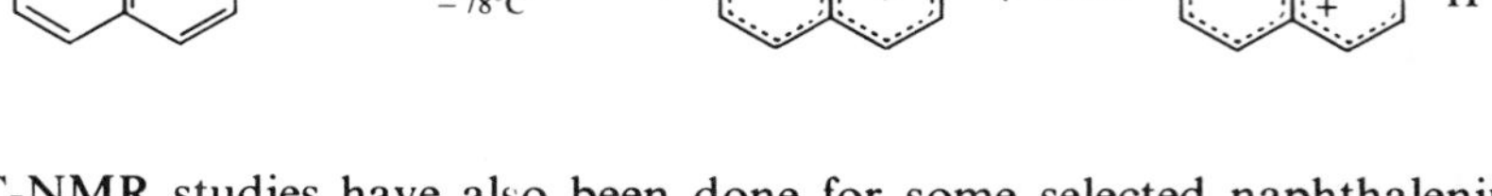

$$\xrightarrow[-78^{\circ}C]{HSO_3F\text{-}SbF_5\text{-}SO_2ClF}$$

^{13}C-NMR studies have also been done for some selected naphthalenium ions. Several acenaphthenium ions,[179] excepting the parent 1-

acenaphthenium ion and 1,2,2-trimethyl-1-acenaphthenium ions, undergo a 1,2-hydrogen or methyl shift.

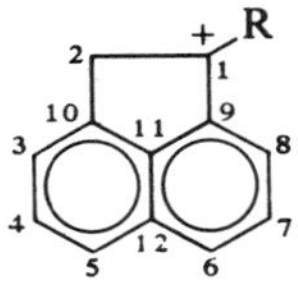

Buck *et al.*[180] reported the formation of the dipositive ions of naphthacene and 1,2-benzanthracene in HSO_3F-SbF_5 and their conversion to the protonated arenes by hydride transfer.

Brouwer and VanDoorn[181,182] found that when perybene is dissolved in SbF_5-SO_2ClF or SbF_5-HSO_3F (> 1/1 m/m) a violet solution was obtained in a markedly exothermic reaction (major bands at 512 nm, log ε = 4.7, with shoulders near 500, 485, and 455 nm, and 315 nm, log ε = 4.3; with a shoulder near 280 nm).

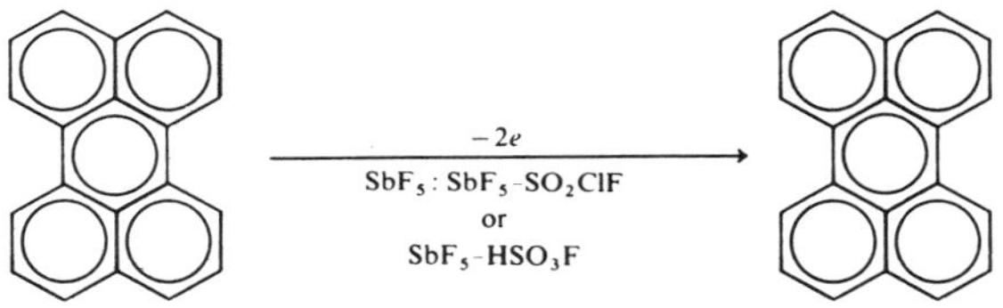

$-2e$

SbF_5: SbF_5-SO_2ClF or SbF_5-HSO_3F

The dipositive ion of naphthacene (green, absorption maximum at 651 nm, log ε = 4.7 and 344 nm, log ε = 5.0, with a shoulder at 330 nm) is formed in SbF_5-SO_2ClF as well as in HSO_3F-SbF_5. It is also obtained on addition of SbF_5 to a solution of the mono-positive ion in H_2SO_4 or HSO_3F.

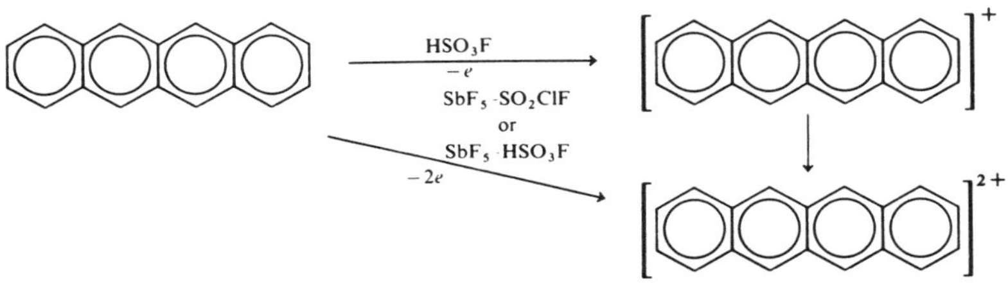

HSO_3F, $-e$

SbF_5-SO_2ClF or SbF_5-HSO_3F, $-2e$

The possible mechanism for the generation of carbonium ions from hydrocarbons in magic acid solution has been discussed by Larsen *et al.*[183] The tropylium and phenalenium ions were generated cleanly from their precursors, cycloheptatriene and phenalene, respectively, in HSO_3F, magic acid, and H_2SO_4.

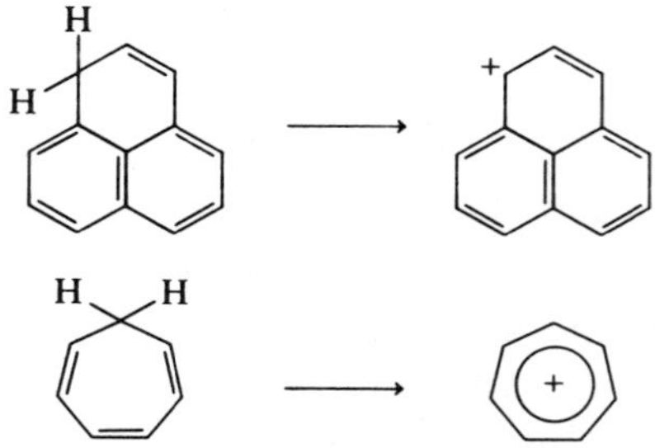

In all cases SO_2 was formed. Isobutane gave the *tert*-butyl cation in magic acid without the formation of either SO_2 or hydrogen. The mechanism described by them is limited to the formation of rather stable cations. The stability of the cation plays a role in determining the process by which it is formed. According to Larsen *et al.*, there are three processes by which carbonium ions can be generated from hydrocarbons in magic acid solutions: (1) oxidation by HSO_3F (or SO_2), (2) oxidation by SbF_5, and (3) hydride abstraction by a proton.

21. Carbonium Ions

A detailed review of alkylcarbonium ions derived from various precursors in different solvents has been reported by Olah *et al.*[106,184,185] A detailed study of HSO_3F–SbF_5–SO_2 and related solvent systems[186] found most suitable for the low-temperature study of carbonium ions permits the assignment of all Raman spectra lines due to the solvent and gegenions and hence, unambiguous assignment of signals for the alkyl carbonium ions.[187]

The fluorescence problem was minimized by using a red He–Ne laser as the exciting light and by recording the Raman spectra of solutions at temperatures as low as $-100°C$. The Raman spectra of a number of alkylcarbonium ions and their deuterated products have been reported.

Olah *et al.*[188] have carried out a detailed UV investigation of 10^{-1} to 10^{-1} *M* solutions of alkyl cations in HSO_3F–SbF_5 solutions at $-60°C$ and were able to confirm that they do not absorb above 210 mμ.

a. *Alkanes.* HSO_3F–SbF_5 acts as a very effective hydride (alkide) ion abstracting agent in the formation of alkyl carbonium ions from alkanes.[189]

They found SO_2ClF to be a more suitable solvent than SO_2, which reacts with carbonium ions to give sulfinic acids or sulfonyl fluorides.

$$R_3CH \xrightarrow{HSO_3F-SbF_5} R_3C^+SbF_5FSO_3^- + H_2$$

The overall reaction of these alkanes in magic acid is complicated. For example, methane forms[190] several intermediate cations like ethyl, isopropyl, and related carbonium ions which themselves can undergo various di-, tri-, and polymerization processes. The higher molecular weight hydrocarbonium ions, in turn, undergo fragmentation. A possible scheme is

$$CH_4 \rightleftharpoons [CH_5^+] \longrightarrow CH_3^+ + H_2$$

$$CH_3^+ \xrightarrow{CH_4} [C_2H_7^+]$$

$$C_2H_6 \rightleftharpoons [C_2H_7^+] \xrightarrow{-H_2} [C_2H_5^+] \underset{H^+}{\overset{-H^+}{\rightleftharpoons}} CH_2{=}CH_2$$

$$[C_2H_5^+] \xrightarrow{CH_4} [C_3H_9^+]$$

$$C_3H_8 \rightleftharpoons [C_3H_9^+] \overset{-H_2}{\rightleftharpoons} [C_3H_7^+]$$

$$[C_3H_7^+] \xrightarrow{CH_4} [C_4H_{11}^+]$$

$$C_4H_{10} \rightleftharpoons [C_4H_{11}^+] \underset{-H_2}{\rightleftharpoons} [C_4H_9^+]$$

$$CH_2{=}CH_2 \longrightarrow [C_4H_9^+]$$

$$[C_4H_9^+] \xrightarrow{CH_4} \text{etc.}$$

Ethane and higher alkanes show similar reactions in magic acid relating to both hydrogen exchange and polycondensation reactions. Because of the high chemical reactivity of these alkanes, Olah *et al.* recommended that these saturated hydrocarbons should be referred to as alkanes and not as "paraffins." The ease with which these alkanes undergo ionization follows the order primary $<$ secondary $<$ tertiary hydrogen atom.

All alkyl carbonium ions ultimately convert on heating through isomerization, fragmentation, and dimerization to the *tert*-butyl cation, which was found to be stable in the acid solution up to 150°C. In branched alkanes, the tendency to cleave increases with the length and degree of branching of the alkanes.

The most surprising observation of their work is that the salts of the cations *tert*-butyl and *tert*-amyl are stable at least at room temperature. X-ray crystallographic studies are being undertaken.

Cycloalkanes and polycycloalkanes in the superacid medium form cycloalkyl cations at low temperatures by hydride ion abstraction.[191] At higher temperatures, ring cleavage takes place to give acyclic alkylcarbonium ions.

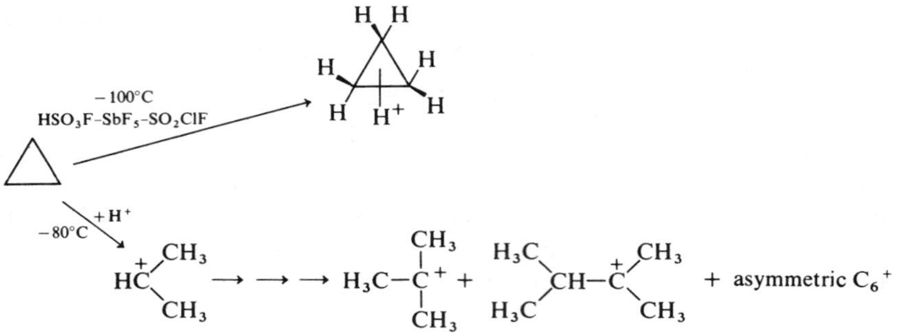

$$\triangle \xrightarrow[HSO_3F\text{-}SbF_5\text{-}SO_2ClF]{-100°C} \text{(protonated cyclopropane, } C_3H_6 \cdot H^+\text{)}$$

$$\triangle \xrightarrow[-80°C]{+H^+} H\overset{+}{C}(CH_3)_2 \longrightarrow \longrightarrow \longrightarrow H_3C\text{—}\overset{+}{C}(CH_3)_2 + (H_3C)_2CH\text{—}\overset{+}{C}(CH_3)_2 + \text{asymmetric } C_6^{\,+}$$

b. *Olefins.* Olefins also form carbonium ions in superacid media.[184,192]

$$RCH{=}CH_2 \xrightarrow[HF\text{-}SbF_5]{HSO_3F\text{-}SbF_5} R\overset{+}{C}H\text{-}CH_3$$

Olah and Halpern[193] found the proper experimental conditions for the preparation of alkyl(aryl)carbenium ions. This nomenclature is suggested to differentiate the trivalent carbo cation from the penta- or tetracoordinated carbonium ion[194] from olefins without concomitant polymerization. They have emphasized the fact that there is not a single general procedure which can be applied to any olefin. Specific conditions have to be used in each case.

Diphenylbenzylcarbenium ion is formed from 1-phenyl-2,2-diphenylethylene.

$$(C_6H_5)_2C{=}CHC_6H_5 \xrightarrow[-80°C]{HSO_3F\text{-}(SO_2ClF)} (C_6H_5)_2C^+CH_2C_6H_5$$

pale red solution

The corresponding methyl olefin forms several cations in the super-acid media.

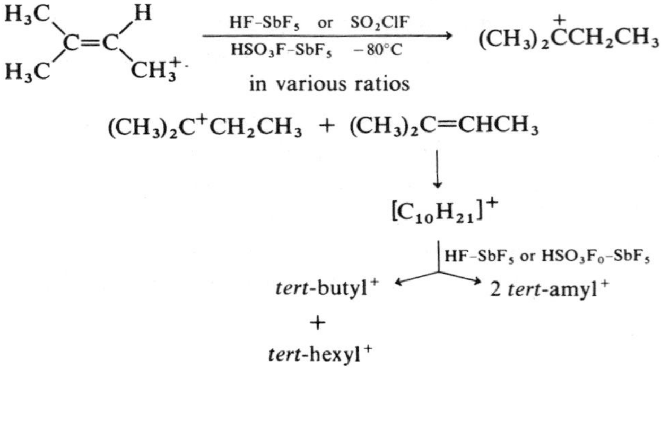

$$(H_3C)_2C{=}C(H)CH_3^{+\cdot} \xrightarrow[HSO_3F\text{-}SbF_5 \quad -80°C]{HF\text{-}SbF_5 \quad \text{or} \quad SO_2ClF} (CH_3)_2\overset{+}{C}CH_2CH_3$$

in various ratios

$$(CH_3)_2C^+CH_2CH_3 + (CH_3)_2C{=}CHCH_3$$

$$\downarrow$$

$$[C_{10}H_{21}]^+$$

HF–SbF$_5$ or HSO$_3$F$_0$–SbF$_5$

tert-butyl$^+$ ← → 2 *tert*-amyl$^+$

\+

tert-hexyl$^+$

c. *Electrophilic Reactions at Single Bonds.* A survey of the possible reactions of alkanes as substrates and protons or carbonium ions as electrophilic agents has been made by Hogeveen and Bickel.[195] They have suggested that the intermediates in these electrophilic substitutions at the alkane carbon is likely to have a C_s-type geometry. H–D exchange of CH_3D was also found to take place in HSO_3F–SbF_5 (4 : 1). They found that this electrophilic substitution reaction in HSO_3F–SbF_5 is about 100 times less reactive than that in HF–SbF_5. This has been ascribed to the differences in the acidity between the two media.

The formation of alkyl carbonium ions from alkanes has been independently discovered both by Hogeveen *et al.* and Olah *et al.* The reversible reactions between carbonium ions and hydrogen have been carried out by Hogeveen and his co-workers. The mechanism of the reverse reaction, namely, the formation of carbonium ions from alkane, was studied both in HF–SbF_5 and HSO_3F–SbF_5.[196] The mechanism in the two solvent systems are likely to differ. In fact, it has been observed by Olah and Lukas[189] that the reaction between alkanes and protons in HSO_3F–SbF_5 only produces a small amount of hydrogen. This cannot be explained by the reaction sequence of the type given by Hogeveen *et al.*[197] because they found that molecular H_2 is not oxidized by HSO_3F–SbF_5 at 20°C.

$$\text{R–H} + \text{H}^{\oplus}(\text{SbF}_5\text{–SO}_3\text{F})^{\ominus} \rightarrow \text{R}^{\oplus}(\text{SbF}_5\text{–SO}_3\text{F})^{\ominus} + \text{H}_2$$

$$\text{H}_2 + \text{HSO}_3\text{F} \rightarrow \text{H}_2\text{O} + \text{SO}_2 + \text{HF}$$

They also found that the reduction of carbonium ions was much slower in HSO_3F–SbF_5 (1 : 3 : 1) than in HF–SbF_5.

$$tert\text{-C}_4\text{H}_9^{+} + \text{H}_2 \rightleftharpoons i\text{-C}_4\text{H}_{10} + \text{H}^{\oplus}$$

The unusual high stability (obtained by NMR studies[195]) of norbornyl ion was shown by the rate constants (see Table XVI) of the equilibrium reaction

$$\text{R}^{+} + \text{CO} \rightleftharpoons \text{RCO}^{+}$$

TABLE XVI

EQUILIBRIUM CONSTANT
$K = [\text{RCO}^{+}]/[\text{R}^{+}][\text{CO}]$

	K(liter/mole)
tert-Butyl	7×10^2
tert-Pentyl	7×10^2
2-Exonorbornyl	10^4
1-Adamantyl	2×10^4

The rates and activation parameters of the decarbonylation reaction (for R = *tert*-butyl) were equal in $HF–SbF_5$ (1 : 1) and $HSO_3F–SbF_5$ (1 : 1), but the second-order rate constant of carbonylation was about a factor of 10 smaller in the former solvent than in the latter (Table XVII). The rates of carbonylation of some unsaturated carbonium ions have been estimated by Hogeveen and Gaasbeek.[198]

The electrophilic reactivity of a number of alkanes has been studied by Olah and his associates.[199–203] They have presented evidence for substantial reactivity in electrophilic reactions such as protolytic processes (isomerization, $H_2–D_2$ exchange, protolysis), alkylation, nitration, and halogenation.

The protolytic (deuterolytic) behavior of 21 alkanes were studied in both $HSO_3F–SbF_5(SO_2ClF)$ and $HF–SbF_5(SO_2ClF)$ solutions as well as in their corresponding deuterated super acids.[201] Stable carbenium ions were observed by NMR spectroscopy, whereas neutral cleavage products were analyzed by GLC and mass spectrometry. These saturated hydrocarbon reaction studies have shown that intermolecular hydrogen transfers involve not linear[196] but triangular, three-center-bonded, pentacoordinated carbonium ion intermediates.

$$[R_3C\cdots(H)(H)]^+ \equiv [R_3C\cdots(H)(H)]^+$$

The nature of the products of protolysis was dependent on the nature of protolytic agent (solvated proton); for example, in ethane CH_4 and H_2 are formed.

$$H_3C{-}CH_3 \xrightarrow{H^+} [H_3C\cdots H\cdots CH_3]^+ \rightleftharpoons CH_4 + CH_3^+ \xrightleftharpoons{CH_3-CH_3} [CH_3CH_2\cdots(H)(CH_3)]^+ \rightleftharpoons CH_3CH_2^+ + CH_4$$

$$H_3C{-}CH_3 \xrightarrow{H^+} [CH_3CH_2\cdots(H)(H)]^+ \rightleftharpoons H_2 + CH_3CH_2^+$$

$$H_2 + CH_3CH_2^+,\ CH_3CH_2^+ + CH_4 \rightarrow tert\text{-}C_4H_9^+,\ tert\text{-}C_6H_{13}^+,\ \text{etc.}$$

The ratio of $CH_4 : H_2$ was dependent on the solvent system used. In HSO_3F, $H_2SO_3F^+$ is more space demanding than protonated HF. $H_2SO_3F^+$ will, therefore, react preferentially with the more open C–H than

TABLE XVII

RATE CONSTANT ACTIVATION PARAMETER $R^+ + CO \rightleftharpoons RCO^+$

	E_a	log $A(s^{-1})$ R = *tert*-butyl	K_c (20°C)
$HF\text{-}SbF_5$ (1 : 1)	15.6 ± 0.8	13.3 ± 0.6	3×10^3 liter $mole^{-1}$ sec^{-1}
$HSO_3F\text{-}SbF_5$ (1 : 1)	15.6 ± 0.04	13.3 ± 0.2	3×10^4 liter $mole^{-1}$ sec^{-1}

the more crowded C–C bond. A further significant factor which can affect the association and nature of the protolytic agent is the solvent medium. The most frequently used solvents, other than superacids, in the order of decreasing nucleophilicity are SO_2, SO_2ClF, and SO_2F_2. The more nucleophilic SO_2 will, therefore, solvate the ions more than SO_2ClF or SO_2F_2. The entire scheme (see Scheme 1) of the protolytic behavior of alkanes is given by Olah

$$CH_4 \rightleftharpoons H_2 + (CH_3)^+ \xrightarrow{CH_4} (C_2H_7)^+ \rightleftharpoons H_2 + C_2H_5^+ \rightarrow\rightarrow C_4H_9^+, \text{etc.} \quad (1)$$

a, H_2; *b*, $C_4H_9{}^+ > C_5H_{11}{}^+ > C_6H_{13}{}^+$

$$CH_3 \overset{1}{-} CH_2H \xrightarrow{1} CH_4 + (CH_3^+) \xrightarrow{C_2H_6} CH_4 + (C_2H_3)^+ \longrightarrow C_4H_9^+ \quad (2)$$
$$\xrightarrow{2} (C_2H_5^+) + H_2$$

a, $CH_4 > C_4H_{10}$; $H_2 > C_3H_8$; *b*, $C_4H_9{}^+$

$$CH_3\overset{H}{\overset{2|}{C}}H \overset{1}{-} CH_3 \xrightarrow{1} CH_4 + C_2H_5^+ \text{ and } C_2H_6 + CH_3^+ \quad (3)$$
$$\xrightarrow{2} (CH_3)_2\overset{+}{C}H + H_2$$
$$CH_3CH_2CH_3 \xrightarrow{CH_3{}^+} CH_4 + (C_3H_7)^+$$
$$CH_3CH_2CH_3 \xrightarrow{C_2H_5{}^+} C_2H_6 + (C_3H_7)^+$$
$$(C_3H_7)^+ + (C_3H_7)^+ \longrightarrow C_6H_{13}^+$$

a, $C_2H_6 > CH_4 \gg H_2$; *b*, $C_6H_{13}{}^+ \gg C_5H_{11}{}^+ > C_4H_9{}^+$

Scheme 1. Protolytic Cleavage Paths of Alkanes in Superacids

(*a*) Volatile products, determined by glc and mass spectrometry; (*b*) Carbocations in acid solution, determined by nmr spectroscopy; (*c*) −60°C; *tert*-$C_7H_{15}{}^+$ is the only product. *J. Am. Chem. Soc.* **95**, 4960 (1973).

$$CH_3CH_2\overset{1}{-}\underset{\underset{H}{|3}}{C}H\overset{2}{-}CH_3 \begin{cases} \xrightarrow{1} C_2H_6 + (C_2H_5^+) \xrightarrow{C_4H_{10}} C_2H_6 + C_4H_9^+ \\ \xrightarrow{2} CH_4 + (C_3H_7) \xrightarrow{C_4H_{10}} C_3H_8 + C_4H_9^+ \\ \xrightarrow{3} C_4H_9^+ + H_2 \end{cases} \tag{4}$$

a, $CH_4 > C_2H_6 \gg H_2$; *b*, $C_4H_9^+$

$$(CH_3)_2\overset{\overset{H}{2|}}{C}\overset{1}{-}CH_3 \begin{cases} \xrightarrow{1} CH_4 + C_3H_7^+ \text{ and } C_3H_8 + (CH_3^+) \\ \xrightarrow{2} H_2 + C_4H_9^+ \end{cases} \tag{5}$$

a, $H_2 > CH_4 \gtrsim C_3H_8$; *b*, $C_4H_9^+$

$$CH_2\overset{3}{C}H_2CH_2\overset{1}{-}CH_2\overset{2}{-}CH_3 \begin{cases} \xrightarrow{1} C_2H_6 + C_3H_7 \xrightarrow{C_5H_{12}} C_3H_8 + C_3H_{11}^+ \\ \xrightarrow{2} CH_4 + C_4H_9^+ \text{ and } C_4H_{10} + CH_3^+ \\ \xrightarrow{3} H_2 + C_3H_{11}^+ \end{cases} \tag{6}$$

a, $C_4H_{10}, C_3H_8 > C_2H_6 \gg CH_4 > H_2$; *b*, $C_4H_9^+ \sim C_5H_{11}^+$

$$CH_3\overset{2}{-}\underset{\underset{CH_3}{|}}{\overset{\overset{H}{4|}}{C}}\overset{1}{-}CH_2\overset{3}{-}CH_3 \begin{cases} \xrightarrow{1} C_2H_6 + C_3H_7^+ \xrightarrow{C_5H_{12}} C_3H_8 + C_3H_{11}^+ \\ \xrightarrow{2,3} CH_4 + C_4H_9^+ \\ \xrightarrow{4} H_2 + C_3H_{11}^+ \end{cases} \tag{7}$$

a, $H_2 \gg CH_4 > C_2H_6$; *b*, $C_4H_9^+ \sim C_5H_{11}^+$

$$CH_3-\underset{\underset{CH_3}{|}}{\overset{\overset{CH_3}{|}}{C}}\overset{1}{-}CH_2\overset{2}{-}H \begin{cases} \xrightarrow{1} CH_4 + C_4H_9^+ \\ \xrightarrow{2} H_2 + C_3H_{11}^+ \end{cases} \tag{8}$$

a, $CH_4 > H_2$; *b*, $C_4H_9^+ > C_5H_{11}^+$

$$CH_3CH_2CH_2\overset{1}{-}CH_2\overset{2}{-}CH_2\overset{3}{-}CH_3 \begin{cases} \xrightarrow{1} C_3H_8 + C_3H_7^+ \xrightarrow{C_6H_{14}} C_3H_8 + C_6H_{13}^+ \\ \xrightarrow{2} C_2H_6 + C_4H_9^+ \\ \xrightarrow{3} CH_4 + C_3H_{11}^+ \end{cases} \tag{9}$$

a, $C_2H_6 > C_4H_{10}$; $C_3H_8 > CH_4 > H_2$; *b*, $C_6H_{13}^+ > C_5H_{11}^+ \sim C_4H_9^+$

$$CH_3\overset{3}{-}\underset{\displaystyle CH_3}{\overset{\displaystyle CH_3}{C}}\overset{1}{-}CH_2\overset{2}{-}CH_3 \xrightarrow{1} C_2H_4 + C_4H_9 \text{ and } C_4H_{10} + C_2H_5 \xrightarrow{C_6H_{14}} C_2H_6 + C_6H_{13}^+ \tag{10}$$

$$\xrightarrow{2,3} CH_4 + C_3H_{11}^+ \text{ and } C_3H_{12} + CH_3 \xrightarrow{C_6H_{14}} CH_4 + C_6H_{13}^+$$

a, C_5H_{12}, C_4H_{10}, C_2H_6, $CH_4 > H_2$; *b*, $C_6H_{13}^+ \gg C_5H_{11}^+ > C_4H_9^+$

$$CH_3\overset{2}{-}\underset{\displaystyle \underset{1}{|}\,H}{\overset{\displaystyle CH_3}{C}}\overset{3}{-}CH_2\overset{4}{-}CH_2\overset{5}{-}CH_3 \xrightarrow{1} H_2 + C_6H_{13}^+ \tag{11}$$

$$\xrightarrow{2,5} CH_4 + C_5H_{11}^+$$

$$\xrightarrow{3} C_3H_8 + C_3H_7^+ \qquad \xrightarrow{4} C_2H_6 + C_4H_9^+$$

a, $H_2 \gg CH_4 > C_2H_6$; *b*, $C_6H_{13}^+ > C_5H_{11}^+ \gg C_4H_9^+$

$$CH_3CH_2CH_2CH_2\overset{1}{-}CH_2\overset{2}{-}CH_2\overset{3}{-}CH_3 \xrightarrow{1} C_3H_8 + C_4H_9^+ \text{ and } C_4H_{10} + C_3H_7^+ \tag{12}$$

$$\xrightarrow{2} C_2H_6 + C_5H_{11}^+$$

$$\xrightarrow{3} CH_4 + C_6H_{13}^+$$

a, $C_4H_{10} > C_3H_8 > C_2H_6 \gg CH_4 > H_2$; *b*, $C_4H_9^+ \gg C_6H_{13}^+ \sim C_5H_{11}^+$

$$CH_3\overset{3}{-}\underset{\displaystyle \underset{4}{|}\,H}{\overset{\displaystyle H_3C}{C}}\overset{1}{-}\underset{\displaystyle CH_3}{\overset{\displaystyle CH_3}{C}}\overset{2}{-}CH_3 \xrightarrow{1} C_3H_8 + C_4H_9^+ \text{ and } C_4H_{10} + C_3H_7^+ \tag{13}$$

$$\xrightarrow{2,3} CH_4 + C_6H_{13}^+$$

$$\xrightarrow{4} H_2 + C_7H_{15}^+ \longrightarrow C_4H_9^+, \text{ etc.}$$

a, C_4H_{10}, $C_3H_8 \gg CH_4$, $H_2 > C_2H_6$ (trace); *b*,[c] $C_4H_9^+ \gg C_6H_{13}^+ > C_5H_{11}^+$

$$CH_3-\underset{\displaystyle H}{\overset{\displaystyle CH_3}{C}}-CH_2\overset{1}{-}\underset{\displaystyle \underset{3}{|}\,H}{\overset{\displaystyle CH_3}{C}}\overset{2}{-}CH_3 \xrightarrow{1} C_3H_8 + C_4H_9^+ \text{ and } C_4H_{10} + C_3H_7^+ \tag{14}$$

$$\xrightarrow{2} CH_4 + C_6H_{13}^+$$

$$\xrightarrow{3} H_2 + C_7H_{15}^+ \longrightarrow C_4H_9^+, \text{ etc.}$$

a, C_4H_{10}, $C_3H_8 \gg CH_4$, $H_2 > C_2H_6$ (trace); *b*,[c] $C_4H_9^+ > C_6H_{13}^+ > C_5H_{11}^+$

$$\mathrm{CH_3\overset{4}{C}H_2-\overset{\displaystyle CH_3}{\underset{\displaystyle 2|\,CH_3}{C}}\overset{1}{-}CH_2\overset{3}{-}CH_3}\left\{\begin{array}{l}\xrightarrow{1}\mathrm{C_2H_6 + C_5H_{11}^+ \text{ and } C_5H_{12} + C_2H_5^+}\\ \xrightarrow{2,3}\mathrm{CH_4 + C_6H_{13}^+}\\ \xrightarrow{4}\mathrm{H_2 + C_7H_{15}^+ \longrightarrow C_4H_9^+,\ etc.}\end{array}\right. \tag{15}$$

a, $C_2H_6 > C_4H_{10}, C_2H_3, CH_4 > C_5H_{12}, H_2$; *b*,[c] $C_4H_9^+ \gg C_5H_{11}^+ > C_6H_{13}^+$

$$\mathrm{CH_3CH_2CH_2CH_2\overset{1}{-}CH_2\overset{2}{-}CH_2\overset{3}{-}CH_2\overset{4}{-}CH_3}\left\{\begin{array}{l}\xrightarrow{1}\mathrm{C_4H_{10} + C_4H_9^+}\\ \xrightarrow{2}\mathrm{C_2H_5 + C_2H_{11}^+}\\ \xrightarrow{3}\mathrm{C_2H_6 + C_6H_{13}^+}\\ \xrightarrow{4}\mathrm{CH_4 + C_7H_{15}^+ \longrightarrow C_4H_9^+;\ C_5H_{11}^+}\end{array}\right. \tag{16}$$

a, $C_4H_{10} > C_3H_8 > C_2H_6 \gg CH_4 > H_2$; *b*, $C_4H_9^+ \gg C_5H_{11}^+$

$$\mathrm{CH_3-\overset{\displaystyle H_3C}{\underset{\displaystyle H_3C}{C}}\overset{1}{-}\overset{\displaystyle CH_3}{\underset{\displaystyle CH_3}{C}}\overset{2}{-}\overset{3}{C}H_3}\left\{\begin{array}{l}\xrightarrow{1}\mathrm{\mathit{i}\text{-}C_4H_{10} + C_4H_9^+}\\ \xrightarrow{2}\mathrm{CH_4 + C_7H_{15}^+ \longrightarrow C_4H_9^+}\\ \xrightarrow{3}\mathrm{H_2 + (C_8H_{17}^+) \longrightarrow C_4H_9^+}\end{array}\right. \tag{17}$$

a, $CH_4 \gg H_2$; *b*, $C_4H_9^+ > C_7H_{15}^+$

$$\mathrm{CH_3-\overset{\displaystyle CH_3}{\underset{\displaystyle CH_3}{C}}-\overset{4}{C}H_2\overset{1}{-}\overset{\displaystyle CH_3}{\underset{\displaystyle CH_3}{C}}\overset{2}{-}\overset{3}{C}H_3}\left\{\begin{array}{l}\xrightarrow{1}\mathrm{\mathit{i}\text{-}C_4H_{10} + C_5H_{11}^+ \text{ and } \mathit{n}\text{-}C_5H_{12} + C_4H_{11}^+}\\ \xrightarrow{2}\mathrm{CH_4 + C_8H_{17}^+ \longrightarrow C_4H_9^+;\ C_5H_{11}^+}\\ \xrightarrow{3,\,4}\mathrm{H_2 + (C_9H_{18}^+) \longrightarrow C_4H_9^+ + C_5H_{11}^+}\end{array}\right. \tag{18}$$

a, $CH_4 \gg H_2 > C_5H_{12}, C_4H_{10}, C_2H_6$; *b*, $C_4H_9^+ \gg C_5H_{11}^+$

$$\mathrm{CH_3-\overset{\displaystyle H_3C}{\underset{\displaystyle 4|\,H_3C}{C}}\overset{1}{-}\overset{\displaystyle CH_3}{\underset{\displaystyle CH_3}{C}}\overset{2}{-}CH_2\overset{3}{-}CH_3}\left\{\begin{array}{l}\xrightarrow{1}\mathrm{\mathit{i}\text{-}C_4H_{10} + C_5H_{11}^+ \text{ and } C_5H_{12} + C_4H_9^+}\\ \xrightarrow{2}\mathrm{C_2H_6 + C_7H_{15}^+ \longrightarrow C_4H_9^+;\ C_5H_{11}^+}\\ \xrightarrow{3,\,4}\mathrm{CH_4 + C_8H_{17}^+ \longrightarrow C_4H_9^+;\ C_5H_{11}^+}\end{array}\right. \tag{19}$$

a, $C_4H_{10} > C_2H_6 > CH_4 > H_2$; *b*, $C_4H_9^+ \sim C_5H_{11}^+$

$$\mathrm{CH_3CH_2\overset{1}{-}\underset{\underset{CH_3\overset{2}{-}CH_3}{|}}{\overset{\overset{CH_2CH_3}{|}}{C}}-CH_2CH_3} \begin{array}{l} \xrightarrow{1} \mathrm{C_2H_6 + C_7H_{15}^+ \longrightarrow C_4H_9^+; C_5H_{11}^+} \\ \xrightarrow{2} \mathrm{CH_4 + C_8H_{17}^+ \longrightarrow C_4H_9^+; C_5H_{11}^+} \end{array} \tag{20}$$

a, C_4H_{10}, $C_2H_6 > C_5H_{12}$, $C_3H_8 > CH_4$, H_2; *b*, $C_4H_9^+ > C_5H_{11}^+ \gg C_6H_{13}^+$

$$\mathrm{[(CH_3)_3C]_3CH \longrightarrow (CH_3)_3C\ CH_2\ C(CH_3)_3 + C_4H_9^+} \\ \hookrightarrow n\text{-}\mathrm{C_5H_{12} + C_4H_9^+} \tag{21}$$

a, C_4H_{10}, $C_3H_8 > CH_4 \gg C_2H_6$, H_2; *b*, $C_4H_9^+ \gg C_5H_{11}^+$

---

et al.[202] They[203] suggested that hydride ion abstraction (or transfer) should be more properly named "hydrogen transfer."

Molecular H_2 and D_2 were found to undergo exchange reaction in superacids at room temperature, as indicated by HD formation.[204] The following systems were studied:

H_2 + DF–SbF_5	H_2 + D_2SO_4–SbF_5
H_2 + DSO_3F–SbF_5	D_2 + HF–SbF_5
D_2 + H_2SO_4–SbF_5	D_2 + HSO_3F–SbF_5
H_2 + D_2 + HF–SbF_5	

The exchange reactions indicated the intermediacy of isomeric, trigonal $(H,D)_3^+$ ions in superacid solution.

$$\left[\begin{array}{c} \mathrm{H} \\ \mathrm{D}\cdots\mathrm{D} \end{array}\right]^+ \qquad \left[\begin{array}{c} \mathrm{D} \\ \mathrm{H}\cdots\mathrm{H} \end{array}\right]^+$$

Gillespie and Pez[205] concluded from cryoscopic and solubility measurements in HSO_3F that D_3^+ or H_3^+ exist only as metastable species.

The first observation of an intermediate σ complex in electrophilic aromatic substitution reactions other than protonation and alkylation was the formation of benzenium ions in hexamethylbenzene and 2,4,6-trifluoromesitylene.[206]

When hexamethylbenzene in SO_2 was added, with vigorous stirring, to a solution containing $NO_2^+BF_4^-$–HSO_2F–SO_2 at $-70°C$, nitrohexamethyl benzenium ions were formed, as indicated by PMR.

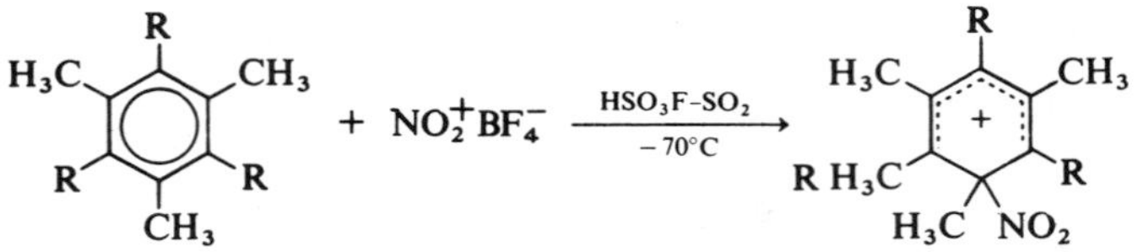

$R = CH_3F$ + $NO_2^+BF_4^-$ $\xrightarrow[-70°C]{HSO_3F\text{-}SO_2}$ (XXXV) $R = CH_3$; (XXXVI) $R = F$

Chlorohexamethyl benzenium ion **(XXXVII)** was obtained by adding a solution of hexamethyl benzene in SO_2ClF to a solution of Cl_2 in magic acid–SO_2ClF solution at $-70°C$.

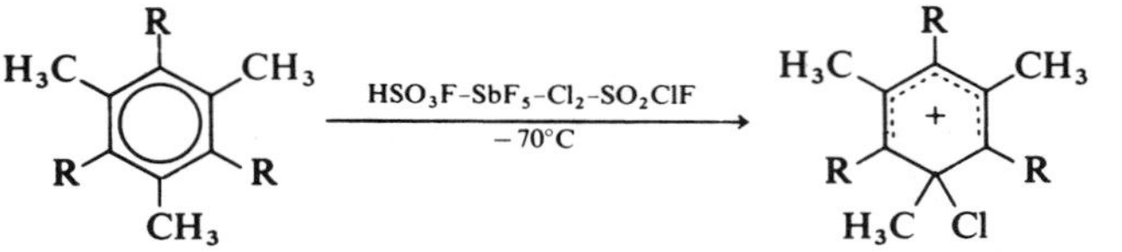

$R = CH_3, F$ $\xrightarrow[-70°C]{HSO_3F\text{-}SbF_5\text{-}Cl_2\text{-}SO_2ClF}$ (XXXVII) $R = CH_3$; (XXXVIII) $R = F$

Temperature-dependent PMR of **(XXXV)** indicated the migration of NO_2 which is assumed to proceed via 1,2-nitroshifts involving a three-center-bond benzonium ion intermediate.

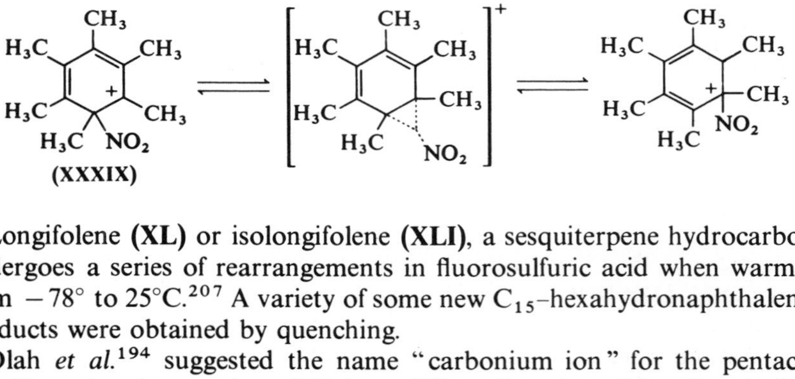

(XXXIX)

Longifolene **(XL)** or isolongifolene **(XLI)**, a sesquiterpene hydrocarbon, undergoes a series of rearrangements in fluorosulfuric acid when warmed from $-78°$ to $25°C$.[207] A variety of some new C_{15}–hexahydronaphthalenes products were obtained by quenching.

Olah *et al.*[194] suggested the name "carbonium ion" for the pentacoordinated cations such as CH_5^+, to differentiate them from the trivalent carbocations. Farnum and his associates felt that changing a well-established nomenclature should be done only if it proves to be unwieldy by comparison with its replacement, or is clearly logically inconsistent in essentials, or is ambiguous, or is unable to accommodate new structures. None of

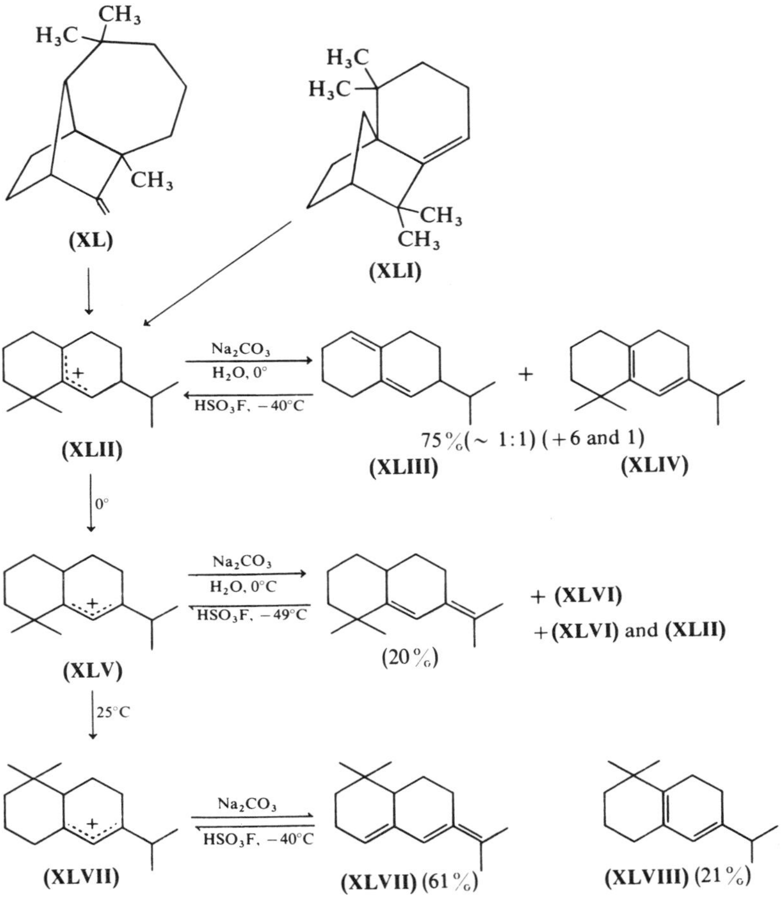

(XL) (XLI)

(XLII) $\xrightleftharpoons[\text{HSO}_3\text{F, } -40°\text{C}]{\text{Na}_2\text{CO}_3,\ \text{H}_2\text{O, } 0°}$ (XLIII) + (XLIV)

75%(~ 1:1) (+6 and 1)

(XLII) $\xrightarrow{0°}$ (XLV)

(XLV) $\xrightleftharpoons[\text{HSO}_3\text{F, } -49°\text{C}]{\text{Na}_2\text{CO}_3,\ \text{H}_2\text{O, } 0°\text{C}}$ (20%) + (XLVI) +(XLVI) and (XLII)

(XLV) $\xrightarrow{25°\text{C}}$ (XLVII)

(XLVII) $\xrightleftharpoons[\text{HSO}_3\text{F, } -40°\text{C}]{\text{Na}_2\text{CO}_3}$ (XLVII) (61%) (XLVIII) (21%)

these criteria seemed to them to apply to the term "carbonium ion" when properly used. They found it at least as "wieldy" as the alternative carbenium ion system proposed by Olah. Recently Olah and his associates[208,209] have carried out electrophilic reactions at multiple bonds also.

d. *Cyclic Systems.* (i) Cycloalkyl cations. It is interesting to note that the 1-methylcyclopentyl cation is the most stable cation formed by 1,2-hydride shift from both cyclopentane and cyclohexane systems.[210] The following scheme shows the formation of 1-methylcyclopentyl cation from a variety of precursors in $HSO_3F–SbF_5$ as well as in other superacids.

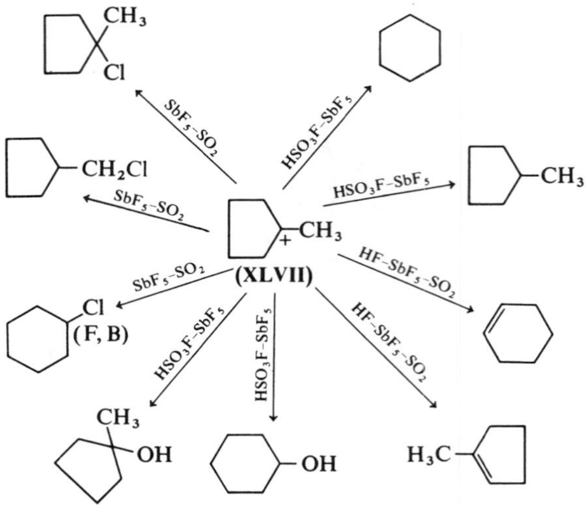

(XLVII)

(XLIX)

1-methyl cyclopentyl cation

Although the exact mechanism is not yet known, Olah *et al.*[210] favor the mechanism involving a protonated cyclopropane intermediate.

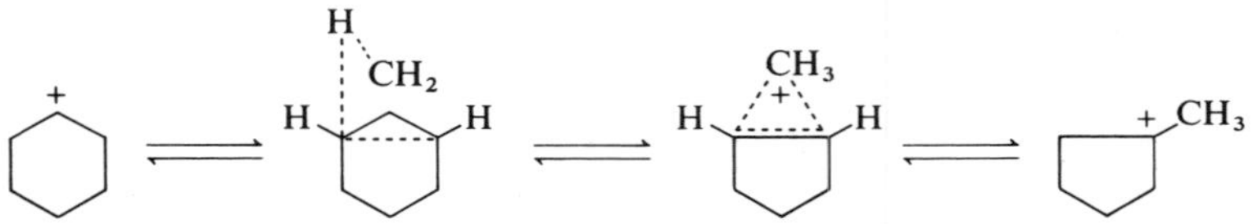

(ii) Cyclopropyl carbonium ions. Pittman and Olah[211] observed several mono- and dicyclopropyl carbonium ions in HSO_3F–SbF_5–SO_2 at low temperatures. The most interesting result was the nonequivalence of the methyl groups in dimethylcyclopropyl carbonium ion, proving the bisected structure.

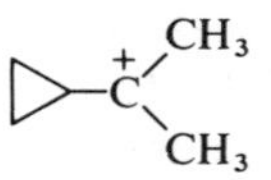

The cyclopropane ring lies in a plane perpendicular to the plane of the

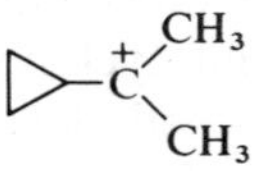

system. Other cyclopropyl carbonium ions of the type

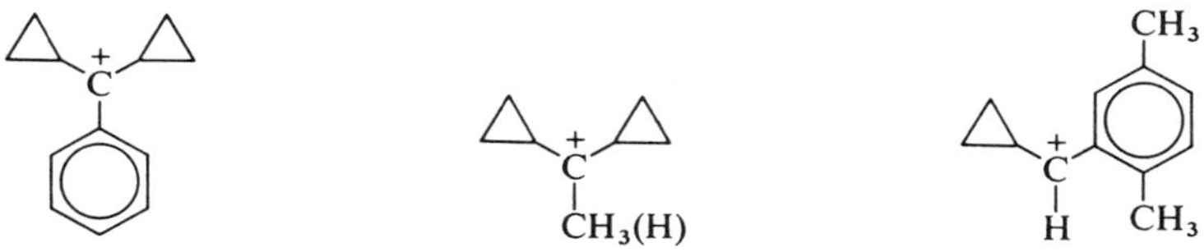

$CH_3(H)$ H CH_3 CH_3

were also studied by the same authors in $HSO_3F–SbF_5$.

e. *Alkylaryl Carbonium Ions.* Several mono- and dialkylaryl carbonium ions are readily formed from their corresponding alcohols both in HSO_3Cl and in HSO_3F[212] or $HSO_3F–SbF_5$.[213] Representative ions are:

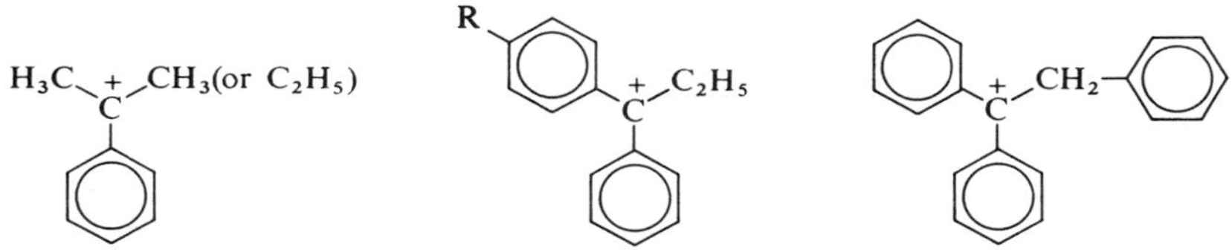

H_3C $\overset{+}{C}$ CH_3(or C_2H_5)   R $\overset{+}{C}$ C_2H_5 $\overset{+}{C}$ CH_2

R = H, OCH_3, CH_3

Because of the high stability of the tertiary ions, these are preferentially formed in the strong acid systems from tertiary, secondary, and even primary precursors:

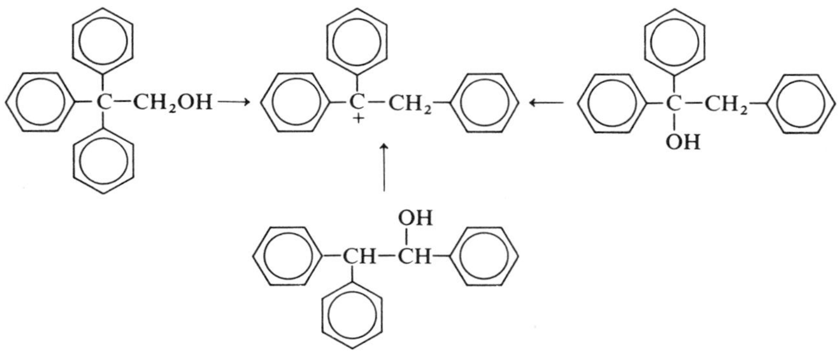

C–CH_2OH → $\overset{+}{C}$–CH_2 ← C(OH)–CH_2

↑

OH

CH–CH

If, however, the tertiary carbonium ion is nonbenzylic, rearrangement to a secondary, but benzylic, ion can be observed:

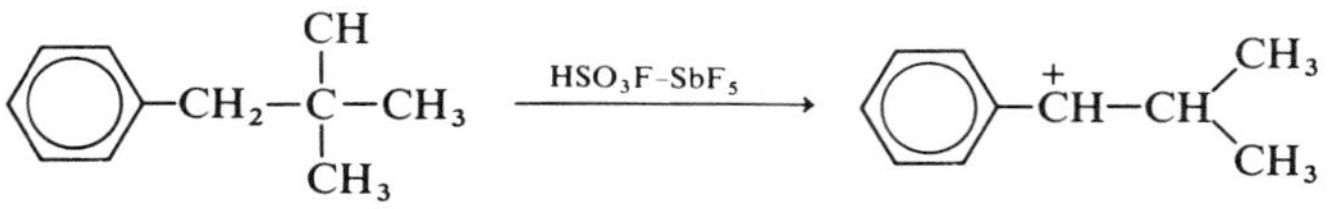

CH

CH_2–C–CH_3 $\xrightarrow{HSO_3F–SbF_5}$ $\overset{+}{C}H$–CH(CH_3)(CH_3)

CH_3

A comparative study of the cyclopentenyl, -hexenyl, -heptenyl, -octenyl, and -nonenyl cations was made in HSO_3F–SbF_5 and SbF_5 diluted with SO_2ClF (SO_2) by both ^1H- and ^{13}C-NMR spectroscopy.[214,215]

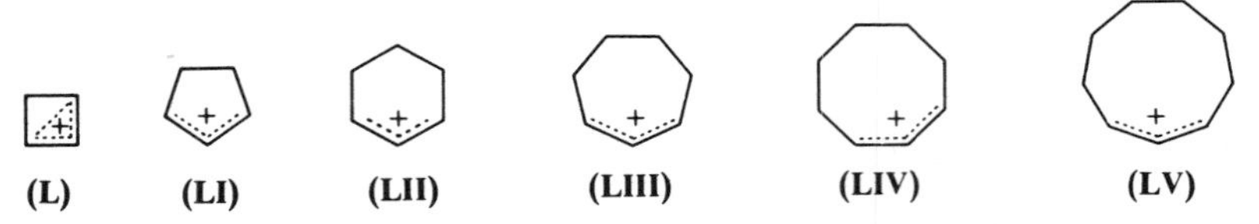

Of these, **(L)**, the homocyclopropenyl cation, is the simplest 2 π-homoaromatic species prepared from 3-acetoxyclobutene.[216]

$$\text{OAc} \xrightarrow[\text{SO}_2\text{ClF(SO}_2\text{)},\ -78^\circ\text{C}]{\text{HSO}_3\text{F}-\text{SbF}_5}$$

$$-CH_3CO_2H_2^+$$

The order of stability of these allylic cycloalkenyl cations, based on their reactivity in superacid solution, is: cyclopentenyl > cyclohexenyl > cycloheptenyl > cyclooctenyl > cyclononenyl. These stable cycloalkenyl cations are believed to be the most thermodynamically stable intermediates in superacid solvents (low-nucleophilicity media).

^{13}C-NMR studies revealed no significant 1, 3-orbital interactions in cyclopentenyl cations. The ring contraction of cyclohexenyl and its methyl derivatives to the corresponding cyclopentenyl cations were also studied.[217]

22. Norbornyl, Adamantyl, and Related Cations

A detailed account of degenerate carbonium ions of the type $(CH)_n^+$ ($n = 3$ to 9), including from cyclopropenyl to homocyclononenyl cations and several other degenerate ions such as butyl, amyl, cycloalkyl, aryl, norbornyl, and adamantyl cations in general is given by Leone, Barborak, and Schleyer.[218] In their review the major emphasis is placed on ions capable of multiple degeneracy. These degenerate ions have been generated under solvolytic conditions as well as in highly acid solvents.

The thermal conversion of the 7-norbornadienyl ion **(LVI)** into the tropylium ion **(LVII)** in HSO_3F solution ($K = 6.2 \times 10^{-4}$/sec at 47°C; 25% yield) is reported by Brookhart *et al.*[219] The photochemically induced conversion of the tropylium ion into 7-norbornadienyl ion was observed by Hogeveen and Gaasbeck.[220]

Irradiation of a 0.5 *M* solution of tropylium ion **(LVII)** in HSO_3F at −70°C in an NMR tube, with a Hanau S-89 high pre-Hg arc, afforded the 7-norbornadienyl ion **(LVI)** in a yield of about 75%.

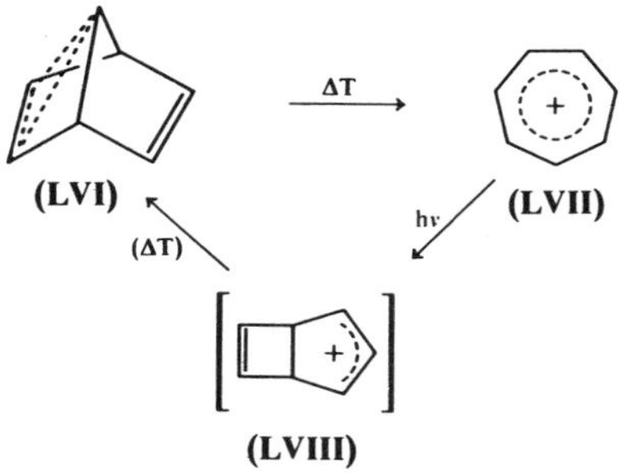

(LVI) —ΔT→ (LVII)

(LVII) —hν→ (LVIII)

(LVIII) —(ΔT)→ (LVI)

(LVIII)

Ion **(LVI)** proved to be photochemically stable during the irradiation at low temperatures. Conversion of **(LVII)** → **(LVI)** probably proceeds via the bicyclo[3 · 2 · 0]heptadienyl ion **(LVIII)**.

Olah *et al.*[221] generated the norbornyl cation in strong acid solutions from the precursors shown in the following scheme:

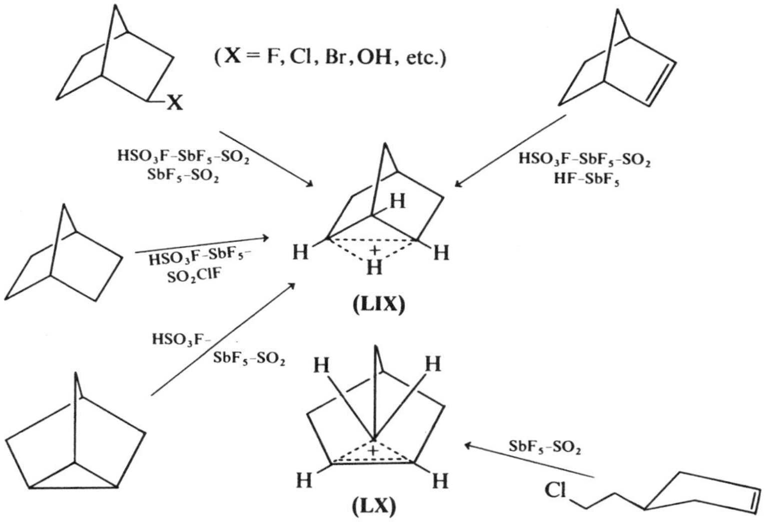

(X = F, Cl, Br, OH, etc.)

HSO_3F-SbF_5-SO_2
SbF_5-SO_2

HSO_3F-SbF_5-SO_2
HF-SbF_5

HSO_3F-SbF_5-
SO_2ClF

HSO_3F-
SbF_5-SO_2

(LIX)

SbF_5-SO_2

Cl

(LX)

These workers obtained a 100 MHz NMR resonance spectrum of this cation. This showed the fine structure reported by Jensen and Beck. They also obtained a Raman spectrum of the cation in a strong acid solution and concluded that the ion was best represented as a protonated nortricyclene. These authors felt that either a corner protonated nortricyclonium ion **(LIX)** or an edge-protonated nortricyclonium ion **(LX)** was most compatible with the NMR and Raman spectra.

An example of a carbonium ion system which is neither an equilibrating classical ion nor a fully delocalized nonclassical one was studied by Olah *et al.*[222,223] The cation obtained in SbF_5–SO_2 or HSO_3F–SbF_5–SO_2 from 1,2-dimethyl-2-endo-norbornanol **(LXI)** or endo- and exo-2-chloride **(LXII)** was probed with combined techniques of ^{13}C- and ^{1}H-NMR and Raman spectroscopy. The ion was assigned to the degenerately rearranging unsymmetrically bridged structures **(LXIII)** and **(LXIV)**

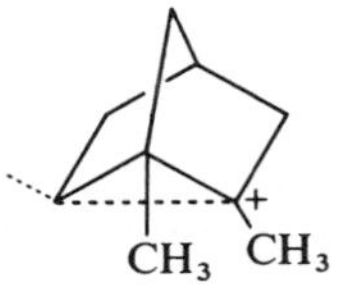

(LXIV) 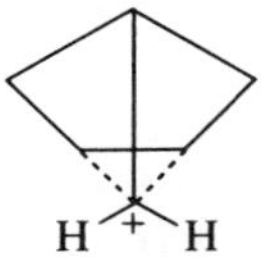 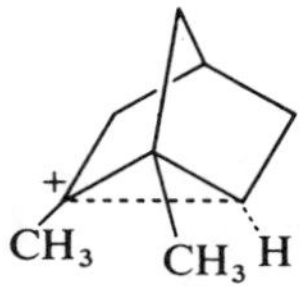 **(LXIII)** 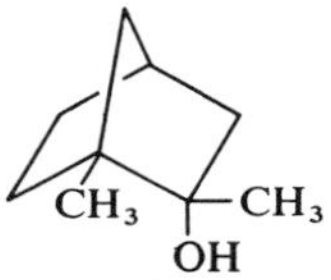 **(LXI)** 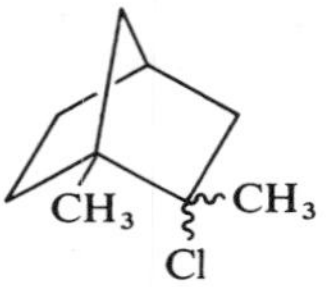 **(LXII)**

The result demonstrates the existence of a large variety of possible intermediates, from classical to nonclassical, that can convincingly be called rearranging cationic species.

The interconversions of the 2-exo- and 2-endonorbornyloxocarbonium ion via the norbornyl cation was studied kinetically by Hogeveen and Roobeek[224] in HSO_3F–SbF_5 (1 : 1) in the temperature range 75°–102°C by NMR integration. The system obeys first-order kinetics to about 80–90% conversion. The free enthalpy diagram is shown in Fig. 3. Rate constants and activation parameters are given in Table XVIII.

TABLE XVIII

RATE CONSTANTS AND ACTIVATION PARAMETERS OF THE INTERCONVERSION 2-EXONORBORNYLOXOCARBONIUM ION ⇌ 2-ENDONORBORNYLOXOCARBONIUM ION[224]

Temperature (°C)	$k^a \times 10^4$/sec	
75	1.53	$\Delta H_D^{\ddagger}$(endo) = 31.3 kcal/mole
87.5	7.8, 9.0	$\Delta S_D^{\ddagger}$(endo) = 13 cal · degree^{-1} · mole^{-1}
94.5	16	$\Delta G_D^{\ddagger}$(endo) = 27.5 kcal/mole at 20°C
102	38	

$k^a = k_D^{endo} = k_D^{exo} \cdot k_C^{endo}/k_C^{exo}$. [224]

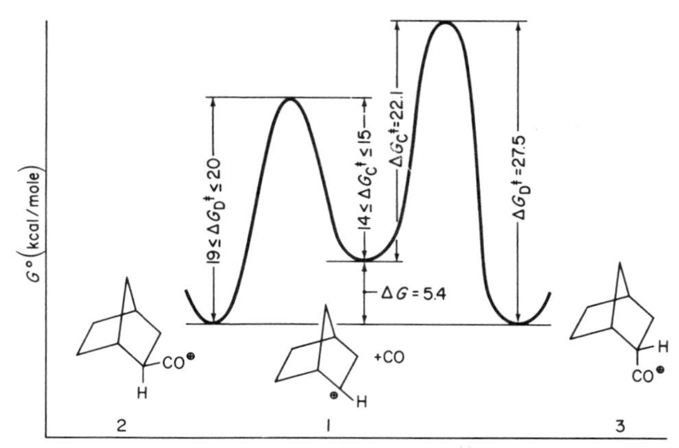

FIG. 3. Free-enthalpy diagram at 20°C and [CO] in mole/liter for the system 2-exonorbornyloxocarbonium ion ⇌ 2-norbornyl ion + CO ⇌ 2-endonorbornyloxocarbonium ion in $FHSO_3/SbF_5$ (1 : 1 m/m). H. Hogeveen and R. E. Roobeek, *Tetrahedron Lett.* **56**, 4941 (1969).

$\overset{+}{CO}$ H **(LXV)** ⇌ + H + CO ⇌ H CO^+ **(LXVI)**

A comparison of σ and π delocalization in phenylnorbornyl and related cations were made by Olah and Liang.[225] The proton and ^{13}C-NMR shows ion 1-H as a rapidly equilibrating carbenium ion undergoing fast Wagner–Meerwin 1,2-shift. The degree of σ-delocalization in 1-H is found to be very similar to that in the 1,2-dimethyl-2-norbornyl cation, 2-CH_3. σ Delocalization in the 2-methyl-2-norbornyl cation is found to be more significant than in the 2-phenyl-2-norbornyl cation in which π delocalization dominates. 1-H is 1,2-diphenyl-2-norbornyl cation. 1-OCH_3 is 1,2-dianisyl-2-norbornyl cation.

room temp; (OCH_3) or (H); (OCH_3); 1-(OCH_3) (1-H); (OCH_3) or (H); (OCH_3) (or H); 1—(OCH_3) 1—H; −78°C; CH_3 CH_3; CH_3 CH_3 (2-CH_3)

Adamantyl cation: A number of bridgehead cations[226,227] have been prepared in the magic-acid system of Olah.[191] Of particular interest is the transformation of the second equation. This rearrangement, when the usual aluminum chloride or aluminum bromide catalyst systems are used, is considered to proceed by repeated formation and quenching of the carbonium ion. In HSO_3F–SbF_5 solutions, such hydride transfers cannot occur[191]; the reaction must therefore involve a substantial number of either intermolecular hydride transfers, or intramolecular hydride migrations. A considerable driving force for the formation of the adamantane skeleton is apparent. Rearrangement of tricyclo[5 · 2 · 1 · 0]decylalcohol to 1-adamantol has been observed.[228]

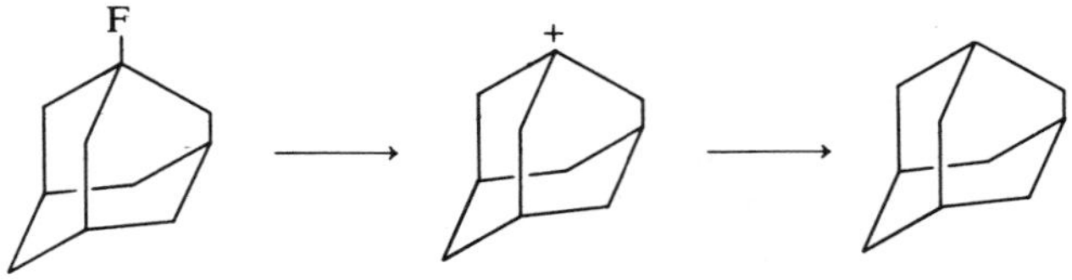

F → + →

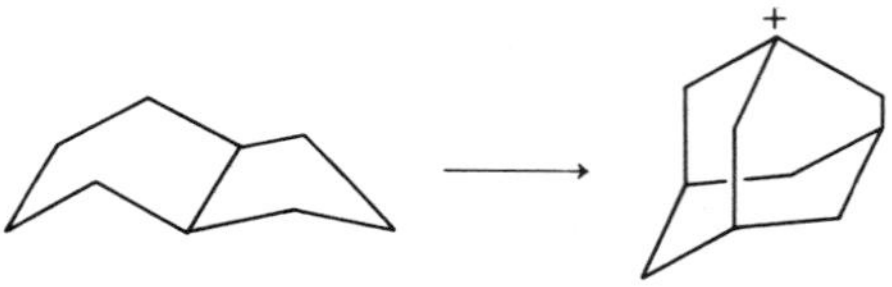

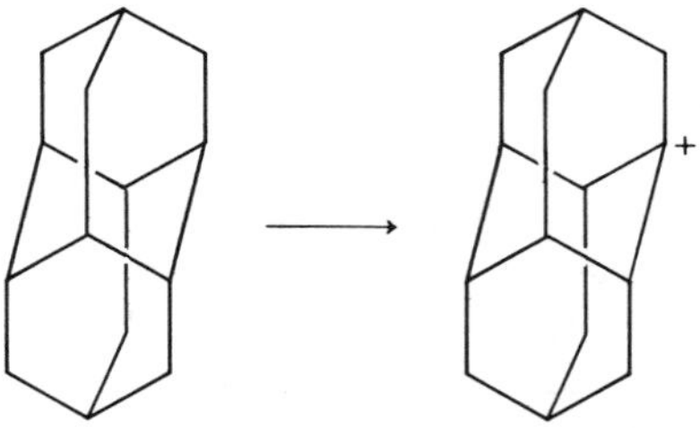

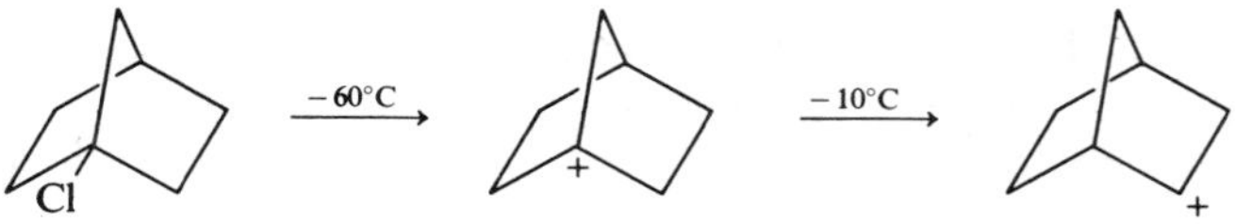

→ +

→ +

Cl —60°C→ + —10°C→ +

A series of 2-alkyl, 2-phenyl-, and 2-halo-substituted 2-adamantyl cations were obtained in HSO_3F, HSO_3F–SbF_5, and SbF_5 (SO_2ClF or SO_2) solu-

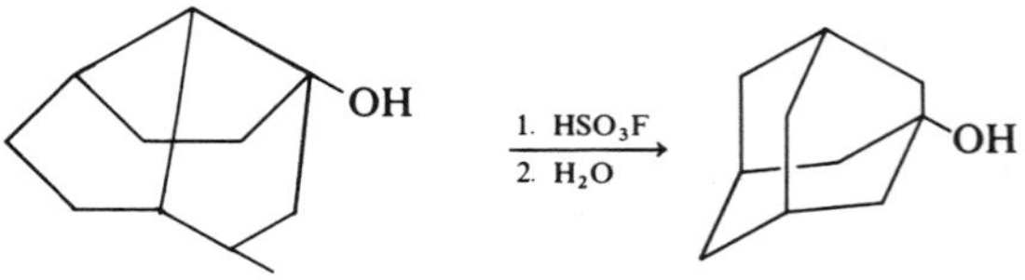

OH —1. HSO_3F 2. H_2O→ OH

tions at $-78°C$.[229] The PMR and FT (Fourier transform) ^{13}C-NMR spectra of 2-methyl and 2-ethyladamantyl cation are considerably different from that of the 1-adamantyl cation.[191,230,231] These two ions are formed from their respective alcohols. Both are stable and show no rearrangement in the temperature range studied ($-78°$ to $+25°C$).

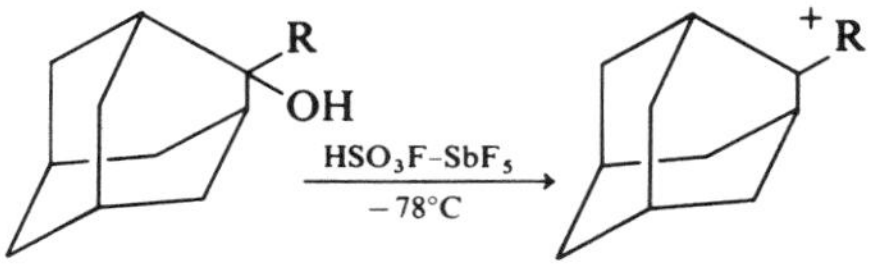

HSO_3F–SbF_5, $-78°C$

$R = CH_3, CH_2CH_3$

2-Halo-2-adamantyl cations are formed directly from their respective 2,2-dihaloadamantanes.

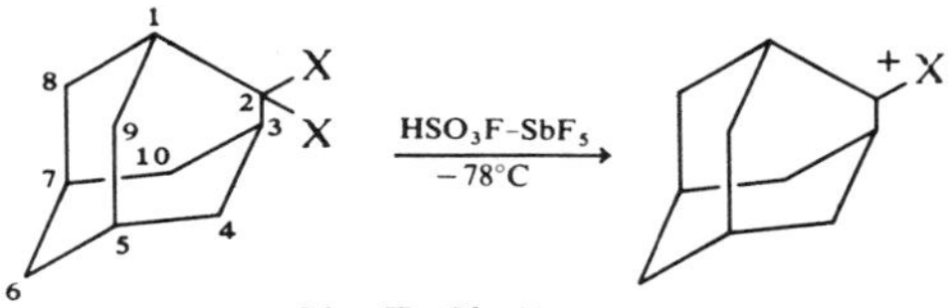

HSO_3F–SbF_5, $-78°C$

X = F, Cl, Br

The results have been interpreted in terms of (1) X = F, Cl, and Br with the different anisotropic effects of the halogen atom, (2) decreasing *n*-*p* conjugation between the halogen atoms and the empty *p* orbital in the order F > Cl > Br. Consideration of the Fourier transform ^{13}C-NMR data reveals several interesting points: (1) the magnitude of the deshielding of the carbenium carbon increases in the order 1-F < 1-Cl < 1-Br, (2) the two bridgehead carbons (C_1 and C_3) show increasing shielding effects according to the order 1-Br < 1-Cl < 1-F, and (3) the methylene C_6 carbons farthest from the positive charge are more deshielding than the bridgehead carbons (C_5 and C_7) which are closer to the positive charge. ^1H- and ^{13}C-NMR data of both secondary **(LXVII)** and tertiary **(LXVIII)** 8,9-dehydro-2-adamantyl cations[232] indicate the carbenium ion nature with the charge delocalization into the cyclopropane ring.

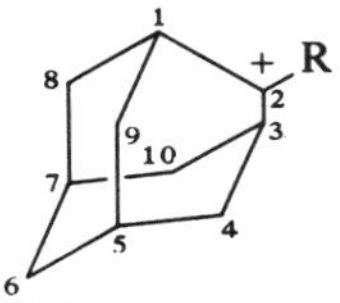

(LXVII) R = H
(LXVIII) $R = CH_3$

The parent secondary and the tertiary ion were prepared from the corresponding alcohols in HSO_3F–SO_2ClF (SO_2) solutions at $-120°$ and $-78°C$ respectively.[233]

23. Hexamethyl(Dewar)Benzene and Related Compounds

The protonation of hexamethylbenzene and hexamethylprismane has been studied in HSO_3F–SbF_5–SO_2.[234]

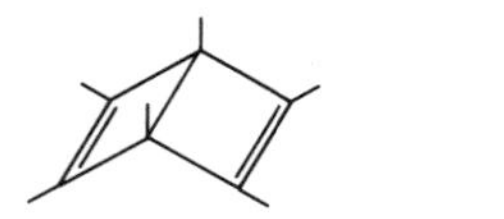

Hexamethyl(Dewar)benzene

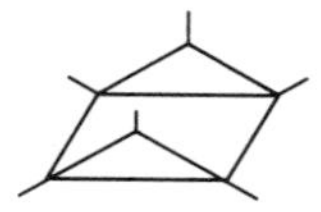

Hexamethylprismane

The presence of two isomeric bicyclo[2 · 1 · 1]hexenyl cations were shown by NMR.

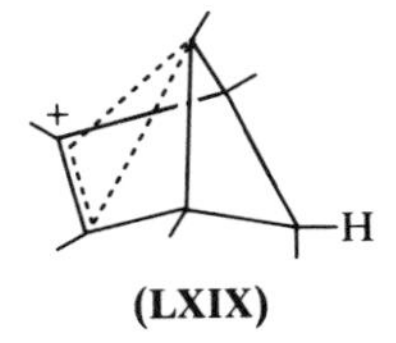

(LXIX)

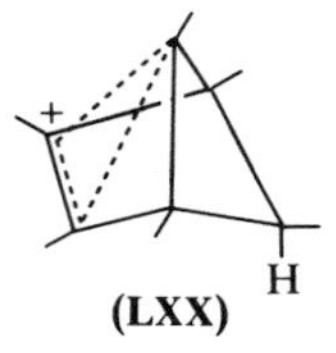

(LXX)

The dissolution of hexamethylprismane in these media at $-78°C$ gave a spectra similar to that obtained from **(LXIX)** except that a significant concentration of methylbenzenium ion was present in solution.

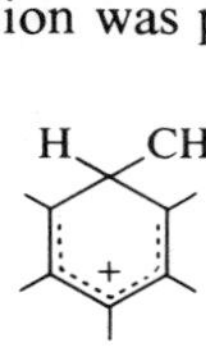

H CH$_3$
+

The idea that in some cations the carbon atom might have a coordination number higher than 4 was suggested by Wilson *et al.*[235] and was later developed by Winstein.[236] Nonclassical ions with a pentacoordinated C atom as norbornyl and norbornenyl cations have become common phenomena reported in the literature since then. Examples of ions with an electronic structure strongly related to that of the norbornenyl cations are the nonclassical 1,2,3,4,5,6-hexamethylbicyclo[2 · 1 · 1]hexenyl cations. These ions are important precursors and intermediates in the reactions of an unusual dication $[CCH_3]_6^{2+}$ reported by Hogeveen and Kwant.[237] This dication can be generated in strongly acidic solutions from a variety of precursors.

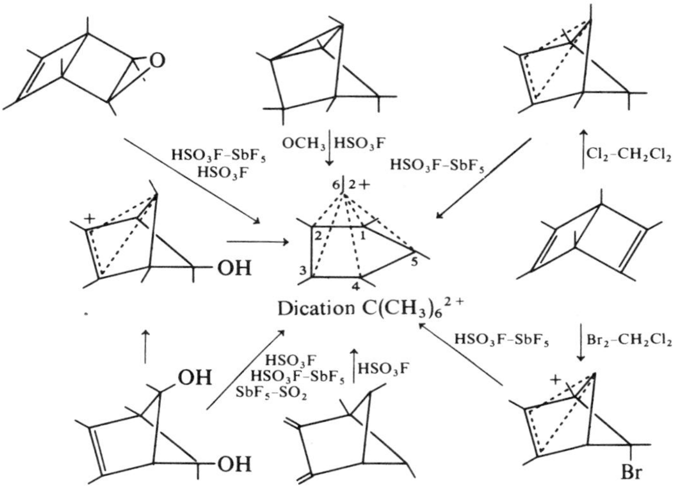

Both the PMR and CMR of dication ($C(CH_3)_6^{2+}$) appeared to be temperature independent in the range from $-140°$ to 100°C. This dication finally rearranges intramolecularly to hexamethylbenzenium ion.

Childs and Winstein[238] studied heptamethylbicyclo[3 · 1 · 0]hexenyl cation **(LXXII)** and observed fivefold degenerate scrambling as predicted by Swatton and Hart. Cation **(LXXII)** was prepared by the irradiation of a solution of the heptamethylbenzenonium ion **(LXXI)** in HSO_3F at $-78°C$. The reverse process, **(LXXII)** to **(LXXI)**, occurs smoothly on warming the solution of ion **(LXXII)** up to $-9°C$.

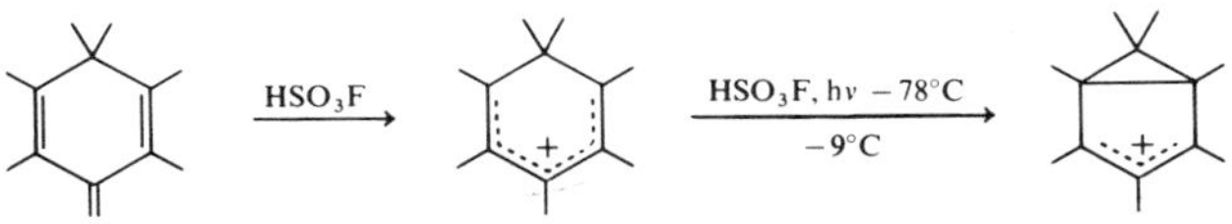

(LXXI) **(LXXII)**

$\Delta G = 19.8$ kcal/mole
$k = 2.2 \times 10^{-4}$ sec^{-1}

Cation **(LXXII)** displays a temperature-dependent NMR spectrum.

The possible mechanism for the photochemical conversion of durene into a bicyclo[3 · 1 · 0]hexenyl cation has been discussed by Childs and Parrington.[239] Protonation of durene in fluorosulfuric acid[175] at $-78°C$ gave the tetramethylbenzenonium cation **(LXXIII)**. Irradiation of **(LXXIII)** at $-90°C$ gave a product **(LXXIV)** which had a striking resemblance to that of the pentamethylbicyclo[3 · 1 · 0]hexenyl cation **(LXXII)**.[238] Upon increasing the length of the irradiation time, a photostationary state was reached between **(LXXIII)** and **(LXXIV)**. The position of the photostationary state was dependent upon the type of lamp and filters used, for example, **(LXIV)** (60 %) and **(LXIII)** (40 %) with a pyrex filter and Philips SP 500 W high-pressure mercury source.

Heating the fluorosulfuric acid solution of the photoproduct (57) to $-31°C$ caused it to undergo a clean isomerization back to protonated durene (56), with a first-order constant rate of $1.46 \times 10^{-3} \text{sec}^{-1}$ ($\Delta F = 17.1$ kcal/mole). They[239] indicated that the preferred path of photoisomerization of the benzenonium cation to a bicyclohexenyl cation is by a disrotatory electrocyclic ring closure rather than by a $[\sigma 2a + \pi 2a]$ cyclo addition reaction.[239]

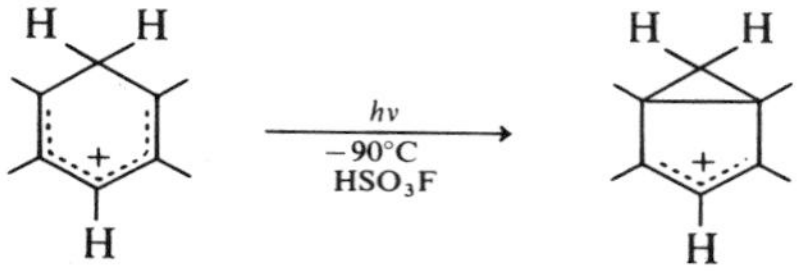

H H
$\xrightarrow[-90°C,\ HSO_3F]{h\nu}$
H H
H
(LXXIII)
H
(LXXII)

(LXXIII) $\xrightarrow[HSO_3F]{h\nu\ -90°C}$ **(LXXII)**

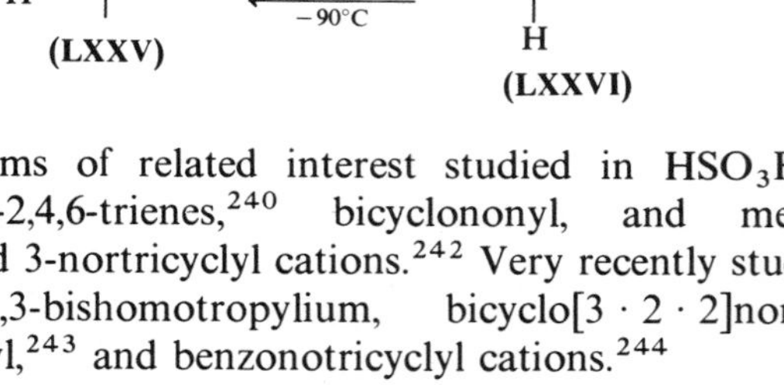

$-31°C$, $k = 1.46 \times 10^{-3}$/sec, $\Delta F^{\ddagger} = 17.1$ kcal/mole
$-31°C$ | $\Delta F^{\ddagger} > 18.6$ kcal/mole, $k \leqslant 7.3 \times 10^{-5}$/sec
$[\sigma 2a + \pi 2a]$
$[\sigma 2a + \pi 2a]$

H H
H
H
H
$h\nu\ -90°C$
$HSO_3F\ -90°C$
H
(LXXV)
(LXXVI)

Other systems of related interest studied in HSO_3F are bicyclo × [6 · 1 · 0]nona-2,4,6-trienes,[240] bicyclononyl, and methylbicyclooctyl cations,[241] and 3-nortricyclyl cations.[242] Very recently studied systems are monomethyl-1,3-bishomotropylium, bicyclo[3 · 2 · 2]nonatrienyl, and norbornadienyl,[243] and benzonotricyclyl cations.[244]

24. Alkoxysilanes and Disiloxanes

Unlike the corresponding carbon systems, silanes and siloxanes do not form siliconium ions similar to carbonium ions. Instead they form fluorosilanes. Several alkoxy silanes and disiloxanes were examined by Olah, O'Brien, and Lui[245] in HSO_3F–SbF_5 at low temperatures by NMR methods. They discussed the mechanism of cleavage of trimethoxy silane and hexamethyl disiloxane. No experimental evidence was found for the intermediacy of the siliconium ions. The siloxanes studied were $(CH_3)_3SiOCH_3$, $(C_2H_5)_3SiOCH_3$, $(CH_3)_2SiO(C_2H_5)_2$, and

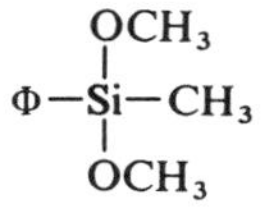

$$\Phi-\overset{\overset{\displaystyle OCH_3}{|}}{\underset{\underset{\displaystyle OCH_3}{|}}{Si}}-CH_3$$

Typical cleavage of alkoxysilanes and disiloxanes is shown below.

$$CH_3Si(OR)_3 \xrightarrow[-60°C]{HSO_3F\ SbF_5\ SO_2} CH_3SiF_3 + 3ROH_2^+$$

$$R = CH_3,\ C_2H_5$$

$$Si(OCH_3)_4 \xrightarrow[-60°C]{HSO_3F\ SbF_5\ SO_2} SiF_4 + 4CH_3OH_2^+$$

For the cleavage of hexamethyl disiloxanes

$$(CH_3)_3Si-O-Si(CH_3)_3 \xrightarrow[-78°C]{HSO_3F\text{-}SbF_5\text{-}SO_2} [(CH_3)_3Si]_2O^+H \xrightarrow{-60°C} 2(CH_3)_3SiF + H_3O^+$$

$$\downarrow > -30°C$$

$$2(CH_3)_2SiF_2 + 2CH_4\uparrow$$

$$(CH_3)_2SiO-Si(C_6H_5)_3 \longrightarrow (CH_3)_3SiF + (C_6H_5)_3SiO^+H_2$$

$$\downarrow$$

$$(C_6H_5)_3SiF$$

The facile formation of fluorosilanes in the superacid medium is explained by an Si(V) intermediate arising from a nucleophilic attack by a fluoride ion at Si prior to the loss of methanol.

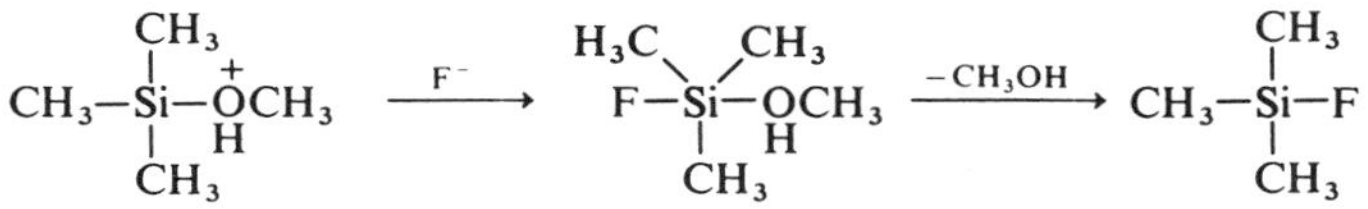

$$CH_3-\overset{\overset{\displaystyle CH_3}{|}}{\underset{\underset{\displaystyle CH_3}{|}}{Si}}-\overset{+}{\underset{H}{O}}CH_3 \xrightarrow{F^-} (H_3C)(CH_3)F-\underset{\underset{\displaystyle CH_3}{|}}{Si}-\underset{H}{O}CH_3 \xrightarrow{-CH_3OH} CH_3-\overset{\overset{\displaystyle CH_3}{|}}{\underset{\underset{\displaystyle CH_3}{|}}{Si}}-F$$

25. Selenonium and Telluronium Ions

As an extension of the study of onium ions, Olah *et al.*[246] prepared and studied selenonium and telluronium ions R_2Se (Te) H^+ in a superacid solution. They also prepared and isolated a series of trialkyl selenonium and telluronium ions as well as trialkylsulfonium ions as their fluorosulfate salts. A comparative study of all these onium ions has been made.

Unlike hydrogen selenide, which is easily oxidized to elemental Se by magic acid, alkyl selenides are much more stable to oxidation than hydrogen selenide and can be protonated in $HSO_3F–SbF_5–SO_2$ solutions.

$$R_2Se \xrightarrow[SO_2,\ -60^\circ C]{HSO_3F-SbF_5} R_2SeH^+$$

dialkyl selenonium ion

The acidic, secondary alkylselenonium ions are remarkably stable. The PMR spectra showed no significant change from $-60°$ to 65°C. The alkyltellurides are also protonated in $HSO_3F–SbF_5–SO_2$ at -60°C.

The protons on Se in selenonium ions and that on Te in telluronium ions are considerably more shielded than the protons on oxygen in the related onium ions. (δ 7.88–9.21) and the proton on S in the corresponding sulfonium ions (δ 5.80–6.52). There is a consistent trend of increasing shielding going from related onium to sulfonium to telluronium ions. Charge delocalization and shielding by increasingly heavier atoms is thus indicated.

26. Mercurinium Ions

The observation of the first stable, long-lived mercurinium ions and the evidence for their structure were observed by Olah and Clifford.[247] They have demonstrated that these stable, long-lived mercurium ions can be prepared both by σ and π routes in superacid media of low nucleophilicity.

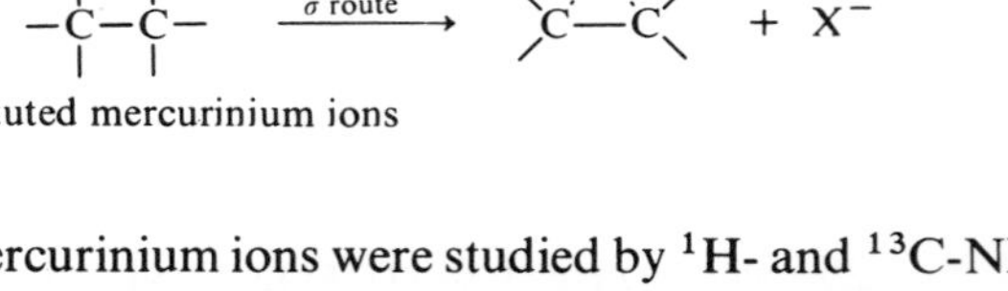

$$>C=C< + Hg^{2+} \xrightarrow{\pi \text{ route}} >C\overset{Hg^{2+}}{\triangle}C<$$

olefins

$$-\overset{X}{\underset{|}{\overset{|}{C}}}-\overset{Hg^+}{\underset{|}{\overset{|}{C}}}- \xrightarrow{\sigma \text{ route}} >C\overset{Hg^{2+}}{\triangle}C< + X^-$$

β-substituted mercurinium ions

The following mercurinium ions were studied by ^1H- and ^{13}C-NMR quenching experiments: ethylene-, *cis*- and *trans*-dimethylethylene-, propylene-, cyclohexene-, cyclohexenemethyl-, norbornylene-, and norbornylenemethyl mercurinium ions.

Cyclohexene mercurinium ions were made by both π and σ routes.

$$\text{C}_6\text{H}_{10} + Hg(OCOCF_3)_2 \xrightarrow[\pi \text{ route}]{HSO_3F\text{-}SO_2} \text{cyclohexenemercurinium ion } (\cdots Hg^{2+}) \longleftarrow \text{2-hydroxycyclohexyl-HgOCOCH}_3$$

cyclohexenemercurinium ion

Norbornylenemercurinium was formed by exo-*cis*-3-hydroxy-2-norbornylmercuric acetate.

$$\text{(OH)(HgOAc)norbornane} \xrightarrow[SO_2]{HSO_3F\text{-}SbF_5} \text{norbornylene}\cdots Hg^{2+} + CH_3CO_2H_2^+ + H_3O^+$$

27. Ferrocenes

The site of protonation in acylferrocenes was found to be at the carbonyl oxygen and not at the metal atom.[248] Even protonation of ferrocene itself in HSO_3F–SO_2ClF (SO_2) at $-80°C$ showed no evidence of Fe protonation.

$$\text{Fc–C(=O)R} \xrightarrow[-60°C]{HSO_3F\text{-}SO_2ClF} \text{Fc–C(R)=O}^+\text{–H}$$

$(R = H, CH_3, C_2H_5, C_6H_5, OCH_3)$

Protonated acylferrocenes undergo intermolecular hydrogen exchange with the solvent systems at higher temperature:

$$\text{Fe}\overset{*}{\text{C}}\text{HO} + \text{Fe CHOH}^+ \quad \text{Fe}\overset{*}{\text{C}}\text{HOH} + \text{Fe CHO}$$

The PMR of acylferrocenes were also dependent on the acid concentration.

28. Organophosphorus Compounds

The protonation and cleavage of trialkyl (aryl) phosphates and phosphites, dialkylphosphonates, and phosphorus oxyacids,[249] in HSO_3F and HSO_3F–SbF_5 were studied by Olah and McFarland.[250] In alkoxy (aryloxy) and hydroxy phosphonium ions NMR data showed substantial d_π–p_π bond formation in protonated intermediates.

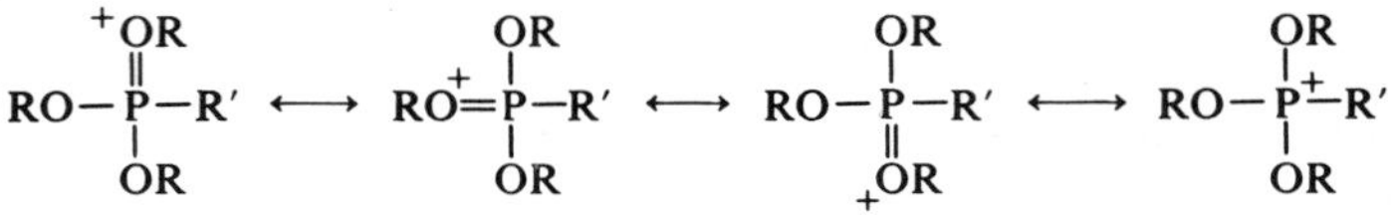

$$RO-\overset{\overset{+}{OR}}{\underset{OR}{\overset{\|}{\underset{|}{P}}}}-R' \longleftrightarrow R\overset{+}{O}=\overset{OR}{\underset{OR}{\overset{|}{\underset{|}{P}}}}-R' \longleftrightarrow RO-\overset{OR}{\underset{\underset{+}{OR}}{\overset{|}{\underset{\|}{P}}}}-R' \longleftrightarrow RO-\overset{OR}{\underset{OR}{\overset{|}{\underset{|}{P^{+}}}}}-R'$$

R = H, alkyl, aryl; R′ = H, alkoxy, aryloxy, hydroxy

At higher temperatures the protonated compounds were subjected to decomposition reactions, including C–O bond cleavage and fluorination.

29. Azo, Azoxy, Hydrazo, and Related Compounds

A new mechanism for the benzidine and Wallach rearrangement was proposed based on the observation of intermediate compounds.[251] Protonation of azobenzenes, azoxybenzenes, pyridine *N*-oxides, hydrazobenzene, phenylhydrazine, and arylamines were studied in HSO_3F–SbF_5–SO_2 and HF–SbF_5–SO_2 solutions. Only monoprotonation of azoxy and 4,4-dichloroazoxybenzene was observed in HSO_3F. In the stronger acid, HF–SbF_5–SO_2, diprotonation was observed. Both azobenzene and benzidine were diprotonated in HSO_3F.

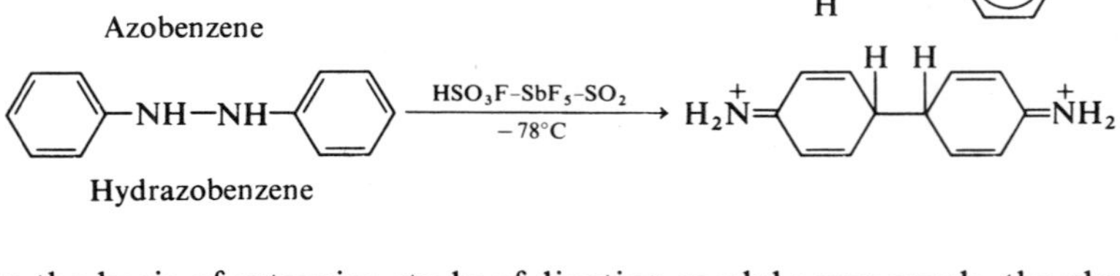

R–C₆H₄–N(→O)=N–C₆H₄–R $\xrightarrow[-78°C]{HSO_3F\text{–}SO_2}$ C₆H₅–$\overset{+}{N}$(–O–H)=N–C₆H₅

R = H, Cl

Azobenzene: C₆H₅–N=N–C₆H₅ $\xrightarrow[-80°C]{HSO_3F\text{–}SbF_5\text{–}SO_2}$ C₆H₅–NH=NH–C₆H₅

Hydrazobenzene: C₆H₅–NH–NH–C₆H₅ $\xrightarrow[-78°C]{HSO_3F\text{–}SbF_5\text{–}SO_2}$ $H_2\overset{+}{N}$=C₆H₄(H)–(H)C₆H₄=$\overset{+}{N}H_2$

On the basis of extensive study of dication model compounds, the absence of H_2–D_2 incorporation into the benzidine rings during rearrangements in deuterated acids, and general consideration of benzenium ions in alkylative

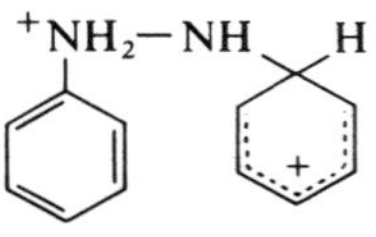

$^{+}NH_2$–NH (C₆H₅ and H-substituted benzenium ion, +)

systems, a new mechanism for the benzidine rearrangement is proposed involving N,C-diprotonated hydrazobenzene as the key intermediate.

Azoxybenzenes substituted in 4,4′ position were reduced to substituted 4,4′ azobenzene both in HSO_3Cl[252] and HSO_3F[253] by the Wallach transformation. Unsubstituted azoxybenzenes with an open 4 or 4′ position are converted to their corresponding 4-chlorosulfato- or 4-fluorosulfateazobenzenes in chlorosulfonic and fluorosulfonic acids, respectively.

30. Halonium and Related Ions

Halonium ions are now recognized as an important class of onium compounds in their own right and their chemistry is rapidly being explored. They are important reaction intermediates in halogenation reactions. With the development of highly acidic and low-nucleophilicity solvent systems such as SbF_5–SO_2 or HSO_3F–SbF_5 solutions the preparation and direct observation, even isolation of stable halonium ions have become possible.[254–259a] A comprehensive study of the structure of cyclic and acyclic halonium ions was carried out by CMR.[260] Halonium ions are classified into acyclic and cyclic halonium ions. Acyclic halonium ions include dialkylhalonium ions,[254] alkylarylhalonium ions,[255] and diarylhalonium[256] di- and trihalonium ions.[257,258,258a] All these ions have been prepared and characterized by NMR spectroscopy. Cyclic halonium ions include the three-membered ring ethylene halonium ion,[258,258a] five-membered ring tetramethylene halonium ions[259,259a] and six-membered ring pentamethylene halonium ions,[261] which was recently reported in SbF_5–SO_2. Attempted preparation of four-membered-ring trimethylene halonium ions under similar conditions was generally unsuccessful,[262] although a recent report has appeared only on four-membered-ring fluorinated halonium ions.[263]

An indirect evaluation of the relative stability of the β-halogenocarbonium ion was obtained by following the fate[258a] of 1-halogeno-2-methyl-2-methoxypropanes treated with HSO_3F–SbF_5 in SO_2 at -60°C.

For **(LXVII)** when X = Cl, the protonated ether **(LXXVIII)** slowly undergoes loss of methanol to give the open cation **(LXXIX)**; for 1, when X = Br, the loss of methanol is faster and the cyclic ion **(LXXX)** is formed. In the case of the iodo compound [**(LXXVII)**, X = I] the protonated ether cannot be observed because of very fast formation of the cyclic iodonium ion **(LXXXI)**.

The examples reported above stress the different abilities of the halogens to give bridged ions. It is also shown that the importance of cyclic structures increases with decreasing stability of the open carbonium ion.

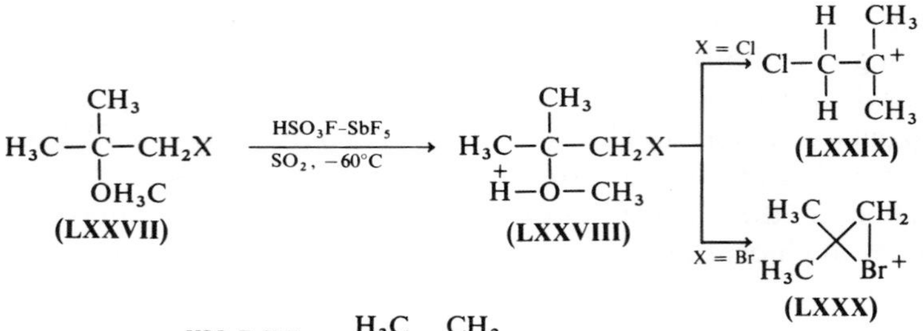

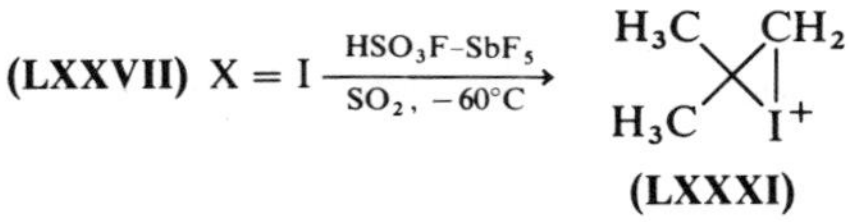

$$H_3C-C(CH_3)(OCH_3)-CH_2X \xrightarrow[SO_2,\ -60°C]{HSO_3F\text{-}SbF_5} H_3C-C(CH_3)(\overset{+}{O}H-CH_3)-CH_2X$$

(LXXVII) → **(LXXVIII)**

X = Cl → $Cl-CH_2-\overset{+}{C}(CH_3)_2$ **(LXXIX)**

X = Br → $(H_3C)_2C(CH_2)Br^+$ (three-membered bromonium ring) **(LXXX)**

$$\textbf{(LXXVII)}\ X = I \xrightarrow[SO_2,\ -60°C]{HSO_3F\text{-}SbF_5} (H_3C)_2C(CH_2)I^+$$

(LXXXI)

2-Methyl-, 2,2-dimethyl-, and 2,5-dimethyltetramethylene halonium (X = Cl, Br, I) ions have also been prepared according to the reactions shown below:

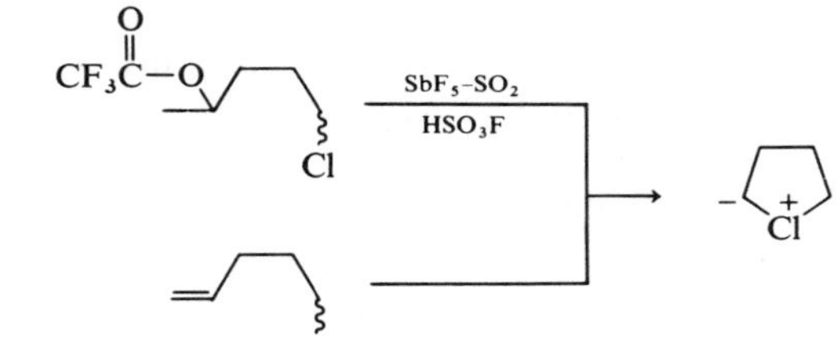

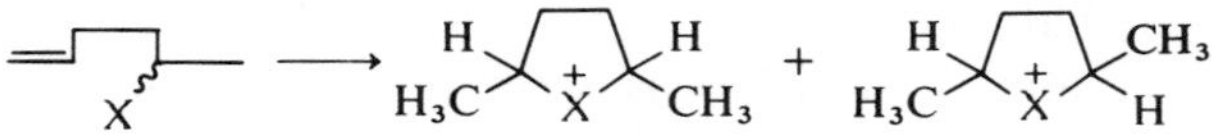

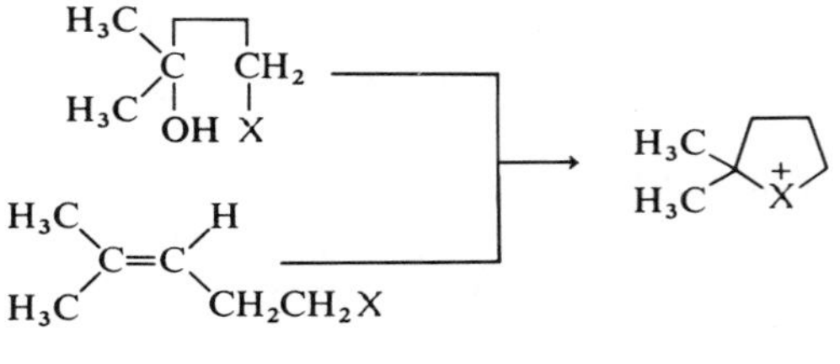

$CF_3C(=O)-O-CH(CH_3)CH_2CH_2CH_2Cl \xrightarrow[HSO_3F]{SbF_5\text{-}SO_2}$ 2-methyltetramethylenechloronium ion ($\overset{+}{Cl}$ five-membered ring)

$CH_2=CHCH_2CH_2CH_2Cl$ → same product

$CH_2=CHCH_2CH_2CH(CH_3)X$ → cis- and trans-2,5-dimethyltetramethylenehalonium ions ($\overset{+}{X}$ ring with H, CH_3 at C-2 and C-5)

$(H_3C)_2C(OH)CH_2CH_2CH_2X$ and $(H_3C)_2C{=}CH-CH_2CH_2X$ → 2,2-dimethyltetramethylenehalonium ion ($\overset{+}{X}$ ring with $(H_3C)_2$ at C-2)

Protonation of 5-iodo-1-pentyne **(LXXXII)** in magic acid gives an NMR spectrum indicating the formation of a five-membered-ring iodonium ion.[259a]

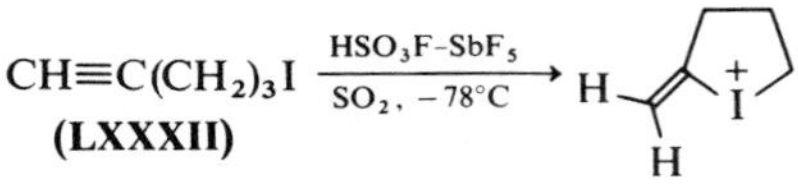

$$CH{\equiv}C(CH_2)_3I \xrightarrow[SO_2,\ -78°C]{HSO_3F\text{-}SbF_5} H_2C{=}\text{(five-membered ring with } \overset{+}{I})$$

(LXXXII)

Reactions of 5-chloro and 5-fluoro-1-pentyne in the same solutions gave uninterpretable spectra.

In order to understand better the bonding and chemistry of the intermediate halonium ions, Larsen and Metzner[264] did some calorimetric measurements of their heats of formation from dihalides in 11.5 mole% HSO_3F–SbF_5 at $-60°C$. The tetramethylenebromonium ions are about as stable as the *tert*-butyl cation and their heats of formation are about 10 kcal/mole more stable than a correspondingly substituted ethylene bromonium ion. The tetramethylenechloronium ion is approximately 7.5 kcal/mole less stable than its bromo analog. The interaction of a methyl group with a three-membered ring seemed to be unusually favorable.

The behavior of chloroalkenes in both H_2SO_4–SO_3 and HSO_3F has been studied by Deno *et al.*[265]

Another aspect of how the halogens may affect the stability of a positive center is offered by the studies of halobenzenonium ions which have been obtained by protonation of halobenzenes in HSO_3F–SbF_5[266] or HF–SbF_5.[267] Olah and Kiovsky[266] reported the spectra of protonated fluorobenzene in HSO_3F–SbF_5, *m*- and *p*-difluorobenzene, 1,3,5- and 1,2,4-trifluorobenzene, 1,2,3,5-tetrafluorobenzene, *o*-, *m*-, and *p*-fluorotoluene, 2,4- and 2,5-difluorotoluene, 2,3,5,6-tetrafluorotoluene, and of monofluoro, difluoro, and trifluoromesitylene. Arenonium ions have been reviewed by Brouwer *et al.*[268] The importance of resonance forms and high degree of charge delocalization are shown in both bis- and tris-pentafluorophenyl carbonium ions.[269]

The observation of the first stable fluorocarbonium ions, the phenyl and diphenylfluorocarbonium ions was reported by Cupas, Comisarow, and Olah.[270]

$$C_6H_5-\overset{+}{C}F_2 \qquad C_6H_5-\overset{+}{C}F-C_6H_5$$

The acidities of a number of perfluorinated alcohols of the type $C_nF_{2n}CH_2OH$ have been compared. In a number of cases these alcohols are only protonated in HSO_3F–SbF_5–SO_2 solutions and no ionization was observed. The basicities of a number of fluorinated ketones have also been compared in HSO_3F–SbF_5.

$$C_6H_5-C(CF_3)_2-\overset{+}{O}H_2, \quad F_3C-CH(\overset{+}{O}H_2)-CH_3, \quad H_3C-C(CF_3)_2-\overset{+}{O}H_2, \quad CF_3-\overset{\overset{+}{O}H_2}{\overset{\|}{C}H}-CF_3$$

31. Miscellaneous Systems

The use of HSO_3F–SbF_5 or HF–SbF_5 in steroid chemistry has been shown by Jacquesy *et al.*[271] These hyperacidic media provide a novel route to the formation of 13α-steroids.

5α-Pregnane-3,20-dione **(LXXXIII)** with HSO_3F–SbF_5 or HF–SbF_5 gave the corresponding 13α-pregnane dione **(LXXXIV)**.

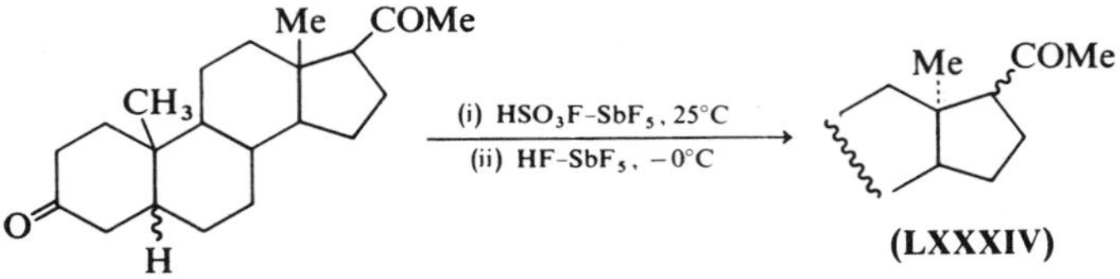

Me COMe
CH_3
O
H
(LXXXIII)

(i) HSO_3F–SbF_5, 25°C
(ii) HF–SbF_5, −0°C

Me COMe
(LXXXIV)

Other interesting systems studied in HSO_3F and its superacids include charge distribution in aldehydes and ketones,[272] rearrangements of *tert*-hexyl cations,[273] acetyl and benzoyl pyridinium ions and its related ions,[274] solvolysis of ethyl tosylates.[275,276] The most recently studied systems are hydride transfer reactions,[277] the relative ability of charge delocalization by phenyl, cyclopropyl, and methyl groups,[278,279] chain elongation in the rearrangement of 2,3-dimethyl-4-penten-2-ol to 2-methyl-3-hexen-2-yl cations,[280] hydridohalonium ions and methylmethylenehalonium ylides,[281] and double bond vs. cyclopropane ring reactivity toward different acids.[282]

II. Chlorosulfuric Acid

A. Introduction

The chemistry of chlorosulfuric acid as a nonaqueous solvent has not been as well explored as fluorosulfuric acid. It was known much earlier than fluorosulfuric acid; it was made in 1857 by treating SO_3 with dry HCl.[283] In 1902 Walden[284] first measured its specific conductivity and established that it is a good ionizing solvent. Later its solvent properties were investigated by several workers.[285–288] It was shown that chlorosulfuric acid is much less acidic than HSO_3F but that its acidity is higher than that of H_2SO_4. The greater acidity of chlorosulfuric acid, as compared with H_2SO_4, its relatively high dielectric constant ($\varepsilon = 60$), and other properties point to its potential use as a solvent for a number of inorganic and organic solutes.

B. Physical Properties

Chlorosulfuric acid is a colorless fuming liquid that decomposes violently with water to form HCl and H_2SO_4. Some of the important properties of HSO_3Cl are summarized in Table XIX.[12,21,284,286–292]

C. Solvent System

Palm[287] and Gillespie *et al.*[12] have determined the Hammett acidity (H_0) function. They found that H_0 value was more negative than that of pure H_2SO_4 showing that HSO_3Cl is a stronger acid than H_2SO_4. This result was confirmed by Barr, Gillespie, and Robinson,[286a] who found that HSO_3Cl behaves as a very weak acid in 100% H_2SO_4. Waddington and Klanberg[288] found that HSO_3Cl also behaves as an acid in liquid HCl

$$HSO_3Cl + HCl \rightleftharpoons H_2Cl^+ + SO_3Cl^-$$

Paul and his co-workers[285] have studied the electrical conductivities of solutions of a few tertiary bases and alkali chlorosulfates and concluded that HSO_3Cl ionizes according to

$$HSO_3Cl \rightleftharpoons HCl^+ + SO_3Cl^-$$

Like fluorosulfuric acid, chlorosulfuric acid also undergoes autoprotolysis.

$$2HSO_3Cl \rightleftharpoons H_2SO_3Cl^+ + SO_3Cl^-$$

Acids in the solvent give the chlorosulfuric acidium ion $H_2SO_3Cl^+$ and bases the chlorosulfate ion SO_3Cl^-. Because of its high acidity, bases are the much larger class of electrolytes.

TABLE XIX

PHYSICAL PROPERTIES OF CHLOROSULFURIC ACID

		Reference
Freezing point	$-80°C$	289
	$-77.62°C$	288
Boiling point	152°C	289
Viscosity (cP)	2.4	290
Density (g/ml)	1.741	286
Dielectric constant (25°C)	60 ± 10	291
Specific conductivity ($ohm^{-1}\ cm^{-1}$)	1.72×10^{-4}	284
	4×10^{-4}	286a
Heat of formation (kcal/mole)	142.9	292
Cryoscopic constant ($deg\ mole^{-1}\ kg$)	5.5	288
Hammett acidity function	-12.78	287

D. Inorganic Solutes

1. Alkali and Alkaline Metals

Pure HSO_3Cl has a small electrical conductivity ($k = 4 \times 10^{-4}$ ohm^{-1} cm^{-1} at 25°C).[286] However, alkali and alkaline earth metal chlorosulfates dissolve in HSO_3Cl to give highly conducting solutions. Solutions of $LiSO_3Cl$, $NaSO_3Cl$, and KSO_3Cl have very similar molar conductances and it has been concluded that these salts are fully dissociated in chlorosulfuric acid and that a large proportion of the current is carried by SO_3Cl^-.

$$MSO_3Cl \rightarrow M^+ + SO_3Cl^-$$

The specific conductivities of alkali metal chlorides are all very similar but are slightly higher than those of the corresponding chlorosulfates. This is due to the presence of HCl in solution which behaves as a very weak base.

The lower conductivities of alkaline earth chlorosulfates has been explained on the basis of relatively low dielectric constant of HSO_3Cl ($\varepsilon = 60$) as compared with H_2SO_4 ($\varepsilon = 100$) and HSO_3F ($\varepsilon = 120$) which would encourage ion pair formation especially in the case of divalent cations with smaller radii.

From density measurements and hence the apparent molal volumes of the chlorosulfates, Robinson and Ciruna[286] have shown that the extent of solvation increases in the order $K^+ < Na^+ < Ba^{2+}$. This is also the order that would be expected on the basis of increasing polarizing power of cations and is exactly the order of the specific conductivities of the chlorofulfates of group IA and IIA.

Heubel and Wartel[293] have reviewed some reactions between HSO_3Cl, $NaSO_3Cl$, and $NOSO_3Cl$ and 42 different chemical agents. They found four fundamental reaction schemes (where C_1 is the cation),

$$C_1SO_3Cl \rightleftharpoons C_1^+ + SO_3Cl^-$$

$$SO_3Cl^- \rightleftharpoons SO_3 + Cl^-$$

$$SO_3Cl^- + O^{2-} \rightleftharpoons SO_4^{2-} + Cl^-$$

$$SO_3Cl^- \rightleftharpoons SO_2Cl^+ + O^{2-}$$

The chlorination of acid sulfates with thionyl chloride was greatly improved by using pure HSO_3Cl as solvent.[294,295] Thus, yields of $\simeq 75\%$ MSO_3Cl were quickly obtained for M = K and Na when excess $MHSO_4$ was dissolved in HSO_3Cl and small amounts of $SOCl_2$ were added. MHS_2O_7 is formed as an intermediate.

$$MHSO_4 + HSO_3Cl \rightarrow MHS_2O_7 + HCl;$$

$$MHS_2O_7 + SOCl_2 \rightarrow MSO_3Cl + HSO_3Cl + SO_2$$

Yves, Wartel, and Heubel,[294] also studied the mechanism of the chlorination of acid sulfates ($MHSO_4$ $M = Na^+$, NO^+, $NH_4{}^+$) and pyrosulfates in a chlorosulfuric acid medium.

Heubel and his co-workers[296] studied the reactions of chlorosulfuric acid with MCl_2 ($M = Ca$, Sr, and Br). They found solvates of the type $Sr(SO_3Cl)_2 \cdot 2HSO_3Cl$, and $Ba(SO_3Cl)_2 n \cdot HSO_3Cl$ ($n = 1$, 2, and 3). These salts were characterized by X-ray diffraction, IR spectra, and thermogravimetry. They proposed a mechanism for the thermal decomposition of these salts.

Paul and his co-workers,[26,27,32,297,298] using conductivity measurements in various nonaqueous solvents like Ac_2O, EtOAc, CH_3SO_3H, and ethyl formate, found the following acid strengths: $H_2S_2O_7 > HSO_3F > HSO_3Cl > H_2SO_4 > HBr > HCl$. They also found 1 : 1 adduct formation in these solvents. The same order of acid strengths of these protonic acids were found by measuring the heats of reaction of these protonic acids with anthrone in nitrobenzene.[299] The use of chlorosulfuric acid as an acidic titrant in acetic acid was shown by the same workers.[285] They also indicated[285] that HSO_3Cl could act as dibasic or tribasic acid. This was explained by dissociation.

$$HSO_3Cl \rightleftharpoons HCl + SO_3$$

Rao and Naidu[300] demonstrated the advantage of HSO_3Cl over $HClO_4$ as an acidic titrant in the mixed solvent methylethylketone and acetic acid (20 : 1) for a number of metal acetates by potentiometric, conductometric, visual, and photometric methods. They showed that the inflection values in the potentiometric titrations of metal acetates with HSO_3Cl increases with the increase in the basicity of the cations in the order $K^+ \simeq NH_4{}^+ > Na^+ > Li^+$.

Bessière[301] established the acid–base scale in CF_3COOH by spectrometric measurements. The following acid strength order was found:

$$HClO_4 < H_2SO_3H < p\text{-}CH_3C_6H_5SO_3H < HBr < HBF_4 < HI < HCl < HNO_3 < H_3PO_4$$

The same author[302] compared the classification of the relative strengths of 25 uncharged bases in CF_3COOH with those established in HOAC and H_2O–H_2SO_4 mixtures.

Rumeau[303] established a general acidity scale in $CHCl_3$ solutions extending over 13–14 units for a number of common acids and amines. This scale was based on the equilibrium constants determined by spectrophotometric methods using various indicators. According to them the acid strength order is $HSO_3Cl < H_2SO_4 < HBr$.

The relative strengths of HCl, HIO_3, HSO_3Cl, *p*-toluenesulfonic acid, picric acid, and CCl_3COOH in 1,4-dioxane and THF were determined using

2-(*p*-*N*,*N*-dimethylaminophenylazo) pyridine as an indicator by Smagowski.[304] He determined the equilibrium constants, $\Delta G°$, $\Delta H°$, and $\Delta S°$ at 20°, 25°, and 30°C. He found these equilibrium constants were higher in dioxane than in THF except for HSO_3Cl and picric acids where the reverse was observed. Fischer and Wartel[305] found an increasing order in the acid strengths $HCl < H_2SO_4 < HSO_3Cl < H_2S_2O_7$ in nitromethane from measurements of their Hammett acidity functions. Another method used for measuring the acid strengths is the comparison of SO_2 frequencies, as carried out by Robinson and Ciruna.[306] They correlated the ν SO_2 symmetric frequencies with acid dissociation constants in H_2SO_4 (Table XX). Lead tetraacetate was found to be soluble in HSO_3Cl to give the acid $H_2Pb(ClSO_3)_6$.[307]

2. Nitrogen and Phosphorus

The mechanism of the reaction between nitryl chloride and HSO_3Cl has been discussed by Wartel, Noel, and Heubel.[308] The reaction of N_2O_4 with HSO_3Cl gives HS_2O_7NO and NO_2Cl. Some reaction of NO_2Cl with HSO_3Cl also occurs to give HS_2O_7NO. HSO_4NO is shown to be an intermediate in the formation of HS_2O_7NO.

$$N_2O_4 + HSO_3Cl \rightarrow NOHSO_4 + NO_2Cl$$

The behavior of some inorganic phosphates in HSO_3F has been studied by Dillon and Waddington[309] by ^{31}P-NMR. The ^{31}P-NMR spectrum has been explained by protonation. The lower field peaks in orthophosphates are assigned to the $P(OH)_4^+$-protonated pyrophosphate equilibrium system as with concentrated phosphate solutions in 100% H_2SO_4. Rapid equilibrium between these two

$$2P(OH)_4^+ \rightarrow (HO)_3POP(O)(OH)_2^+ + H_3O^+$$

$$2P(OH)_4^+ \rightarrow (HO)_3POP(OH)_3^{2+} + H_2O$$

The higher field resonances are attributed to trimetaphosphoric acid.

TABLE XX

Acid Strengths

Acid	ν SO_2 (sym)/cm	k_a (moles/kg) (25°C)
Disulfuric	1240	0.014
Fluorosulfuric	1230	0.002
Chlorosulfuric	1209	0.001
Sulfuric	1195	—

$$P(OH)_4^+ + H_5P_2O_7^+ \qquad H_4P_3O_9^+ + H_3O^+ + H_2O$$

$$P(OH)_4^+ + H_5P_2O_7^{2+} \qquad H_5P_3O_9^{2+} + H_3O^+ + H_2O$$

Metaphosphates indicated other than these a branched phosphate group, i.e., $(PO)_3P{=}O$, which are presumably caused by a condensation between two trimetaphosphate rings. P_2O_5 showed a very complex spectrum in HSO_3Cl. The NMR results were explained by the following possible hydrolysis scheme (Scheme 2).

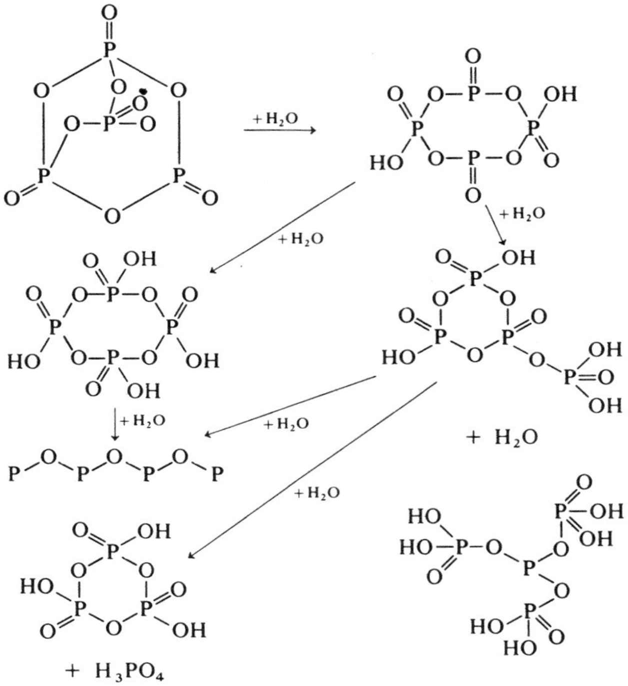

$+ H_3PO_4$

Scheme 2. Hydrolysis of P_2O_5 in HSO_3Cl

Ledson[290] concluded from cryoscopic, conductometric, and spectroscopic measurements that both PCl_5 and PCl_3 form PCl_4^+ ion in HSO_3Cl. $POCl_3$ is not chlorinated as in Se oxychloride but is protonated as a simple weak base.

3. As AND Sb

Lehman[310] investigated the reactions of $SbCl_3$ and $AsCl_3$ with SO_3. The reaction products were considered to be either chlorosulfates or SO_3 adducts.

$$SbCl_3 + SO_3 \rightarrow SbCl_3 \cdot OSO_2 \text{ or } SbCl_3 \cdot 2(OSO_2)$$

(LXXXV) **(LXXXVI)**

(LXXXV) and **(LXXXVI)** are oxidized by chlorine to $SbCl_4 \cdot OSO_2Cl$ and $SbCl_4 \cdot OSO_2Cl \cdot OSO_2$ or $SbCl_4S_2O_6Cl$. $SbCl_4 \cdot OSO_2Cl$ is isomeric with $SbCl_5 \cdot OSO_2$. Similarly,

$$ASCl_3 + SO_3 \rightarrow AsCl_3 \cdot SO_3 \rightleftharpoons ASCl_2 \cdot OSO_2Cl$$

4. S, Se, AND Te

From cryoscopic, conductometric, and spectroscopic measurements of oxides of Te and Se in HSO_3Cl, Ledson[290] showed that these compounds reacted with acid to produce an MCl_3^+ ion.

Robinson and Ciruna[286] have also studied the solutes $SeCl_4$ and $TeCl_4$ in HSO_3Cl by conductivity and Raman spectra. It was shown that both $SeCl_4$ and $TeCl_4$ behave as strong bases according to $MCl_4 + HSO_3Cl \rightarrow MCl_3^+ + SO_3Cl^- + HCl$. They compared the electrical conductivity measurements of $TeCl_4$ and $SeCl_4$ in HSO_3Cl with those of strong bases KSO_3Cl and KCl.

$$KSO_3Cl \rightarrow K^+ + SO_3Cl^-$$

$$KCl + HSO_3Cl \rightarrow K^+ + SO_3Cl^- + HCl$$

Raman spectra also showed that $TeCl_4$ and $SeCl_4$ ionize quantitatively in HSO_3Cl to give the $SeCl_3^+$ and $TeCl_3^+$ cations.

Gillespie and Robinson measured the Raman spectra of SO_3 in HSO_3Cl. They found that the spectra are consistent with the formation of the chlorosulfuric acids HS_2O_6Cl, HS_3O_9Cl, probably $HS_4O_{12}Cl$, and of higher polyacids. As was the case for the corresponding fluoroacids, the spectra indicated the presence of some free SO_3 even at very low stoichiometric concentrations of SO_3 and a probable rapid exchange according to the equation

$$H(SO_3)_nCl + SO_3 \rightleftharpoons H(SO_3)_{n+1}Cl$$

The addition compounds of simple ionic chlorides such as NaCl with SO_3, e.g., $NaCl \cdot SO_3$, $NaCl \cdot 2SO_3$, and $NaCl \cdot 3SO_3$, have been formulated as the salts of HSO_3Cl and the chloropolysulfuric acids, i.e., $NaClSO_3$, $NaClS_2O_6$, and $NaClS_3O_9$.

Schmidt and Talsky[311,312] carried out the reactions of HSO_3Cl with H_2S in ether at $-78°C$.

$$HSO_3Cl + H_2S \xrightarrow[-78°C]{Et_2O} H_2S_2O_3 + HCl$$

The thiosulfuric acid then reacts with HSO_3Cl to form trithionic acid $H_2S_3O_6$. At higher temperatures, $-40°$ to $0°C$, $H_2S_3O_6$ and H_2S again form thiosulfuric acid. In the absence of solvent, HSO_3Cl and H_2S form only thiosulfuric acid at $-78°C$.

5. Ti

Interaction of $TiCl_4$ with HSO_3Cl[313] gave the series of compounds $TiCl_{4-n}(OSO_2Cl)_n$ ($n = 1, 2, 3$) and HCl. However, with excess HSO_3Cl, $Ti(SO_4)_2$ was formed.

$$TiCl_4 + nHOSO_2Cl \rightarrow Cl_{4-n}Ti(OSO_2Cl)_n + nHCl$$

The compound with $n = 2$ prepared from SO_3 had a very similar IR spectra to that of the compound prepared from either HSO_3Cl or CH_3SO_3Cl.

$$TiCl_4 + 2CH_3O \cdot SO_2Cl \text{ (or } 2HSO_3Cl) \rightarrow TiCl_2(OSO_2Cl)_2 + 2CH_3Cl \text{ (2HCl)}$$

E. Organic Solutes

Only very few organic systems have been studied in an HSO_3Cl solvent.

Benzoic and acetic acids are fully protonated in HSO_3Cl. Their conductances are similar to those of alkali metal salts.

$$R'COOH + HSO_3Cl \rightarrow R'COOH_2 + ClSO_3^-$$

Mesitoyl chloride dissolves in $ClSO_3H$ to give the well-known mesitoyl cation.[314]

$$(CH_3)_3C_6H_2COCl + HSO_3Cl \rightarrow (CH_3)_3C_6H_2CO^+ + HCl$$

This ion was readily identified by its NMR and UV absorption spectra. The electrical conductivity of the solution was slightly greater than those of simple bases. The difference was attributed to the weakly basic behavior of HCl in HSO_3Cl.

Solutions of trichloromethylmesitylene in HSO_3Cl were rather more stable than those in H_2SO_4[315] and had conductances similar to those of solutions of mesitoyl chloride. This is consistent with the formation of monopositive chlorocarbonium ion

$$RCCl_3 + HSO_3Cl \rightarrow RCCl_2^+ + HCl + ClSO_3^-$$

together with the partial protonation of the HCl formed. The NMR and UV spectra of solutions of trichloromesityleme in HSO_3Cl were similar to those observed in 100% H_2SO_4. Solutions of trichloropentamethylbenzene **(LXXXV)** and trichloromethylprehnitene **(LXXXVII)** in HSO_3F or HSO_3Cl gave NMR spectra identical to those observed in H_2SO_4. NMR spectra were assigned to the monopositive chlorocarbonium ions **(LXXXVIII)** and **(XC)**.

Berezin[316] studied the solubilities of various metal phthalocyanines(pc) like H_2Pc, NiPcCl, PdPcCl, PtPcCl, CoPcCl, CuPc, $CuPcCl_{15}$, and ZnPc in HSO_3Cl and oleum. Available thermodynamic data were used to calculate

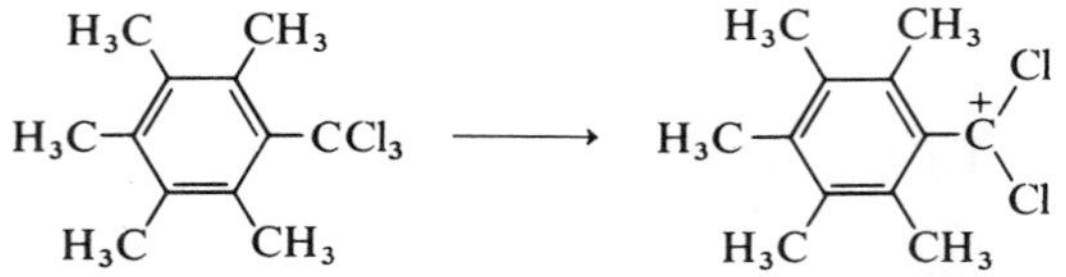

(LXXXVII) → (LXXXVIII)

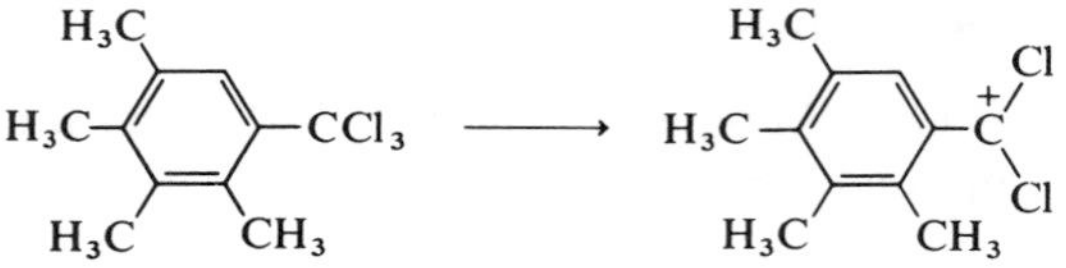

(LXXXIX) → (XC)

the equilibrium constants, K, of the reactions of phthalocyanines, MPc's, with H_2SO_4, HSO_3Cl, and $H_2S_2O_7$ (where M = metal):

$$\mathrm{MPc} + \mathrm{H_2SO_4} \rightleftharpoons \mathrm{MPcH^+} + \mathrm{HSO_4^-} \qquad -K_1$$

$$\mathrm{MPc} + \mathrm{HSO_3Cl} \rightleftharpoons \mathrm{MPcH^+} + \mathrm{OSO_2Cl^-} \qquad -K_2$$

$$\mathrm{MPc} + \mathrm{H_2S_2O_7} \rightleftharpoons \mathrm{MPcH^+} + \mathrm{HS_2O_7^-} \qquad -K_3$$

For the first reaction, $pS_1 = pK_1 + H_0 - pK_{H_2SO_4}$, where S is the solubility of MPc, H_0 is the acidity function, and $K_{H_2SO_4}$ is the acidity constant of H_2SO_4. Similarly, $pS_2 = pK_2 + H_0 - pK_{HSO_3Cl}$ and $pS_3 = pK_3 + H_0 - pK_{H_2S_2O_7}$. Assuming the H_0 value for all three acids are equal then

$$pS_{(1)} = pS_{(2)} = pS_{(3)}$$

$$pK_2 = pK_1 + pK_{HSO_3Cl} - pK_{H_2SO_4}$$

$$pK_3 = pK_1 + pK_{H_2S_2O_7} - pK_{H_2SO_4}$$

The values of pK_1 and pK_2 at 25° and 50°C were calculated for the above phthalocyanine and metal phthalocyanines. The solubility in HSO_3Cl and $H_2S_2O_7$ calculated in the first equation showed that it was very high, approaching complete miscibility. The solubilities and ultraviolet spectra of isoviolanthrene and isoviolanthrone have been studied in various strong acids like 98% H_2SO_4, HSO_3Cl, HSO_3F, and CH_3SO_3H.[317]

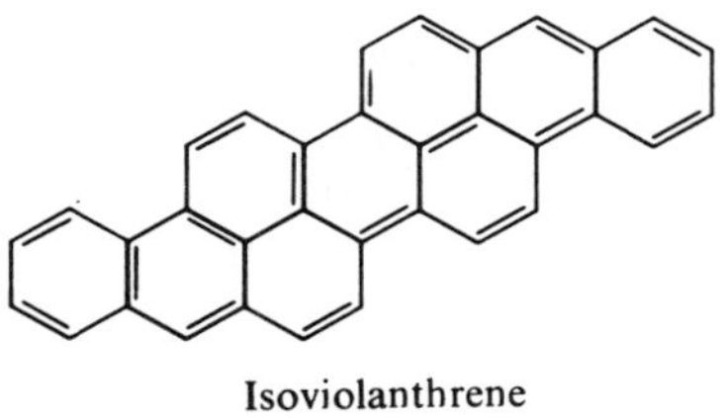

Isoviolanthrene

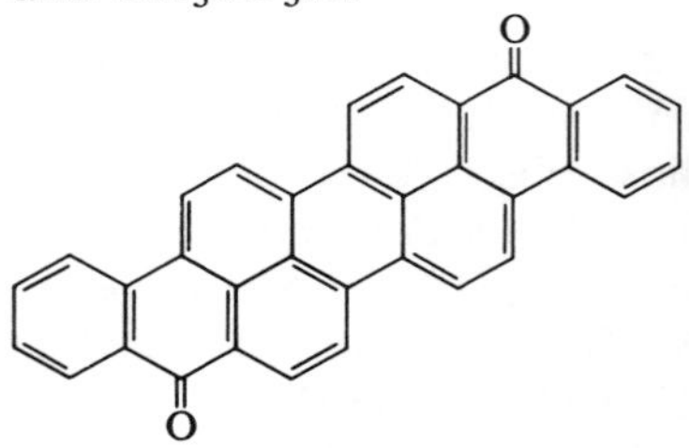

Isoviolanthrone

Isoviolanthrene (R′) and isoviolanthrone (R^2) dissolve readily in both HSO_3Cl and HSO_3F to give the protonated adducts $R'H^+$ and R^2H^+.

Isoviolanthrene gives a green solution in HSO_3Cl and a transient blue solution which turns green in HSO_3F. The dissolution of these compounds are reversed on dilution with H_2O, the original compound usually being recovered. This can be represented by the equation

$$R' + H^+ \rightleftharpoons R'H^+$$

$$R^2 + H^+ \rightleftharpoons R^2H^+$$

In mixtures of CH_3COCl and HSO_3X ($X = CF_3$, Cl) a Raman line at 2309 cm^{-1} associated with acetylium ion (CH_3CO^+) is found by Corriu, Dabosi, and Germain.[318] It increases as the ratio of acid to acid chloride is increased, decreases when a base is added, and is constant when an inert solvent, $CH_2Cl–CH_2Cl$, is added. The species in solution and the equilibrium between them is given by:

$$RCO^+ + B^- + HX \rightleftharpoons \left[R{:}-C(=O)X + HB\right] \rightleftharpoons R-C(OH)(X)-B$$

$$RCO^+ + B^- + HX \rightleftharpoons \left[R-C(=OH)X\right]^+ \rightleftharpoons R-C(OH)(X)-B; \qquad \left[R{:}-C(=O)X + HB\right] \rightleftharpoons \left[R-C(=OH)X\right]^+$$

Infrared and Raman spectra[306,319–321] of HSO_3Cl indicate that it is a tetrahedral molecule with a structure analogous to that of H_2SO_4, with one OH replaced by a Cl atom. The opportunities for molecular association through hydrogen bonding are fewer with chlorosulfuric acid and fluorosulfuric acid than with sulfuric acid since there is one less hydroxyl group.

References

1. T. F. Thorpe and W. Kirman, *J. Chem. Soc.* **61**, 921 (1892).
2. W. Lange, *Fluorine Chem.* **1**, 126 (1950).
3. G. H. Cady, *Adv. Inorg. Chem. Radiochem.* **2**, 105 (1960).
4. S. M. Williamson, *Prog. Inorg. Chem.* **7**, 39 (1968).
5. R. C. Thompson, *in* "Inorganic Sulfur Chemistry" (G. Nickless, ed.), Chapter 17. Elsevier, Amsterdam, 1968.
6. R. J. Gillespie, *Acc. Chem. Res.* **1**, 202 (1968).
7. A. W. Jache, *Adv. Inorg. Chem. Radiochem.* **16**, 177 (1975).
8. R. J. Gillespie, J. B. Milne, and R. C. Thompson, *Inorg. Chem.* **5**, 4668 (1966).
9. J. Barr, R. J. Gillespie, and R. C. Thompson, *Inorg. Chem.* **3**, 1149 (1964).
10. G. W. Richards and A. A. Woolf, *J. Chem. Soc. A* p. 1118 (1967).
11. A. A. Woolf, *J. Inorg. Nucl. Chem.* **14**, 21 (1969).
12. J. Barr, R. I. Gillespie, and E. A. Robinson, *J. Am. Chem. Soc.* **93**, 5083 (1971).
13. A. A. Woolf, *J. Chem. Soc.* p. 433 (1955).
14. L. P. Hammett and A. J. Deyrup, *J. Am. Chem. Soc.* **54**, 2721 (1932).

15. M. A. Paul and F. A. Long, *Chem. Rev.* **57**, 1 (1957).
16. J. C. D. Brand, W. C. Horning, and M. B. Thornby, *J. Chem. Soc.* p. 997 (1950).
17. V. A. Palm, *Proc. Acad. Sci. USSR, Chem. Sect.* **198**, 249 (1956).
18. P. Tickle, A. G. Briggs, and J. M. Wilson, *J. Chem. Soc. B* p. 65 (1970).
19. C. D. Johnson, A. R. Katritzky, and S. A. Shapiro, *J. Am. Chem. Soc.* **91**, 6594 (1969).
20. M. J. Jorgenson and D. R. Harter, *J. Am. Chem. Soc.* **85**, 876 (1963).
21. R. J. Gillespie, T. E. Peel, and E. A. Robinson, *J. Am. Chem. Soc.* **95**, 5173 (1973).
22. E. N. Arnett, R. P. Quirk, and J. J. Burke, *J. Am. Chem. Soc.* **92**, 1260 (1970).
23. E. M. Arnett, R. P. Quirk, and J. W. Larsen, *J. Am. Chem. Soc.* **92**, 3977 (1970).
24. R. C. Paul, K. S. Dhindsa, S. C. Ahluvalia, and S. P. Narula, *Indian J. Chem.* **9**, 700 (1971).
25. R. C. Paul and B. R. Sreenathan, *Indian J. Chem.* **4**, 382 (1966).
26. R. C. Paul, D. Singh, and K. C. Malhotra, *Z. Anorg. Allg. Chem.* **321**, 56 (1963); R. C. Paul, S. K. Rehani, S. S. Pahil, and S. C. Ahluvalia, *Indian J. Chem.* **7**, 715 (1969).
27. R. C. Paul and J. L. Vashisht, *Indian J. Chem.* **7**, 1243 and 1246 (1969).
28. R. C. Paul, J. Singh, S. C. Ahluvalia, and S. S. Pahil, *Indian J. Chem.* **2**, 134 (1964).
29. R. C. Paul, R. Kaushal, K. S. Dhindsa, S. S. Pahil, and S. C. Ahluvalia, *J. Indian Chem. Soc.* **44**, 964 (1967), and previous papers in this series.
30. R. C. Paul, K. S. Dhindsa, S. C. Ahluvalia, and S. P. Narula, *J. Indian Chem. Soc.* **48**, 381 (1971).
31. R. C. Paul and R. Dev, *Res. Bull. Panjab Univ., Sci.* **20**, Part 1-2, 139–48 (1969); *Indian J. Chem.* **7**, 392 (1969); R. C. Paul, J. P. Singla, and R. Dev, *ibid.* p. 170.
32. R. C. Paul, K. K. Paul, and K. C. Malhotra, *J. Chem. Soc. A* p. 2712 (1970).
33. W. Reed, D. W. Secret, and R. C. Thompson, *Can. J. Chem.* **47**, 4275 (1969).
34. R. C. Paul, K. Romesh, K. S. Dhindsa, S. C. Ahluvalia, and S. P. Narula, *Indian J. Chem.* **6**, 641 (1968).
34a. R. C. Paul, H. M. Kapil, S. S. Pahil, and S. C. Ahluvalia, *Indian J. Chem.* **6**, 720 (1968).
34b. R. C. Paul, K. S. Dhindsa, S. C. Ahluvalia, and S. P. Narula, *Indian J. Chem.* **11**, 277 (1973).
35. R. L. Benoit, C. Buisson, and G. Chou, *Can. J. Chem.* **48**, 2353 (1970).
36. A. Engelbrecht and B. M. Rode, *Monatsh. Chem.* **103**, 1315 (1972).
37. A. A. Woolf, *J. Chem. Soc.* p. 433 (1955).
38. R. C. Paul, K. Krishnan, and K. C. Malhotra, *Proc. Chem. Symp., 1st, 1969*, Vol. 2, p. 64 (1970).
39. R. C. Paul, K. K. Paul, and K. C. Malhotra, *Inorg. Nucl. Chem. Lett.* **5**, 689 (1969).
40. R. J. Gillespie and J. B. Milne, *Inorg. Chem.* **5**, 1233 (1966).
41. R. Seely and A. W. Jache, *J. Fluorine Chem.* **2**, 225 (1972).
42. J. N. Brazier and A. A. Woolf, *J. Chem. Soc. A* p. 97 (1967).
43. H. A. Carter, C. A. Milne, and F. Aubke, *J. Inorg. Nucl. Chem.* **37**, 282 (1975).
44. R. J. Gillespie and E. A. Robinson, *Proc. Chem. Soc.* p. 145 (1957).
45. R. J. Gillespie, R. Kapoor, and E. A. Robinson, *Can. J. Chem.* **44**, 1197 (1966).
46. K. O. Mailer, *Diss Abstr., Int. B* **31,** 7168 (1971); *C.A.* **76,** 39738r (1972).
47. R. C. Paul, S. K. Sharma, R. D. Sharma, K. K. Paul, and K. C. Malhotra, *J. Inorg. Nucl. Chem.* **33**, 2905 (1971).
48. E. Hayek and A. Engelbrecht, *Monatsh. Chem.* **86**, 735 (1955).
49. R. C. Paul, K. K. Paul, and K. C. Malhotra, *J. Inorg. Nucl. Chem.* **34**, 2523 (1972).
50. R. J. Gillespie and O. C. Vaidya, *Chem. Commun.* p. 40 (1972).
51. R. C. Paul, K. K. Paul, and K. C. Malhotra, *Chem. Commun.* p. 453 (1970).
52. R. C. Thompson, J. Barr, R. J. Gillespie, J. B. Milne, and R. A. Rothenburg, *Inorg. Chem.* **4**, 1641 (1965).

53. A. Commeyras and G. A. Olah, *J. Am. Chem. Soc.* **91**, 2929 (1969).
54. J. Badoz-Lambling, M. Herlem, A. Thiebault, and G. Adhami, *Anal. Lett.* **5**, 305 (1972).
55. R. J. Gillespie, K. Ouchi, and G. P. Pez, *Inorg. Chem.* **8**, 63 (1969).
56. G. A. Olah and C. W. McFarland, *Inorg. Chem.* **11**, 845 (1972); *J. Org. Chem.* **36**, 1374 (1971).
57. E. M. Arnett and J. F. Wolf, *J. Am. Chem. Soc.* **95,** 978 (1973).
58. P. A. W. Dean and R. J. Gillespie, *J. Am. Chem. Soc.* **92**, 2362 (1970).
59. R. J. Gillespie and E. A. Robinson, *Can. J. Chem.* **40**, 675 (1962).
60. A. A. Carter and F. Aubke, *Inorg. Nucl. Chem. Lett.* **5**, 999 (1969).
61. J. Barr, R. J. Gillespie, R. Kapoor, and K. C. Malhotra, *Can. J. Chem.* **46**, 149 (1968).
62. M. Herlem, A. Thiebault, and G. Adhami, *Anal. Lett.* **5**, 309 (1972).
63. I. D. Brown, D. B. Crump, R. J. Gillespie, and D. D. Santry, *Chem. Commun.* p. 853 (1968).
64. J. Barr, R. J. Gillespie, R. Kapoor, and P. K. Ummat, *Can. J. Chem.* **46**, 3607 (1968).
65. R. J. Gillespie and W. A. Whitla, *Can. J. Chem.* **47**, 4153 (1969).
66. R. J. Gillespie and J. Passmore, *Chem. Commun.* p. 1339 (1969).
67. J. Barr, R. J. Gillespie, G. P. Pez, P. K. Ummat, and O. C. Vaidya, *Inorg. Chem.* **10**, 362 (1971).
68. F. Seel, V. Hartmann, I. Molnar, R. Budenz, and W. Gombler, *Angew. Chem., Int. Ed. Engl.* **10**, 186 (1973).
69. R. J. Gillespie and J. B. Milne, *Inorg. Chem.* **5**, 1236 (1966).
70. R. J. Gillespie and J. B. Milne, *Inorg. Chem.* **5**, 1577 (1966).
71. R. J. Gillespie and M. J. Morton, *Chem. Commun.* p. 1565 (1968); *Inorg. Chem.* **11**, 586 (1972).
72. R. J. Gillespie and M. J. Morton, *Inorg. Chem.* **11**, 591 (1972).
73. G. A. Olah and M. B. Comisarow, *J. Am. Chem. Soc.* **90**, 5033 (1968).
74. R. S. Eachaus, T. P. Sleight, and M. C. R. Symons, *Nature (London)* **222**, 769 (1969).
75. H. A. Carter, A. M. Quereshi, and F. Aubke, *Chem. Commun.* p. 1461 (1968).
76. H. A. Lehmann and G. Krueger, *Z. Anorg. Allg. Chem.* **274**, 141 (1953); A. A. Woolf, *J. Chem. Soc.* p. 4113 (1954); M. Schmeisser and F. L. Ebenloch, *Angew. Chem.* **66**, 230 (1954).
77. R. C. Paul, S. K. Sharma, K. K. Paul, and K. C. Malhotra, *J. Inorg. Nucl. Chem.* **34**, 2535 (1972).
78. R. J. Gillespie and G. P. Pez, *Inorg. Chem.* **8,** 1233 (1972).
79. R. J. Gillespie and G. J. Schrobilgen, *Inorg. Chem.* **13**, 1964 (1974).
79a. R. J. Gillespie, B. Landa, and G. J. Schrobilgen, *Winter Fluorine Conf., 1st, 1972*; *55th Chem. Inst. Can. Meet., 1972.*
79b. H. Eisenberg and D. D. Des Marteau, *J. Am. Chem. Soc.* **92**, 4759 (1970).
79c. D. D. Des Marteau and M. Eisenberg, *Inorg. Chem.* **11**, 2641 (1972).
80. G. A. Olah, A. M. White, and D. H. O'Brien, *Carbonium Ions* **4**, 1697 (1973).
81. G. A. Olah and J. A. Olah, *Carbonium Ions* **2**, 715 (1970).
82. G. A. Olah and C. V. Pittman, *Adv. Phys. Org. Chem.* **4**, 305 (1966).
83. R. J. Gillespie, *Friedel-Crafts Relat. React.* **1**, 169 (1963).
84. G. A. Olah, C. V. Pittman, and H. C. Symons, *Carbonium Ions* **1**, 153 (1968).
85. G. A. Olah and E. Namanworth, *J. Am. Chem. Soc.* **88**, 5327 (1966).
86. G. A. Olah, J. Sommer, and E. Namanworth, *J. Am. Chem. Soc.* **89**, 3576 (1967).
87. G. A. Olah and J. Sommer, *J. Am. Chem. Soc.* **90**, 927 (1968).
88. G. A. Olah, D. H. O'Brien, and C. U. Pittman, Jr., *J. Am. Chem. Soc.* **89**, 2996 (1967).
89. G. A. Olah and D. H. O'Brien, *J. Am. Chem. Soc.* **89**, 1725 (1967).
90. G. A. Olah and P. J. Szilagyi, *J. Org. Chem.* **36**, 1121 (1971).

91. T. Birchall, A. N. Bourns, R. J. Gillespie, and P. J. Smith, *Can. J. Chem.* **42**, 1433 (1964).
92. D. M. Brouwer, E. L. Mackor, and C. MacLean, *Recl. Trav. Chim. Pays-Bas* **85**, 109 and 114 (1966).
93. R. W. Alder and F. J. Taylor, *J. Chem. Soc. B* p. 845 (1970).
94. M. P. Hartshorn, K. E. Richards, J. Vaughn, and G. J. Wright, *J. Chem. Soc. B* p. 1624 (1971); J. W. Larsen and M. Eckert-Maksic, *Croat. Chem. Acta* **45**, 503 (1973).
95. G. A. Olah and Y. K. Mo, *J. Am. Chem. Soc.* **94**, 5341 (1972).
96. G. A. Olah and Y. K. Mo, *J. Org. Chem.* **38**, 353 (1973).
97. G. A. Olah and Y. K. Mo, *J. Org. Chem.* **38**, 2212 (1973).
98. D. M. Brouwer, E. L. Mackor, and C. MacLean, *Recl. Trav. Chim. Pays-Bas* **85**, 109 and 114 (1966); T. Birchall, A. N. Bourns, R. J. Gillespie, and P. J. Smith, *Can. J. Chem.* **42**, 1433 (1966); T. Birchall and R. J. Gillespie, *ibid.* p. 502; C. MacLean and E. L. Mackor, *Discuss. Faraday Soc.* **34**, 165 (1962); D. M. Brouwer, C. MacLean, and E. L. Mackor, *ibid.* **39**, 121 (1965); M. P. Hartshorn, K. E. Richards, J. Vaughn, and G. J. Wright, *J. Chem. Soc. B* p. 1624 (1971).
99. D. M. Brouwer, E. L. Mackor, and C. MacLean, *Carbonium Ions* **2**, 865 (1970).
100. D. M. Brouwer, *Recl. Trav. Chim. Pays-Bas* **87**, 335 and 342 (1968).
101. G. A. Olah, S. Kobayashi, and Y. K. Mo, *J. Org. Chem.* **38**, 4057 (1973); G. A. Olah, G. D. Mateescu, and Y. K. Mo, *J. Am. Chem. Soc.* **95**, 1865 (1973).
102. T. Birchall and R. J. Gillespie, *Can. J. Chem.* **43**, 1045 (1965).
103. M. Brookhart, G. C. Levy, and S. Winstein, *J. Am. Chem. Soc.* **89**, 1735 (1967).
104. G. A. Olah and A. M. White, *J. Am. Chem. Soc.* **89**, 3591 (1967).
105. G. A. Olah, M. Calin, and D. H. O'Brien, *J. Am. Chem. Soc.* **89**, 3586 (1967).
106. G. A. Olah and A. M. White, *Chem. Rev.* **70**, 561 (1970).
107. H. Hogeveen, *Recl. Trav. Chim. Pays-Bas* **87**, 1313 (1968).
108. D. M. Brouwer and A. A. Kiffen, *Recl. Trav. Chim. Pays-Bas* **92**, 1053 (1973).
109. H. Hogeveen, *Recl. Trav. Chim. Pays-Bas* **86**, 809 (1967).
110. H. Hogeveen and C. F. Roobeek, *Recl. Trav. Chim. Pays-Bas* **89**, 1121 (1970).
111. G. A. Olah and A. M. White, *J. Am. Chem. Soc.* **89**, 4752 (1967).
112. G. A. Olah and A. T. Ku, *J. Org. Chem.* **35**, 3912 (1970).
113. G. A. Olah and W. Westerman, *J. Org. Chem.* **39**, 1307 (1974).
114. G. A. Olah and M. Calin, *J. Am. Chem. Soc.* **90**, 405 (1965).
115. G. A. Olah and M. B. Comisarow, *J. Am. Chem. Soc.* **89**, 2964 (1967).
116. G. A. Olah, A. T. Ku, and J. Sommer, *J. Org. Chem.* **35**, 2159 (1970).
117. G. A. Olah, K. Dunne, Y. K. Mo, and P. J. Szilagyi, *J. Am. Chem. Soc.* **94**, 4200 (1972).
118. G. A. Olah, Y. K. Mo, and J. L. Grant, *J. Org. Chem.* **38**, 3207 (1973).
119. G. A. Olah, D. H. O'Brien, and A. M. White, *J. Am. Chem. Soc.* **89**, 5694 (1967).
120. G. A. Olah, A. T. Ku, and A. M. White, *J. Org. Chem.* **34**, 1827 (1969).
121. G. A. Olah and A. T. Ku, *J. Org. Chem.* **35**, 331 (1970).
122. G. A. Olah and M. Calin, *J. Am. Chem. Soc.* **90**, 401 (1968).
123. V. C. Armstrong, D. W. Farlow, and R. B. Moodie, *Chem. Commun.* p. 1362 (1968).
124. G. A. Olah, A. M. White, and A. T. Ku, *J. Org. Chem.* **36**, 3585 (1971).
125. G. A. Olah, J. A. Olah, and R. H. Schlosberg, *J. Org. Chem.* **35**, 328 (1970).
126. E. Spinner, *Spectrochim. Acta* **15**, 95 (1959).
127. E. Spinner, *J. Phys. Chem.* **64**, 275 (1960).
128. M. Daries and L. Hopkins, *Trans. Faraday Soc.* **53**, 1563 (1957).
129. J. T. Edward, H. S. Chang, K. Yates, and R. Stewart, *Can. J. Chem.* **38**, 1518 (1960).
130. M. Liler, *Spectrochim. Acta, Part A* **23**, 139 (1967).
131. A. R. Katritzky and R. A. Y. Jones, *Chem. Ind.* (*London*) p. 722 (1961).
132. R. J. Gillespie and T. Birchall, *Can. J. Chem.* **41**, 148 (1963).

133. T. Birchall and R. J. Gillespie, *Can. J. Chem.* **41**, 2642 (1963).
134. G. Fraenkel and C. Franconi, *J. Am. Chem. Soc.* **82**, 4478 (1960).
135. G. A. Olah and A. M. White, *J. Am. Chem. Soc.* **90**, 6087 (1968).
136. G. A. Olah, D. L. Brydon, and R. D. Porter, *J. Org. Chem.* **35**, 317 (1970).
137. G. A. Olah and D. L. Brydon, *J. Org. Chem.* **35**, 313 (1970).
138. G. A. Olah and P. Kreienbühl, *J. Am. Chem.* **35**, 313 (1970).
139. G. A. Olah, D. H. O'Brien, and M. Calin, *J. Am. Chem. Soc.* **89**, 3582 and 3586 (1967).
140. G. A. Olah, J. M. Bollinger, and J. M. Brinich, *J. Am. Chem. Soc.* **90**, 2587 (1968).
141. H. Hogeveen, *Recl. Trav. Chim. Pays-Bas* **86**, 696 (1967).
142. D. M. Brouwer, *Recl. Trav. Chim. Pays-Bas* **86**, 879 (1967).
143. G. A. Olah and M. Calin, *J. Am. Chem. Soc.* **90**, 938 (1968).
144. D. M. Brouwer, *Chem. Commun.* p. 515 (1967).
145. G. A. Olah and C. U. Pittman, Jr., *J. Am. Chem. Soc.* **88**, 3310 (1966).
146. G. A. Olah and A. M. White, *J. Am. Chem. Soc.* **91**, 5801 (1969).
147. R. W. Taft, E. Price, I. R. Fox, I. C. Lewis, K. K. Andersen, and G. T. Davis, *J. Am. Chem. Soc.*, **85**, 709 and 3146 (1963).
148. S. Patai, ed., "The Chemistry of the Carbonyl Group," Chapter 9. Wiley, New York, 1966.
149. M. Brookhart, G. C. Levy, and S. Winstein, *J. Am. Chem. Soc.* **89**, 1735 (1967).
150. T. J. Sekkur and P. Kranenburg, *Tetrahedron Lett.* p. 4793 (1966).
151. G. A. Olah and A. T. Ku, *J. Org. Chem.* **35**, 3922 (1970).
152. H. Hogeveen, *Recl. Trav. Chim. Pays-Bas* **87**, 1295 (1968).
153. D. M. Brouwer, *Recl. Trav. Chim. Pays-Bas* **88**, 530 (1969).
154. D. M. Brouwer and J. A. VanDoorn, *Recl. Trav. Chim. Pays-Bas* **91**, 895 (1972).
155. D. M. Brouwer and J. A. VanDoorn, *Recl. Trav. Chim. Pays-Bas* **90**, 1010 (1971).
156. D. M. Brouwer and J. A. VanDoorn, *Recl. Trav. Chim. Pays-Bas* **91**, 261 (1972).
157. D. M. Brouwer and J. A. VanDoorn, *Recl. Trav. Chim. Pays-Bas* **90**, 535 (1971).
158. D. M. Brouwer and J. A. VanDoorn, *Recl. Trav. Chim. Pays-Bas* **89**, 553 and 896 (1970).
159. G. A. Olah and A. T. Ku, *J. Org. Chem.* **35**, 3916 (1970).
160. G. A. Olah, A. T. Ku, and J. A. Olah, *J. Org. Chem.* **35**, 3904 (1970).
161. G. A. Olah, A. T. Ku, and J. A. Olah, *J. Org. Chem.* **35**, 3908 (1970).
162. G. A. Olah and A. M. White, *J. Am. Chem. Soc.* **90**, 1884 (1968).
163. L. P. Hammett and A. J. Deyrup, *J. Am. Chem. Soc.* **54**, 2721 (1932).
164. G. A. Olah, Y. Halpern, P. W. Westerman, and J. L. Grant, *J. Org. Chem.* **39**, 2390 (1974).
165. G. A. Olah, P. Schilling, J. M. Bollinger, and J. Nishimura, *J. Am. Chem. Soc.* **96**, 2221 (1974).
166. G. A. Olah, A. T. Ku, and J. A. Olah, *J. Org. Chem.* **36**, 3582 (1971).
167. G. A. Olah, J. Nishimura, and P. Kreienbühl, *J. Am. Chem. Soc.* **95**, 7672 (1973).
168. G. A. Olah and G. D. Mateescu, *J. Am. Chem. Soc.* **93**, 781 (1971).
169. G. A. Olah and T. E. Kiovsky, *J. Am. Chem. Soc.* **90**, 4666 (1968).
170. H. Hogeveen, *Recl. Trav. Chim. Pays-Bas* **86**, 1288 (1967).
171. G. A. Olah and T. E. Kiovsky, *J. Am. Chem. Soc.* **90**, 4666 (1968).
172. G. A. Olah and T. E. Kiovsky, *J. Am. Chem. Soc.* **90**, 6461 (1968).
173. E. M. Arnett and J. W. Larsen, *J. Am. Chem. Soc.* **91**, 1438 (1969).
174. E. M. Arnett and J. W. Larsen, *Carbonium Ions* **2**, 445 (1970).
175. D. M. Brouwer and J. A. VanDoorn, *Recl. Trav. Chim. Pays-Bas* **89**, 88 (1970).
176. G. A. Olah, R. H. Schlosberg, R. D. Porter, Y. K. Mo, D. P. Kelly, and G. D. Mateescu, *J. Am. Chem. Soc.* **94**, 2034 (1972).
177. T. Birchall and R. J. Gillespie, *Can. J. Chem.* **42**, 502 (1964).
178. G. A. Olah, G. D. Mateescu, and Y. K. Mo, *J. Am. Chem. Soc.* **95**, 1865 (1973).
179. G. A. Olah, G. Liang, and P. W. Westerman, *J. Am. Chem. Soc.* **95**, 3678 (1973).

180. H. M. Buck, G. Holtrust, L.-J. Oosterhaff, and M. J. Van der Sluys, *Ind. Chim. Belge* **32**, Spec. Issue, Part III, 158 (1967); W. T. A. M. Van der Luft, H. M. Buck, and L.-J. Oosterhaff, *Tetrahedron* **24**, 4941 (1968).
181. D. M. Brouwer and J. A. VanDoorn, *Recl. Trav. Chim. Pays-Bas* **91**, 1110 (1972).
182. J. Lukas, unpublished results quoted by D. M. Brouwer and H. Hogeveen, in *Prog. Phys. Org. Chem.* **9**, 179 (1972).
183. J. W. Larsen, P. A. Bouis, C. R. Watson, Jr., and R. M. Pagni, *J. Am. Chem. Soc.* **96**, 2284 (1974).
184. G. A. Olah, *Chem. Eng. News* **45**, 77 (1967).
185. G. A. Olah, *Science* **168**, 1298 (1970).
186. A. Commeyras and G. A. Olah, *J. Am. Chem. Soc.* **91**, 2929 (1969).
187. G. A. Olah, J. R. Demember, A. Commeyras, and T. L. Bribes, unpublished.
188. G. A. Olah, C. U. Pittman, Jr., R. Waack, and M. A. Doran, *J. Am. Chem. Soc.* **88**, 1488 (1966).
189. G. A. Olah and J. Lukas, *J. Am. Chem. Soc.* **89**, 2227 and 4739 (1967).
190. G. A. Olah and R. H. Schlosberg, *J. Am. Chem. Soc.* **90**, 2726 (1968).
191. G. A. Olah and J. Lukas, *J. Am. Chem. Soc.* **90**, 933 (1968).
192. D. M. Brouwer and E. L. Mackor, *Proc. Chem. Soc., London* p. 147 (1964); D. M. Brouwer, *Recl. Trav. Chim. Pays-Bas* **87**, 210 (1968).
193. G. A. Olah and Y. Halpern, *J. Org. Chem.* **36**, 2354 (1971).
194. G. A. Olah, *J. Am. Chem. Soc.* **94**, 808 (1972); G. A. Olah, J. M. Bollinger, Y. K. Mo, and J. M. Brinich, *ibid.* p. 1164.
195. H. Hogeveen and A. F. Bickel, *Recl. Trav. Chim. Pays-Bas* **88**, 371 (1969).
196. H. Hogeveen, C. J. Gaasbeek, and A. F. Bickel, *Recl. Trav. Chim. Pays-Bas* **88**, 703 (1969).
197. H. Hogeveen, F. Baardman, and R. F. Roobeek, *Recl. Trav. Chim. Pays-Bas* **89**, 227 (1970).
198. H. Hogeveen and C. J. Gaasbeek, *Recl. Trav. Chim. Pays-Bas* **89**, 395 (1970).
199. G. A. Olah, Y. K. Mo, and J. A. Olah, *J. Am. Chem. Soc.* **95**, 4939 (1973).
200. G. A. Olah and J. A. Olah, *J. Am. Chem. Soc.* **93**, 1256 (1971), and related preview papers by Olah *et al.*
201. G. A. Olah, J. R. DeMember, and J. Shen, *J. Am. Chem. Soc.* **95**, 4952 (1973).
202. G. A. Olah, Y. Halpern, J. Shen, and Y. K. Mo, *J. Am. Chem. Soc.* **95**, 4960 (1973).
203. G. A. Olah, Y. Halpern, J. Shen, and Y. K. Mo, *J. Am. Chem. Soc.* **93**, 1251 (1971).
204. G. A. Olah, J. Shen, and R. H. Schlosberg, *J. Am. Chem. Soc.* **95**, 4957 (1973); **92**, 3831 (1970).
205. R. J. Gillespie and G. P. Pez, *Inorg. Chem.* **8**, 1233 (1969).
206. G. A. Olah, H. C. Lin, and Y. K. Mo, *J. Am. Chem. Soc.* **94**, 3667 (1972).
207. D. G. Farnum, R. A. Mader, and G. Mehta, *J. Am. Chem. Soc.* **95**, 8692 (1973); D. G. Farnum and G. Mehta, *Chem. Commun.* p. 1643 (1968).
208. G. A. Olah, P. Schilling, P. W. Westerman, and H. C. Lin, **96**, 3581 (1974).
209. G. A. Olah and R. J. Spear, *J. Am. Chem. Soc.* **97**, 1845 (1975).
210. G. A. Olah, J. M. Bollinger, C. A. Cupas, and J. Lukas, *J. Am. Chem. Soc.* **89**, 2692 (1967).
211. C. U. Pittman, Jr. and G. A. Olah, *J. Am. Chem. Soc.* **87**, 2998 (1965).
212. D. G. Farnum, *J. Am. Chem. Soc.* **86**, 934 (1964).
213. G. A. Olah, C. U. Pittman, Jr., E. Namanworth, and M. B. Comisarow, *J. Am. Chem. Soc.* **88**, 5571 (1966); G. A. Olah, *ibid.* **86**, 932 (1964).
214. G. A. Olah, J. S. Staral, and G. Liang, *J. Am. Chem. Soc.* **96**, 6233 (1974).
215. G. A. Olah and G. Liang, *J. Am. Chem. Soc.* **94**, 6434 (1972).
216. K. B. Wiberg, V. Z. Williams, Jr., and X. E. Friedrich, *J. Am. Chem. Soc.* **92**, 564 (1970).
217. G. A. Olah, G. Liang, and Y. K. Mo, *J. Am. Chem. Soc.* **94**, 3544 (1972).
218. R. E. Leone, J. L. Barborak, and P. v. R. Schleyer, *Carbonium Ions* **4**, 1837 (1973).

219. M. Brookhart, R. K. Lustgarten, and S. Winstein, *J. Am. Chem. Soc.* **89**, 6352 (1967).
220. H. Hogeveen and C. J. Gaasbeek, *Recl. Trav. Chim. Pays-Bas* **89**, 1079 (1970).
221. G. A. Olah, A. Commeyras, and C. Y. Lui, *J. Am. Chem. Soc.* **90**, 3882 (1968).
222. G. A. Olah, J. R. DeMember, C. Y. Lui, and R. D. Porter, *J. Am. Chem. Soc.* **93**, 1442 (1971).
223. G. A. Olah and G. Liang, *J. Am. Chem. Soc.* **96**, 189 (1974).
224. H. Hogeveen and R. F. Roobeek, *Tetrahedron Lett.* **56**, 4941 (1969).
225. G. A. Olah and G. Liang, *J. Am. Chem. Soc.* **96**, 195 (1974).
226. P. v. R. Schleyer, R. C. Fort, Jr., W. E. Watts, M. B. Comisarow, and G. A. Olah, *J. Am. Chem. Soc.* **86**, 4195 (1964).
227. P. v. R. Schleyer, W. E. Watts, R. C. Fort, Jr., M. B. Comisarow, and G. A. Olah, *J. Am. Chem. Soc.* **86**, 5679 (1974).
228. L. A. Paquette, G. V. Meehan, and S. J. Marshall, *J. Am. Chem. Soc.* **91**, 6779 (1969).
229. G. A. Olah, G. Liang, and G. D. Mateescu, *J. Org. Chem.* **39**, 3750 (1974).
230. G. A. Olah, *J. Am. Chem. Soc.* **86**, 4195 (1964).
231. G. A. Olah, *J. Am. Chem. Soc.* **87**, 2997 (1965).
232. G. A. Olah, G. Liang, K. A. Babiak, and R. K. Murray, Jr., *J. Am. Chem. Soc.* **96**, 6796 (1974).
233. R. K. Murray, Jr. and K. A. Babiak, *Tetrahedron Lett.* p. 311 (1974).
234. L. A. Paquitte, G. R. Krow, J. M. Bollinger, and G. A. Olah, *J. Am. Chem. Soc.* **90**, 7147 (1968).
235. T. P. Nevell, E. de-Salas, and C. L. Wilson, *J. Chem. Soc.* p. 1188 (1939).
236. S. Winstein and D. J. Trifan, *J. Am. Chem. Soc.* **71**, 2953 (1949); S. Winstein, *Q. Rev., Chem. Soc.* **23**, 1411 (1969).
237a. H. Hogeveen and P. W. Kwant, *J. Am. Chem. Soc.* **96**, 2208 (1974).
237b. H. Hogeveen and P. W. Kwant, *Tetrahedron Lett.* pp. 1665 and 3747 (1973).
238. R. F. Childs and S. Winstein, *J. Am. Chem. Soc.* **90**, 7146 (1968).
239. R. F. Childs and B. Parrington, *J. Chem. Soc. D* p. 1540 (1970).
240. L. A. Paquette, M. J. Broadhurst, P. Warner, G. A. Olah, and G. Liang, *J. Am. Chem. Soc.* **95**, 3386 (1973).
241. G. A. Olah, G. Liang, J. R. Wiseman, and J. A. Chong, *J. Am. Chem. Soc.* **94**, 4927 (1972).
242. G. A. Olah and G. Liang, *J. Am. Chem. Soc.* **97**, 1920 (1975).
243. L. A. Paquette, M. Oku, W. Farnham, G. A. Olah, and G. Liang, *J. Org. Chem.* **40**, 700 (1975).
244. G. A. Olah and G. Liang, *J. Am. Chem. Soc.* **97**, 2236 (1975).
245. G. A. Olah, D. H. O'Brien, and C. Y. Lui, *J. Am. Chem. Soc.* **91**, 701 (1969).
246. G. A. Olah, J. J. Svoboda, and A. T. Ku, *J. Org. Chem.* **38**, 4447 (1973).
247. G. R. Olah and P. R. Clifford, *J. Am. Chem. Soc.* **95**, 6067 (1973); **93**, 1266 and 2320 (1971).
248. G. A. Olah and Y. K. Mo, *J. Organomet. Chem.* **60**, 311 (1973).
249. G. A. Olah and C. W. McFarland, *J. Org. Chem.* **36**, 1374 (1971).
250. G. A. Olah and C. W. McFarland, *J. Org. Chem.* **34**, 1832 (1969).
251. G. A. Olah, K. Dunne, D. P. Kelly, and Y. K. Mo, *J. Am. Chem. Soc.* **94**, 7438 (1972).
252. T. E. Stevens, *J. Org. Chem.* **33**, 2667 (1968).
253. V. O. Lukasherich and T. N. Sokolova, *C. R. Acad. Sci. USSR* **54**, 693 (1946); *C.A.* **41**, 5472 (1947).
254. G. A. Olah and J. R. DeMember, *J. Am. Chem. Soc.* **91**, 2113 (1969); **92**, 718 (1970).
255. G. A. Olah and E. G. Melby, *J. Am. Chem. Soc.* **94**, 6220 (1972).
256. A. N. Nesmeyanov, L. G. Makarova, and T. P. Tolstaya, *Tetrahedron* **1**, 145 (1957); I. Masson and E. Race, *J. Chem. Soc.* p. 1718 (1937).
257. G. A. Olah, Y. K. Mo, E. G. Melby, and H. C. Lin, *J. Org. Chem.* **38**, 367 (1973).

258. G. A. Olah and J. M. Bollinger, *J. Am. Chem. Soc.* **89**, 4744 (1967).
258a. G. A. Olah and J. M. Bollinger, *J. Am. Chem. Soc.* **90**, 947 and 2587 (1968).
259. G. A. Olah and P. E. Peterson, *J. Am. Chem. Soc.* **90**, 4675 (1968).
259a. G. H. Olah, J. M. Bollinger, and J. M. Brinich, *J. Am. Chem. Soc.* **90**, 6988 (1968).
260. G. A. Olah, P. W. Westerman, E. G. Melby, and Y. K. Mo, *J. Am. Chem. Soc.* **96**, 3565 (1974).
261. P. E. Peterson, B. R. Bonazza, and P. M. Hendricks, *J. Am. Chem. Soc.* **95**, 2222 (1973).
262. G. A. Olah, J. M. Bollinger, Y. K. Mo, and J. M. Brinich, *J. Am. Chem. Soc.* **94**, 1164 (1972).
263. J. H. Exner, L. D. Kershner, and T. E. Evans, *J. Chem. Soc., Chem. Commun.* p. 361 (1973).
264. J. W. Larsen and A. V. Metzner, *J. Am. Chem. Soc.* **94**, 1614 (1972).
265. N. C. Deno, G. W. Holland, Jr., and T. Schulze, *J. Org. Chem.* **32**, 1496 (1967).
266. G. A. Olah and T. Kiovsky, *J. Am. Chem. Soc.* **89**, 5692 (1967); **90**, 2583 (1968).
267. D. M. Brouwer, *Recl. Trav. Chim. Pays-Bas* **87**, 335 and 342 (1968).
268. D. M. Brouwer, E. L. Mackor, and C. MacLean, *Carbonium Ions* **2**, 837 (1970).
269. G. A. Olah and M. B. Comisarow, *J. Am. Chem. Soc.* **89**, 1027 (1967).
270. G. A. Olah, M. B. Comisarow, and C. A. Cupas, *J. Am. Chem. Soc.* **88**, 362 (1966); G. A. Olah and C. U. Pittman, Jr., *ibid.* p. 3310.
271. J. C. Jacquesy, R. Jacquesy, S. Moreau, and J. F. Patoiseau, *J. Chem. Soc. D* p. 785 (1973).
272. G. A. Olah and A. M. White, *J. Am. Chem. Soc.* **90**, 1884 (1968).
273. D. M. Brouwer, *Recl. Trav. Chim. Pays-Bas* **88**, 9 (1969).
274. G. A. Olah and M. Calin, *J. Am. Chem. Soc.* **89**, 4736 (1967); **90**, 943 (1968).
275. P. C. Myhre and E. Evans, *J. Am. Chem. Soc.* **91**, 5641 (1969).
276. A. F. Diaz, I. L. Reich, and S. Winstein, *J. Am. Chem. Soc.* **91**, 5637 (1969).
277. D. M. Brouwer, C. F. Roobeek, J. A. VanDoorn, and A. A. Kiffen, *Recl. Trav. Chim. Pays-Bas* **92**, 563 (1973).
278. G. A. Olah, P. W. Westerman, and J. Nishimura, *J. Am. Chem. Soc.* **96**, 3548 (1974).
279. G. A. Olah and R. J. Spear, *J. Am. Chem. Soc.* **97**, 1539 (1975).
280. N. C. Deno and R. R. Lastomirsky, *J. Org. Chem.* **40**, 514 (1975).
281. G. A. Olah, Y. Yamada, and R. J. Spear, *J. Am. Chem. Soc.* **97**, 680 (1975).
282. H. Hogeveen and P. W. Kwant, *Tetrahedron Lett.* p. 5361 (1972); p. 3747 (1973); *J. Org. Chem.* **39**, 2626 (1974).
283. A. W. Williamson, *J. Chem. Soc.* **10**, 97 (1857).
284. P. Walden, *Z. Anorg. Chem.* **29**, 371 (1902).
285. R. C. Paul, S. K. Vasisht, K. C. Malhotra, and S. S. Pahil, *J. Sci. Ind. Res., Sect. B* **21**, 528 (1962).
286. E. A. Robinson and J. A. Ciruna, *Can. J. Chem.* **46**, 1719 and 3197 (1968).
286a. J. Barr, R. J. Gillespie, and E. A. Robinson, *Can. J. Chem.* **39**, 1266 (1961).
287. V. A. Palm, *Dokl. Chem.* (*Engl. Transl.*) **108**, 270 (1956).
288. T. C. Waddington and F. Klanberg, *J. Chem. Soc.* p. 2332 (1960).
289. C. R. Sanger and E. R. Riegel, *Proc. Am. Acad. Arts Sci.* **47**, 673 (1912).
290. D. L. Ledson, *Diss. Abstr. Int. B* **34**, 4273 (1974), *C.A.* **81**, 30366p. (1974).
291. G. P. Luchinski, *Z. Phys. Chem.* (*Leipzig*) **169**, 269 (1934).
292. R. J. Gillespie and R. F. White, *Trans. Faraday Soc.* **54**, 1846 (1958).
293. J. Heubel and M. Wartel, *Bull. Soc. Chim. Fr.* p. 4357 (1968).
294. A. Yves, M. Wartel, and J. Heubel, *Bull. Soc. Chim. Fr.* pp. 3455 and 3462 (1970).
295. A. Yves, P. Legrand, and J. Heubel, *C.R. Hebd. Seances Acad. Sci., Ser. C* **269**, 493 (1969); *C.A.* **71**, 119158 (1969).
296. G. Mairessee, P. Barbier, and J. Heubel, *Bull. Soc. Chim. Fr.* **7-8**, Part 1, 1297 and 1301 (1974); *C.A.* **81**, 85339 (1974).
297. R. C. Paul, S. Darshan, and K. C. Malhotra, *Z. Anorg. Allg. Chem.* **321**, 70 (1963).

298. R. C. Paul, K. C. Malhotra, and K. C. Khanna, *Indian J. Chem.* **3**, 63 (1965).
299. R. C. Paul, S. C. Ahluvalia, and R. Prakash, *Indian J. Chem.* **7**, 815 (1969).
300. K. C. Rao and P. R. M. Naidu, *J. Electroanal. Chem.* **35**, 429 (1972).
301. J. Bessière, *Bull. Soc. Chim. Fr.* **9**, 3356 (1969).
302. J. Bessière, *Anal. Chim. Acta* **52**, 55 (1970).
303. M. Rumeau, *Ann. Chim.* (*Paris*) [14] **8**, 131 (1973); *C.A.* **79**, 70739c (1973).
304. H. Smagowski, *Rocz. Chem.* **46**, 1599 (1972); *C.A.* **78**, 83686 (1973).
305. J. C. Fischer and M. Wartel, *Bull. Soc. Chim. Fr.* p. 3302 (1972).
306. E. A. Robinson and J. A. Ciruna, *Can. J. Chem.* **39**, 247 (1961).
307. R. J. Gillespie, R. Kapoor, and E. A. Robinson, *Can. J. Chem.* **44**, 1197 (1966).
308. M. Wartel, S. Noel, and J. Heubel, *C. R. Hebd. Seances Acad. Sci., Ser. C* **262**, 962 (1966).
309. K. B. Dillon and T. C. Waddington, *J. Chem. Soc. A* p. 1146 (1970).
310. H. A. Lehmann, *Z. Anorg. Allg. Chem.* **371**, 281 (1969).
311. M. Schmidt and G. Talsky, *Angew. Chem.* **70**, 312 (1958).
312. M. Schmidt and G. Talsky, *Chem. Ber.* **92**, 1526 and 1539 (1959).
313. M. J. Frazer, W. Gerrard, and F. W. Parrett, *J. Chem. Soc.* p. 342 (1965).
314. E. A. Robinson and J. A. Ciruna, *J. Am. Chem. Soc.* **86**, 5677 (1964).
315. J. A. Ciruna, E. A. Robinson, S. A. A. Zaidi, R. J. Gillespie, and J. S. Hartman, *Tetrahedron Lett.* p. 1101 (1965).
316. B. D. Berezin, *Zh. Prikl. Khim.* **36**, 1181 (1963); *C.A.* **59**, 10809d (1963).
317. N. Parkyns and A. R. Ubbelhode, *J. Chem. Soc.* p. 4188 (1960).
318. R. Corriu, G. Dabosi, and A. Germain, *Bull. Soc. Chim. Fr.* p. 1617 (1972).
319. S. M. Chackalackal and F. E. Stafford, *J. Am. Chem. Soc.* **88**, 723 and 4816 (1966).
320. T. C. Waddington and F. Klanberg, *J. Chem. Soc.* p. 2339 (1960).
321. R. J. Gillespie and E. A. Robinson, *Can. J. Chem.* **40**, 644 (1962).

3

The Interhalogens

DOMINIQUE MARTIN, ROGER ROUSSON, AND JEAN-MARC WEULERSSE

Division de Chimie
Centre D'Etudes Nucleaires de Saclay
Gif-sur-Yvette, France

I. INTRODUCTION

Most of the interhalogens had been synthesized long ago, at the same time as the halogen themselves, by the reaction of one halogen with another. Their extreme chemical reactivity has limited investigations for years, but recently they have gained a new popularity. This has been made possible because of new technology, essentially those developed by the nuclear research of the last 20 years.

Fourteen interhalogens have now been characterized. Their general formulas are XY, XY_3, XY_5, and XY_7, where X and Y are the different halogen atoms (see Table I).

Numerous excellent treatises have been published on this subject, especially by Sharpe,[1] Stein,[2] Popov,[3] Downs and Adams,[4] and O'Donnell.[5] Our purpose will be to recall briefly the preparations and properties of the interhalogens, principally in the liquid state. Their uses as nonaqueous solvents will then be emphasized, especially through extensive study of their acid–base reactions. We also want to present the recent spectroscopic data and show their contribution to the understanding of their solution chemistries.

II. PREPARATIONS AND PROBLEMS OF PURITY

Different types of reactions have been used to synthesize the interhalogens.

A. Direct Union between Halogens

Probably the most common preparation is the direct union between two different halogens in the gas phase, according to the following generalized reactions:

$$X_2 + Y_2 \rightarrow 2XY \tag{1}$$

$$X_2 + 3Y_2 \rightarrow 2XY_3 \tag{2}$$

$$X_2 + 5Y_2 \rightarrow 2XY_5 \tag{3}$$

In several cases these reactions occur in a solvent or at low temperature. For instance, iodine monofluoride has been detected by ESR spectroscopy after direct union of iodine and fluorine in $CFCl_3$ at $-45°C$.[6] Bromine monochloride has been characterized after the reaction of bromine and chlorine in CCl_4[7,8] or even in water[9] at normal temperatures. Iodine monochloride may be prepared with liquid or gaseous chlorine,[7,10–13] bromine trifluoride with liquid bromine,[7,11,12,14] iodine trifluoride in a suspension of

TABLE I

THE KNOWN INTERHALOGEN MOLECULES[a]

Formula	Interhalogen					
XY	IF	BrF	ClF	BrCl	*ICl*	IBr
XY_3	IF_3	*BrF_3*	ClF_3	I_2Cl_6		
XY_5	*IF_5*	*BrF_5*	ClF_5			
XY_7	IF_7					

[a] Those used as solvents are italicized.

iodine in $CFCl_3$ at $-45°C$,[2,15,16] and I_2Cl_6 in an excess of liquid chlorine at $-80°C$.[7,10,12,13]

B. From Another Interhalogen

1. X in the interhalogen XY_n may be oxidized by the halogen Y_2 in reactions 4, 5, and 6

$$XY + Y_2 \rightarrow XY_3 \quad (4)$$

$$XY_3 + Y_2 \rightarrow XY_5 \quad (5)$$

where in the special case for IF_7

$$IF_5 + F_2 \rightarrow IF_7 \quad (6)$$

This is the case for the preparation of ClF_3, ClF_5, and BrF_5, generally in the gas phase near 200°C[7,17]; for chlorine pentafluoride, a photochemical reaction occurs at room temperature under a pressure of 1 atmosphere.[18]

2. The reaction of XY_3 or XY_5 with the halogen X_2 may also be employed, for instance, to prepare BrF:

$$XY_3 + X_2 \rightarrow 3XY \quad (7)$$

$$XY_5 + 2X_2 \rightarrow 5XY \quad (8)$$

This occurs also in the gas phase and seems to be a very good route for the synthesis of the monofluorides. In the case of iodine monofluoride, as for iodine halides, a solvent such as $CFCl_3$ is used at low temperature.

C. Various Oxidations and Fluorinations

This type of reaction is used mainly to obtain the iodine halides, and the other halogen pentafluorides, especially IF_7.

An analytical procedure to prepare ICl by oxidation of iodine in aqueous acidic solution with potassium permanganate, chlorine, water, and potassium iodate has been developed.[7,10–13] A convenient method of synthesis of I_2Cl_6 is somewhat similar; in a strongly acidic solution chlorate oxidizes iodine below 40°C by the reaction[10]

$$I_2 + 5Cl^- + ClO_3^- + 6H^+ \rightarrow I_2Cl_6 + 3H_2O \quad (9)$$

Xenon difluoride, a smooth oxidizing and fluorinating agent, can convert iodine into its trifluoride. This process is interesting because it is "clean." Gaseous xenon, the only other product of the reaction, is easy to separate. Other powerful oxidizing agents have been used, such as KrF_2, O_2F_2, and PtF_6. Iodine pentafluoride itself may be prepared in small quantities by fluorination under various conditions by several oxidizing agents, like ClF_3, BrF_3, SF_4, and RuF_5.[2,14]

Rogers *et al.*[19] described the electrolytic oxidation of chlorine firstly to chlorine trifluoride and subsequently to chlorine pentafluoride in anhydrous hydrogen fluoride solutions. Another interesting route to ClF_5 is the reaction[20]:

$$MClF_4(s) + F_2 \rightarrow MF(s) + ClF_5 \tag{10}$$

where M is an alkali metal. The reaction occurs between 80° and 150°C. The yield is very good (90%). The problem is that $MClF_4$ is often obtained from ClF_5, so this preparation is not strategically useful.

The oxidation of a metal halide by fluorine gives a very pure interhalogen with a yield generally better than 50% for the following reactions[2,21–23]:

$$NaCl + 3F_2 \xrightarrow{100°-300°C} NaF + ClF_5 \tag{11}$$

$$KBr + 3F_2 \longrightarrow KF + BrF_5 \tag{12}$$

$$MI + 3F_2 \longrightarrow MF + IF_5 \tag{13}$$

$$KI + 4F_2 \xrightarrow{250°C} KF + IF_7 \tag{14}$$

$$MI_2 + 8F_2 \longrightarrow MF_2 + 2IF_7 \tag{15}$$

D. Exchange Reactions

A few fluorine exchange reactions involving one halogen may lead to the corresponding halogen fluoride. This has been used in two cases: (1) to obtain iodine monofluoride[24] (Eq. 16) and (2) to obtain bromine trifluoride, such as when Meinert passed ClF_3 into liquid bromine at 0°C.

$$I_2 + AgF \rightarrow AgI + IF \tag{16}$$

E. Problems of Purification

As a consequence of the methods of preparation and of the significant chemical reactivity of the interhalogens, it is generally difficult to get them pure. The usual impurities are the free halogens themselves, hydrogen fluoride, and the oxyfluorides of the halogen concerned. More than that, the chlorides BrCl, ICl, and I_2Cl_6 are not exceptionally stable; BrF and IF have a tendency to disproportionate. The extremely high dielectric constant of BrF_3 makes it nearly impossible to purify completely.[25] Nevertheless, good results may be obtained with the chlorine fluorides, BrF_5, IF_5, and IF_7.

The handling of these products, which are corrosive toward organic materials and highly toxic,[2,10,12,26] has to be carefully undertaken, as described in detail by several authors.[1,27] Vacuum lines, tanks, and distillation columns are made out of nickel, monel, Kel–F, and teflon. They are inactivated before use by fluorine or chlorine trifluoride. All this technology is now well developed.

The treatments generally used to eliminate the impurities are the removal of HF by an alkali metal fluoride, where the bifluoride which formed (like $NaHF_2$) is nonvolatile, and the trap-to-trap distillation or sublimation under vacuum conditions. In the case of an halogen fluoride, the commercial products are commonly pretreated with fluorine to oxidize any elemental halogen impurity to the halogen fluoride desired.[28]

The purity of the liquids can be checked with the aid of conductometry with a Kel-F cell.[29] For gases, microsublimation and various spectroscopies (infrared, Raman-laser, ^{19}F-NMR) are used.

III. Physical Properties

In Downs and Adams' review article[4] a very exhaustive comparative table lists the physical properties of the interhalogens. For our purpose we choose to show the most useful data for solution chemists, namely, molecular weight, boiling point, melting point, vapor pressures; thermodynamic properties of the gas phase and of the liquid phase; mechanical properties such as density and viscosity; and electrical properties such as specific conductivity, dielectric constant, and dipole moment. The spectroscopic and structural properties are discussed in Section VI.

A. Volatility

The properties shown in Table II are necessary in choosing the correct solvent: the temperature range in which the interhalogens are liquids under reasonable pressure will indicate how they can be handled. From these data it can be seen that at room temperature all the chlorine fluorides are gases at atmospheric pressure. The monofluoride is the most volatile followed by ClF_5 and ClF_3, which may be used as solvents, but obviously only at low temperatures or under high pressure. On the other hand IBr and ICl_3 are solids at 25°C. Iodine monochloride is the only one which has been used thoroughly for this purpose above 27.5°C.

TABLE II

SELECTED PHYSICAL PROPERTIES RELATED TO VOLATILITY

Interhalogen	Molecular weight[a]	Boiling point		Melting point		Vapor pressure p (torr)[m]
		°K	°C	°K	°C	
ClF	54.451	173.0[b]	−100.1[b]	117.5[b]	−155.6[b]	$\log p = 15.738 - \frac{3109}{T} + \frac{1.538 \times 10^5}{T^2}$ (123°–168°K)[b]
BrF	98.902	293[c]	20[c]	240[c]	33[c]	
		disproportionates				
IF	145.9029		disproportionates			
BrCl	115.357	278[c]	5[c]	207	−66[c]	
			dissociates			
ICl	162.357	370–373[d]	97–100[d]	300.53[e]	27.38[e]	Partial pressure[f] liquid: 38.52–135.5 (303.60°–330°K) solid: 0.44–32.62 (250°–300.53°K)
IBr	206.808	389[g]	116[g]	314[g]	41[g]	
						Liquid phase
ClF_3	92.448	284.90[b]	11.75[b]	196.83[b]	−76.32[b]	$\log p = 7.36711 - \frac{1096.917}{t + 232.75}$ (−46.97°–29.55°C)[b]
BrF_3	136.869	398.90[b]	125.75[b]	281.92[b]	8.77[b]	$\log p = 7.74853 - \frac{1685.8}{t + 220.57}$ (38.72°–154.82°C)[b]
IF_3	183.899	yellow solid decomposes at −28°C[h]				
I_2Cl_6	466.527	dissociates		384[g]	101[g] (16 atm)	

ClF_5	130.445	260.05[i]	−13.1[i]	170 ± 4[j]	−103 ± 4[j]	$\log p = 7.4837 - \frac{1197}{T}$ (192.95°–391.05°K)[i]
BrF_5	174.896	314.45[b]	41.30[b]	212.65[b]	−60.5[b]	$\log p = 6.4545 + 0.001101t - \frac{895}{t + 206}$ (t°C)[b]
IF_5	221.8964	377.63[k]	104.48[k]	282.571[k]	9.421[k]	$\log p = 29.02167 - \frac{3090.14}{T} - 6.96834 \log T$[k]
IF_7	259.8933	277.92[b]	4.77[b] sublimation temperature	—	—	$\log p = 7.4967 - \frac{1291.58}{T}$[l]

[a] "Table of Atomic Weights," IUPAC Commission on Atomic Weights, 1969; *Pure Appl. Chem.* **21**, 95 (1970).

[b] L. Stein, *in* "Halogen Chemistry" (V. Gutmann, ed.), Vol. 1, p. 133. Academic Press, New York, 1967.

[c] H. R. Leech, "Mellor's Comprehensive Treatise on Inorganic and Theoretical Chemistry," Vol. II, Suppl. I, p. 158. Longmans, Green, New York, 1956.

[d] A. G. Sharpe, *in* "Non-Aqueous Solvent Systems" (T. C. Waddington, ed.), p. 285. Academic Press, New York, 1965.

[e] G. F. Calder and W. F. Giauque, *J. Phys. Chem.* **69**, 2443 (1965).

[f] R. H. Lamoreaux and W. F. Giauque, *J. Phys. Chem.* **73**, 755 (1969).

[g] H. Meinert, *Z. Chem.* **7**, 41 (1967).

[h] M. Schmeisser, W. Ludovici, D. Naumann, P. Sartori, and E. Scharf, *Chem. Ber.* **101**, 4214 (1968).

[i] H. H. Rogers, M. T. Constantine, J. Quaglino, Jr., H. E. Dubb, and N. N. Ogimachi, *J. Chem. Eng. Data* **13**, 307 (1968); W. R. Bisbee, J. V. Hamilton, J. M. Gerhauser, and R. Rushworth, *ibid.* p. 382.

[j] D. Pilipovich, W. Maya, E. A. Lawton, H. F. Bauer, D. F. Sheehan, N. N. Ogimachi, R. D. Wilson, F. C. Gunderloy, Jr., and V. E. Bedwell, *Inorg. Chem.* **6**, 1918 (1967).

[k] D. W. Osborne, F. Schreiner, and H. Selig, *J. Chem. Phys.* **54**, 3790 (1971).

[l] H. Selig, C. W. Williams, and G. J. Moody, *J. Phys. Chem.* **71**, 2739 (1967).

[m] T = temperature in °K, t = temperature in °C.

Bromine monofluoride, like iodine monofluoride, undergoes disproportionation and is never obtained pure.

$$3BrF \rightleftharpoons Br_2 + BrF_3 \quad (17)$$

The measured values[30] given in Table II may be doubtful, and their use is difficult in solvent chemistry. At room temperature BrF_3, BrF_5, and IF_5 are liquids, under atmospheric pressure. Iodine pentafluoride is less volatile than the heptafluoride, and chlorine and bromine trifluorides are less volatile than the corresponding pentafluorides. For the pentafluorides, the order of volatilities is $IF_5 < BrF_5 < ClF_5$.

B. Thermodynamic Properties

The thermodynamic data of the interhalogens in the gas phase and in the condensed phase (Table III and Table IV) help us understand comparative stabilities of these molecules, and thus, their chemistry. In Fig. 1 the average bond energies of the halogen fluorides XF_n are shown as a function of X and n.[31] Except for ClF and BrF, the energy of the X–F bond increases as X becomes more electropositive, whereas the energy decreases when n increases for the interhalogen XF_n when X remains constant. The values for IF_3 shown have been calculated from those of the other halogen fluorides.

For the monohalides XY, the bond energy can be related to the difference of electronegativity between X and Y. If this were the only contributing factor to the stability of XY then the expected order of stability would be BrCl < IBr < ICl < ClF < BrF < IF. All the interhalogen molecules,

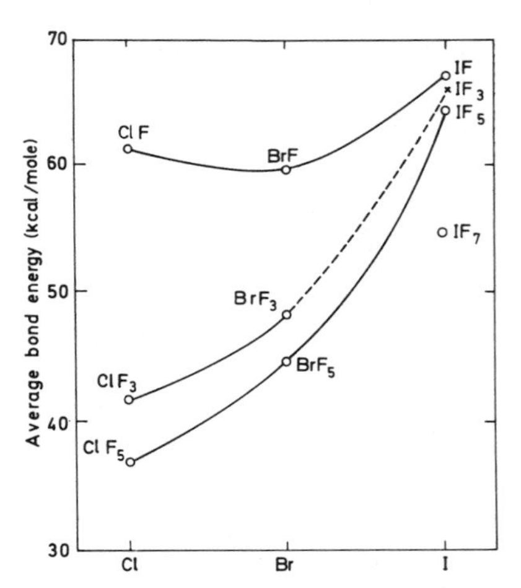

FIG. 1. Average bond energies of halogen fluorides.

TABLE III

THERMODYNAMIC PROPERTIES OF THE INTERHALOGENS IN THE GAS STATE

Interhalogen	ΔH_f (298°K) (kcal mole^{-1})	ΔG_f^0 (298°K)	$S°$ (298°K) (cal deg mole^{-1})	Dissociation energy (kcal)	Mean X–Y bond energy (298°K) (kcal)	C_p (100°–1000°K)[b]
ClF	−13.5[a]	−13.8[a]	52.06[a]	60.35[a]	—	6.646–9.722
BrF	−14.0[a]	−17.6[a]	54.70[a]	59.42[a]	—	6.970–9.704
IF	−22.6[a]	−28.1[a]	56.45[a]	66.2[a]	—	6.983–10.146
BrCl	+3.50[c]	−0.23[c]	57.36[c]	51.4[d]	—	7.101–9.540
ICl	+4.18[e]	−1.37[e]	59.140[c]	49.63[d]	—	7.213–9.580
IBr	+9.76[c]	+0.89[c]	61.822[c]	41.91[d]	—	7.623–9.620
ClF_3	−39.35[f]	−29.75[f]	67.28[c]	—	41.7[j]	9.394–19.857
BrF_3	−61.09[c]	−54.84[c]	69.89[c]	—	48.2[j]	9.347–19.859
IF_3	−116[j]	−110[j]	72[j]	—	66[j]	—
I_2Cl_6	—	—	—	—	—	—
						(250°–1000°K)
ClF_5	−60.9 ± 4.5[g]	−39.0 ± 4.5[g,h]	74.3[h]	—	36.9[j]	20.12–30.85[h]
BrF_5	−102.5[c]	−83.8[c]	76.50[c]	—	44.7[j]	10.839–31.770[b]
IF_5	−200.8[i]	−184.6[i]	80.45[i]	—	64.2[j]	11.830–31.767[b]
IF_7	−229.9[a,i]	−201.3[a,i]	87.40[a,i]	—	55.4[j]	13.213–43.682[b]

[a] L. Stein, *in* "Halogen Chemistry" (V. Gutmann, ed.), Vol. 1, p. 133. Academic Press, New York, 1967.

[b] "JANAF Thermochemical Tables," Dow Chemical Company, Midland, Michigan, 1960–1968.

[c] *Nat. Bur. Stand.* (*U.S.*), *Tech. Note* **270-3** (1968).

[d] A. G. Gaydon, *in* "Dissociation Energies," 3rd ed. Chapman & Hall, London, 1968.

[e] G. F. Calder and W. F. Giauque, *J. Phys. Chem.* **69**, 2443 (1965); R. H. Lamoreaux and W. F. Giaque, *ibid.* **73**, 755 (1969).

[f] R. C. King and G. T. Armstrong, *J. Res. Natl. Bur. Stand., Sect. A* **74**, 769 (1970).

[g] D. Pilipovich, W. Maya, E. A. Lawton, H. F. Bauer, D. F. Sheehan, N. N. Ogimachi, R. D. Wilson, F. C. Gunderloy, Jr., and V. E. Bedwell, *Inorg. Chem.* **6**, 1918 (1967).

[h] R. Bougon, J. Chatelet, and P. Plurien, *C. R. Hebd. Seances Acad. Sci., Ser. C* **264**, 1747 (1967).

[i] D. W. Osborne, F. Schreiner, and H. Selig, *J. Chem. Phys.* **54**, 3790 (1971).

[j] A. J. Downs and C. J. Adams, *in* "Comprehensive Inorganic Chemistry" (J. C. Bailer, H. J. Emeleus, R. Nyholm, and A. F. Trotman-Dickenson, eds.), Vol. 2, pp. 1487–1499. Pergamon Press, 1973.

TABLE IV

THERMODYNAMIC PROPERTIES OF THE INTERHALOGENS IN THE LIQUID STATE

	ΔH_f^0 XY (I) (298°K) (kcal/mole)	ΔG_f^0 (kcal/mole)	S^0 (cal d^{-1} $mole^{-1}$)	C_p^0 (303.68°–317.76°K)	Trouton's constant
ClF	—	—	—	—	28.0[i]
BrF	—	—	—	—	—
IF	—	—	—	—	—
BrCl	—	—	—	—	—
ICl	−5.69[a,b]	−3.314[a,b]	32.3	24.63–24.61[a]	26.5[i]
IBr	—	—	—	—	—
ClF_3	−45.65[c]	—	—	—	23.10[d]
BrF_3	−71.9[e]	−57.5[e]	42.6[e]	—	25.7[d]
IF_3	—	—	—	—	—
ClF_5	—	—	—	—	20.4[f]
BrF_5	−109.6[e]	−84.1[e]	53.8[e]	—	23.3[d]
IF_5	−210.8[g]	−186.6[g]	53.74[g]	41.9241.55[g] (282.57°–350°K)	22.76[g]
IF_7	—	—	—	—	21.1[h]

[a] G. F. Calder and W. F. Giauque, *J. Phys. Chem.* **69**, 2443 (1965).
[b] R. H. Lamoreaux and W. F. Giauque, *J. Phys. Chem.* **73**, 755 (1969).
[c] R. C. King and G. T. Armstrong, *J. Res. Natl. Bur. Stand., Sect. A* **74**, 769 (1970).
[d] L. Stein, *in* "Halogen Chemistry" (V. Gutmann, ed.), Vol. 1, p. 133. Academic Press, New York, 1967.
[e] D. D. Wagman, W. H. Evans, V. B. Parker, I. Halon, S. M. Bailey, and R. H. Schumm, *Natl. Bur. Stand. (U.S.), Tech. Note* **270–3**, 34 and 35 (1968).
[f] H. H. Rogers, M. T. Constantine, J. Quaglino, Jr., H. E. Dubb, and N. N. Ogimachi, *J. Chem. Eng. Data* **13**, 307 (1968); W. R. Bisbee, J. V. Hamilton, J. M. Gerhauser, and R. Rushworth, *ibid.* p. 382.
[g] D. W. Osborne, F. Schreiner, and H. Selig, *J. Chem. Phys.* **54**, 3790 (1971).
[h] H. Selig, C. W. Williams, and G. J. Moody, *J. Phys. Chem.* **71**, 2739 (1967).
[i] A. J. Downs and C. J. Adams, *in* "Comprehensive Inorganic Chemistry" (J. C. Bailar, H. J. Emeleus, R. Nyholm, and A. F. Trotman-Dickenson, eds.), Vol. 2, pp. 1487–1499. Pergamon Press, 1973.

except perhaps BrCl, are thermodynamically stable at 25°C with respect to dissociation into the gaseous elements.

It is also interesting to plot the free energy of formation for each series XF_n as a function of n, at room temperature, as shown in Fig. 2. It is clear that ClF is more stable than BrF and IF, the last two of which, as noted elsewhere, undergo disproportionation, giving the halogen and XF_3 or XF_5. This does not occur with BrCl, ICl, and IBr in the solid state. The iodine pentafluoride is exceptionally stable and dissociates only near 1400°C into IF_3 and fluorine. Similarly ClF_5, BrF_5, and IF_7 dissociate into fluorine and ClF_3, BrF_3, and IF_5, respectively, near 450°C; the reaction exhibits an increase of entropy.

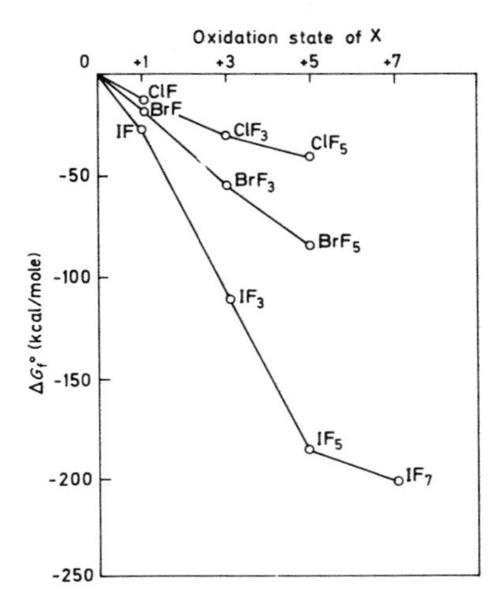

FIG. 2. Free energies of formation (at 298°K) of gaseous halogen fluorides XF_n.

In Table IV the liquid state is described. It is noted that a great part of the work in this field has yet to be done. In fact it is obvious that the liquid state is poorly described. It is only known from qualitative observations, for instance, that IF_3 may disproportionate in the condensed state at low temperature[15,16] according to

$$2IF_3 \rightarrow IF + IF_5 \tag{18}$$

and at 278°K giving iodine and IF_5.

C. Mechanical Properties

Two properties generally known for the interhalogens are density and viscosity. These data have a certain importance because of their consequences either on structure or on solution chemistry and on conductometric measurements and calculations (Table V).

One should note that the density of any interhalogen in the liquid and solid states is near 3 g/cm^3 and that the viscosity is never a constraint for handling.

D. Electrical Properties

The electrical properties (Table VI) give information on the polarity of the interhalogen molecules, on the eventual dissociation into ions in the liquid state, on the association in polymers, and on its ionizing or dissociating power. All these data determine the solvent properties of the liquid interhalogens. Three interhalogens undergo a noticeable self-ionization, namely, iodine monochloride, iodine trichloride, and bromine trifluoride. According

TABLE V

MECHANICAL PROPERTIES OF THE INTERHALOGENS

	Density (g/cm^3)		
	Solid	Liquid	Viscosity (centipoise)
ClF	—	1.62 (173°K)[a]	—
BrF	—	—	—
IF	—	—	—
BrF	—	—	—
ICl	3.85 (α, 0°C)[b] 3.66 (β, 0°C)[b]	$3.18223/[1 + 9.1590 \times 10^{-4} + 8.3296 \times 10^{-7}t^2 + 2.7501 \times 10^{-9}t^3]$ $t°$ in °C[c]	4.19 (28°C)[d]
IBr	4.4157–4.4135 (273°–283°K)[c]	3.7616–3.7345 (315°–523°K)[c]	—
ClF_3	2.53 (153°K)[e]	$t = -5°$ $t°$ 45°C[f] $d = 1.8853–2.942 \times 10^{-3}t - 3.79 \times 10^{-6}t^2$	0.448–0.316 (290.5°–323.2°K)[g]
BrF_3	3.23 (292°K)[f]	2.8030 (298°K) 2.7351 (323°K)[h]	3.036–1.775 (286.4°–312.8°K)[i]
IF_3	—	—	—
I_2Cl_6	3.203–3.1107 (−40, 15°C)[c]	—	—
ClF_5	—	$d = 3.553–1.396 \times 10^{-2}T + 4.565 \times 10^{-5}T^2 - 6.311 \times 10^{-8}T^3$ (193°–372°K)[j]	—

BrF_5	3.09 (212°K)[f]	$2.5509-3.484 \times 10^{-3}t - 3.45 \times 10^{-6}t^2$, $t = -15°-76°C$[f]	0.824–0.590 (275.5°–302.10°K)[i]
IF_5	3.961–3.678 (77.15°–273°K)[k]	3.263–3.031 (283.40°–343.96°K)[k]	2.490–1.686 (292.1°–313.2°K)
IF_7	3.62 (128°K)[l]	$d = 2.7918-0.0049t$ (t = 7.13°–24.32°C)[m]	—

[a] H. R. Leech, "Mellor's Comprehensive Treatise on Inorganic and Theoretical Chemistry," Vol. II, Suppl. I, p. 149. Longmans, Green, New York, 1956.

[b] H. Meinert, *Z. Chem.* **7**, 41 (1967).

[c] N. N. Greenwood, *Rev. Pure Appl. Chem.* **1**, 84 (1951).

[d] A. G. Charpe, *in* "Non-Aqueous Solvent Systems" (T. C. Waddington, ed.), p. 285. Academic Press, New York, 1965.

[e] R. D. Burbank and F. N. Bensey, *J. Chem. Phys.* **21**, 602 (1953).

[f] L. Stein, *in* "Halogen Chemistry" (V. Gutmann, ed.), Vol. 1, p. 133. Academic Press, New York, 1967.

[g] A. A. Banks, A. Davies, and A. J. Rudge, *J. Chem. Soc.* p. 732 (1953).

[h] J. W. Emsley, J. Feeney, and L. H. Sutcliffe, *in* "High Resolution Nuclear Magnetic Resonance Spectroscopy," Vol. 2, p. 871. Pergamon, Oxford, 1966.

[i] M. T. Rogers and E. E. Garver, *J. Phys. Chem.* **62**, 952 (1958).

[j] H. H. Rogers, M. T. Constantine, J. Quaglino, Jr., H. E. Dubb, and N. N. Ogimachi, *J. Chem. Eng. Data* **13**, 307 (1968); W. R. Bisbee, J. V. Hamilton, J. M. Gerhauser, and R. Rushworth, *ibid.* p. 382.

[k] D. W. Osborne, F. Schreiner, and H. Selig, *J. Chem. Phys.* **54**, 3790 (1971).

[l] R. D. Burbank and F. N. Bensey, *J. Chem. Phys.* **27**, 981 (1957); R. D. Burbank, *Acta Crystallogr.* **15**, 1207 (1962); J. Donohue, *ibid.* **18**, 1018 (1965).

[m] H. Selig, C. W. Williáms, and G. J. Moody, *J. Phys. Chem.* **71**, 2739 (1967).

TABLE VI

ELECTRICAL AND MAGNETIC PROPERTIES OF THE INTERHALOGENS

	Magnetic susceptibility (10^6 cgs units)	Specific conductivity in liquid phase (Ω^{-1} cm^{-1})	Dipole moment D (Debye)	Dielectric constant (liquid)
ClF	—	1.9×10^{-7} (145°K)[a]	0.881 (g)[b]	—
BrF	—	—	1.29 (g)[b]	—
IF	—	—	—	—
BrCl	—	—	0.57 (g)[b]	—
ICl	−54.5 (solid at 17°C)[b]	4.403–1.817 $\times 10^{-3}$ [c] (26.75°–122.0°C)	0.65 (g)[d]	—
IBr	—	3.02–4.10 $\times 10^{-4}$ [e,f] (40°–55°C)	1.21 (g)[d]	—
ClF_3	−26.5 (1300°K)[g]	4.9×10^{-9} (298°K)[a]	0.557 (g)[a]	$4.754 - 0.018t$ ($t = 0°$–42°C)[i]
BrF_3	−33.9 (1300°K)[g]	$8.34 \times 10^{-3} - 7.31 \times 10^{-3}$ [c,h] (283.7°–323.2°K)	1.19 (g)[a]	106.8 (25°C)
IF_3	—	—	—	—
I_2Cl_6	−90.2 (ICl_3, 290°K)[b]	$8.60 \times 10^{-3} - 0.36 \times 10^{-3}$ [c] (375°–420°K)	—	—

ClF_5	—	$0.37–1.25 \times 10^{-9}$ [j] (193°–256°K)	1.5 (l)[n]	$3.08 - 0.015t$ [j] ($t = -80°$ to $-17°C$)
BrF_5	−45.1 (298°K)[g]	$7.8–9.91 \times 10^{-8}$ [k] (213°–298°K)	1.51 (g)[a] 1.79 (l)[n]	$8.20 - 0.0117t$ [l] ($t = -11.7°–24.5°C$)
IF_5	−58.1 (298°K)[g]	5.4×10^{-6} (298°K)[a]	2.18 (g)[a]	$41.09 - 0.198t$ [i] ($t = 0°–42°C$)
IF_7	—	$< 10^{-9}$ (298°K)		1.75 (298°K)[m]

[a] L. Stein, *in* "Halogen Chemistry" (V. Gutmann, ed.), Vol. 1, p. 133. Academic Press, New York, 1967.
[b] F. W. Gray and J. Dakers, *Philos. Mag.* [7] **11**, 81 (1931).
[c] N. N. Greenwood, "Mellor's Comprehensive Treatise on Inorganic and Theoretical Chemistry," Vol. II. Suppl. 1, p. 485. Longmans, Green, New York, 1956.
[d] A. L. McClellan, *in* "Tables of Experimental Dipole Moments," Freeman, San Francisco, California, 1963.
[e] A. G. Charpe, *in* "Non-Aqueous Solvent Systems" (T. C. Waddington, ed.), p. 285. Academic Press, New York, 1965.
[f] H. Meinert, *Z. Chem.* **7**, 41 (1967).
[g] M. T. Rogers, M. B. Panish, and J. L. Speirs, *J. Am. Chem. Soc.* **77**, 5292 (1955).
[h] H. H. Hyman, T. Surles, L. A. Quarterman, and A. I. Popov, *J. Phys. Chem.* **74**, 2038 (1970).
[i] M. T. Rogers, H. B. Thompson, and J. L. Speirs, *J. Am. Chem. Soc.* **76**, 4841 (1954).
[j] D. Pilipovich, W. Maya, E. A. Lawton, H. F. Bauer, D. F. Sheehan, N. N. Ogimachi, R. D. Wilson, F. C. Gunderlay, Jr., and V. E. Bedwell, *Inorg. Chem.* **6**, 1918 (1967).
[k] M. T. Rogers, J. L. Speirs, and M. B. Panish, *J. Am. Chem. Soc.* **78**, 3288 (1956).
[l] M. T. Rogers, R. D. Pruett, H. B. Thompson, and J. L. Speirs, *J. Am. Chem. Soc.* **78**, 44 (1956).
[m] H. Selig, C. W. Williams, and G. J. Moody, *J. Phys. Chem.* **71**, 2739 (1967).
[n] C. Neveu, private communication.

to recent work,[25] bromine trifluoride stands apart with its high dielectric constant, which explains its marked complexing power.

IV. Chemical Properties–Halogenation

A. General

The chemistry of the interhalogens is based on a certain analogy with the parent halogen. In fact, interhalogens can behave as a source of halogen and become involved in a whole range of halogenation reactions as well as donor–acceptor interactions, which are described further with the solvent properties.

Halogenation may be an addition type of reaction, particularly with diatomic interhalogen molecules; but, most of the time, substitution occurs, and often an oxidation as well, as in the behavior of the halogen fluorides. Generally for an interhalogen XY_n, the most electropositive halogen X is reduced and the most electronegative Y is added or substituted.

A good example of this is again the chemistry of the halogen fluorides, which are known as fluorinating agents. Meinert has classified them in order of reactivity, IF being a moderate and ClF_3 a strong reagent.[14]

$$IF < IF_3 < BrF < IF_5 < BrF_3 < ClF < IF_7 < BrF_5 < ClF_3$$

ClF_5, now better known,[20,32] is less reactive than ClF_3. It is interesting to note that this experimental order is not in absolute concordance with the order of molar free energies of reduction of the interhalogen relative to reaction[19]

$$XF_n \rightarrow XF_{n-2} + F_2 \qquad (19)$$

and given in Table VII. The kinetic and the surface catalysis, as well as the presence of a Lewis acid, seem to be major influences, but these are not clearly determined. The presence of free radicals, intermediate products of reaction, has also an importance, for instance in the decomposition of ClF_3.[33]

The action of ICl and IBr on chemical elements is always chlorination and bromination, and not iodination, as could be expected.[14] The importance of the environment is major reactions with organic compounds,[34] because in a solvent of high dielectric constant homolytic processes occur and iodination is observed. Iodine monochloride and iodine bromide have a moderate oxidizing power[35] because they do not react with a certain number of elements such as B, C, Cd, Pb, Zr, Nb, Mo, and W. The fluorinating power of ClF is comparable with that of ICl and IBr, taking into account its behavior with respect to S, Se, Mo, W, and other elements.[36–38]

TABLE VII

MOLAR FREE ENERGIES OF REDUCTION OF THE INTERHALOGENS AND CERTAIN OTHER ELEMENTAL FLUORIDES[a]

Reaction	$\Delta G_{298}{}^0$ (kcal/mole)
ClF_5 (g) → ClF_3 (g) + F_2 (g)	+9.25
IF_7 (g) → IF_5 (l) + F_2 (g)	+14.7
ClF_3 (g) → ClF (g) + F_2 (g)	+15.95
2ClF (g) → Cl_2 (g) + F_2 (g)	+27.6
BrF_5 (l) → BrF_3 (l) + F_2 (g)	+27.6
2BrF (g) → Br_2 (l) + F_2 (g)	+35.2
BrF_3 (l) → BrF (g) + F_2 (g)	+39.9
2IF (g) → I_2 (s) + F_2 (g)	+56.2
IF_5 (l) → IF_3 (g) + F_2 (g)	+79
IF_3 (g) → IF (g) + F_2 (g)	+80
XeF_6 (g) → XeF_4 (g) + F_2 (g)	+8.1
XeF_4 (g) → XeF_2 (g) + F_2 (g)	+15.2
XeF_2 (g) → Xe (g) + F_2 (g)	+17.9
$2CoF_3$ (s) → $2CoF_2$ (s) + F_2 (g)	+44
AsF_5 (g) → AsF_3 (l) + F_2 (g)	+62.9
$2AgF_2$ (s) → 2AgF (s) + F_2 (g)	+69
SbF_5 (l) → SbF_3 (s) + F_2 (g)	+90
PF_5 (g) → PF_3 (g) + F_2 (g)	+161.8

[a] A. J. Downs and C. J. Adams, *in* "Comprehensive Inorganic Chemistry" (J. C. Bailar, H. J. Emeleus, R. Nyholm, A. F. Trotman-Dickenson, eds.), Vol. 2, p. 1526. Pergamon Press (1973).

In conclusion, the following rule seems to be observed most often: compared to the parent halogens, the interhalogens are less reactive with respect to homolytic scission, but they are more reactive from the standpoint of heterolytic dissociation.

In a number of cases the reaction of the interhalogens with the elements can give a suggestion of their relative reactivity and oxidizing power. For instance, all the halogen fluorides except IF_5[39] oxidize microquantities of radon dissolved in anhydrous hydrogen fluoride between $-195°C$ and 25°C. Similarly, ClF_3 and IF_7 change xenon into xenon fluorides, but IF_5 does not.[2,40]

Many elements react violently with the interhalogens, even at low temperatures.[2,14,17] Several factors interfere in the kinetics of these reactions as has been mentioned earlier. For instance, nickel, copper, and their alloys are resistant toward halogen fluorides; this resistance is a function of the coating of the corresponding fluoride formed on the interface. If an element that interacts with an interhalogen compound has several valence states, the halogen with the higher atomic number will be reduced, while the element will be oxidized to its highest valence state.

In contrast to the monohalides ICl and IBr, ClF_3 is a powerful oxidizing agent. Hydrogen, potassium, phosphorus, arsenic, antimony, sulfur, selenium, tellurium, molybdenum, tungsten, rhenium, and iridium are ignited. BrF_3 has the same effect on bromine and IF_3 on iodine. The noble metals are only attacked at high temperature.

It appears, then, more or less, that two groups of interhalogens can be distinguished as a result of their reactivities: the less reactive being the monohalides with the general formula XY and ICl_3; then IF_5, and the powerful reactants ClF_3, BrF_3, BrF_5, and even IF_7. This is confirmed by their behavior toward oxides, halides, organic substances, and their reaction with hydrogen-containing compounds.

B. The Monohalides and ICl_3

Their hydrolysis is moderate and in the absence of other halide ions, the first reaction is

$$XY + H_2O \rightarrow H^+ + Y^- + HOX \tag{20}$$

with X = I, Br and Y = Cl, Br. ClF and BrF also react with CO and SO_2,[2,14,41] for example, where they most likely add to the carbon and the sulfur atoms, respectively, because of the free lone pair of electrons.

$$CO + ClF \longrightarrow \text{F(Cl)C=O} \tag{21}$$

$$SO_2 + ClF \longrightarrow \text{F(Cl)S(=O)}_2 \tag{22}$$

But more often, the scheme is different with multiple bonds. For instance, with SO_3[40] ClF gives FSO_2OCl; with $F_4S=O$, F_5SOCl is obtained; and with

$$\text{R}_1\text{R}_2\text{C=O}$$

where R_1 and R_2 are F or CF_3, for instance, the same type of addition occurs leading to

$$\text{R}_1\text{R}_2\text{C(F)–O–Cl}$$

This can be a good synthesis of hypochlorites especially when catalyzed by CsF or HF.[38]

Chlorine monofluoride acting as a mild oxidant transforms $AsCl_3$ into $AsCl_4^+$ AsF_6^-. But $SbCl_5$ is probably changed in $SbCl_4F$. ClF can also be added to SF_4,[36] forming SF_5Cl.

In the case of organic chemistry, BrCl, ICl, and IBr have been used in substitution reactions of the C–H bond. In various solvents, ICl has the property to make organoiodine out of organomercury compounds, except in the case of the existence of secondary reactions. All the diatomic interhalogens can be added to a double or a triple bond.[42–44] The process proposed for these reactions is that the most electropositive halogen atom initiates the electrophilic attack and forms the intermediate cation. Substitution occurs when iodine trichloride reacts with aromatic compounds; but with aryltin or arylmercury compounds, the diaryliodonium derivative is obtained.

C. The Most Reactive Interhalogens: ClF_3, BrF_3, BrF_5, and IF_7

Their reaction with oxides is generally vigorous. For chlorine trifluoride, ignition occurs. For bromine trifluoride and pentafluoride (and their Lewis salts), oxygen is evolved and a whole range of oxides and oxysalts of the central halogen atom are formed.[2] Characteristic of this is the reaction with glass or quartz which is probably initiated by hydrogen fluoride.

$$4HF + SiO_2 \rightarrow SiF_4 + 2H_2O \tag{23}$$

The water thus produced may promote a cycle of reactions such as is shown below for IF_7:

$$IF_7 + H_2O \rightarrow IOF_5 + 2HF \tag{24}$$

On the other hand, IF_7 may directly attack SiO_2:

$$2IF_7 + SiO_2 \rightarrow 2IOF_5 + SiF_4 \tag{25}$$

But if the reactive halogen fluorides, in fact, used are very pure and are manipulated under vacuum conditions, they all can be handled in glass, because they only etch it very slowly.

With water, if no special conditions are used to slow the reaction (like dilution or low temperature), bromine trifluoride and chlorine trifluoride react violently to form a whole range of compounds including, for example, hydrogen fluoride, hydrogen chloride, and oxygen. Only for iodine pentafluoride, where HIO_3 is formed, is the central atom not reduced.

Tantot and Bougon[45] have prepared $KBrO_2F_2$ using bromine pentafluoride to replace one oxygen in $KBrO_3$. This process can also lead to $KBrOF_4$.[45a]

These halogen fluorides fluorinate the metal halides. At 250°C[2] chlorine trifluoride reacts with the chlorides $NiCl_2$, AgCl, and $CoCl_2$ to give the fluorides NiF_2, AgF_2, and CoF_3; at 180°C SF_4 is also fluorinated to SF_6. For bromine trifluoride nearly all the metallic halides are fluorinated and even oxidized to the highest valence state of the metal. Usually the reaction can be complete if the resultant product is volatile (like uranium hexafluoride) or soluble in BrF_3 (like the alkali metal fluorides); with lead(II), thallium(II), and cobalt(II), for example, the reaction is limited by the formation of an insoluble film of fluoride on the surface of the compound.

The reaction with organic compounds is often explosive for the halogen fluorides,[2] and their use in organic chemistry has been well described in several treatises.[2,7,42,46] Most of the time substitution occurs, but in several cases addition reaction is also in competition, especially in the case of ClF_3 with benzene or toluene in the vapor phase.

Generally the reactions with hydrogen-containing compounds are also violent[2] but can be moderated by dilution in an inert gas or by the use of complexes.[47] For instance, the reaction with N_2H_4 is violent

$$2N_2H_4 + 4ClF_3 \rightarrow 3N_2 + 12HF + 2Cl_2 \quad (26)$$

but with NHF_2 it can be moderate

$$3NHF_2 + ClF_3 \rightarrow ClNF_2 + N_2F_4 + 3HF \quad (27)$$

With highly oxidative materials, such as platinum hexafluoride and krypton difluoride, ClF_5 and BrF_5 can be oxidized to the heptavalent fluoride cations ClF_6^+ and BrF_6^+. In the first case the reaction occurs at 25°C exposed to laboratory light after eight days,[48] or under ultraviolet irradiation with a pyrex filter after two weeks[48a]

$$2ClF_5 + 2PtF_6 \rightarrow ClF_4^+PtF_6^- + ClF_6^+PtF_6^- \quad (28)$$

If the UV irradiation is not filtered, after several hours Christe[48a] noted the reaction

$$2ClF_5 + 2PtF_6 \rightarrow ClF_2^+PtF_6^- + ClF_6^+PtF_6^- + F_2 \quad (29)$$

Gillespie and Schrobilgen[49,50] used the ion $Kr_2F_3^+$ to oxidize BrF_5 by dissolving the complex $Kr_2F_3^+SbF_6^-$ in BrF_5.

$$Kr_2F_3^+SbF_6^- + BrF_5 \rightarrow 2Kr + F_2 + BrF_6^+SbF_6^- \quad (30)$$

Christe,[51] too, has prepared and studied the vibrational spectrum of BrF_6^+ cation. From these results the synthesis of ClF_7 and BrF_7 seems hardly probable. For iodine pentafluoride, the least reactive of these four halogen fluorides, many of these reactions can be controlled and occur smoothly.[46]

V. SOLVENT PROPERTIES

A. General

The extreme reactivity of the interhalogens is an obstacle to their use as solvents for several reasons. First of all, they usually react with the solute before it is dissolved. When the solute is dissolved, however, it is either by solvation or ionization. It can be easily deduced that the halogen fluorides might be good reaction media only for the fluorides. Secondly, the usual nonaqueous solvent techniques are difficult or impossible to apply due to the chemical and physical properties of the interhalogens. The careful handling in moisture-free glove boxes, or in nickel or monel vacuum lines, and the use of teflon, Kel-F, and now FEP materials (such as for vacuum lines, tubes, valves, and filters) have made possible the use of some techniques like conductivity, dielectric measurements, and cryoscopy (which provide data on the number and mobilities of ions present), solubilities and pressure determinations, and microsublimations.[52] Spectroscopy has been promoted, especially Raman-laser techniques, the sample being in a Kel-F tube closed by a valve or sealed. With a diamond cell,[53] for example, infrared spectra can now be taken on highly corrosive materials. X-ray diffraction spectroscopy has been performed quite regularly on the crystals of the complexes formed, or at low temperature on the interhalogen themselves, in capillaries which are sealed inside a dry glove box. Fluorine-19 NMR spectroscopy has also been developed with the samples contained in calibrated Kel-F tubes or sealed quartz tubes. Classical electrochemistry, however, including electrolysis, voltommetry, and polarography, which are of fundamental help in other solvent systems, have not been exploited for the interhalogens. The iodine chlorides dissolve gold and platinum, and the identification of products evolving at such electrodes has not been successfully determined.[54] Some attempts have been made by Meinert *et al.* to obtain the limits of the region of electroactivity for BrF_3 and BrF_5[55] but the results to date are inconclusive and the study seems to be interrupted. Now that electrochemistry in anhydrous hydrogen fluoride has been developed,[19,56,57] the way is opened for exploring this field.

For some interhalogens another handling problem can occur. If the vapor pressure at room temperature is too high, the use of Kel-F tubes is excluded and monel bombs become necessary. The necessity for working at sufficiently low temperatures can also be a real hindrance. This is the case for ClF and ClF_5, for instance.

Special mention has to be made to the pionneering work of Emeléus' group at Cambridge University in the 1950's. With a classical glass apparatus they developed practically the whole chemistry of bromine trifluoride.

In spite of all their experimental difficulties, nearly all their results were confirmed afterward with modern technology.[29]

B. Solubilities

The major purpose of a solvent is to dissolve a number of compounds, either by ionization or by solvation. The interhalogens most used as solvents are ICl, BrF_3, BrF_5, and IF_5. For the nonaqueous solvent chemist, solubilities are basic practical data.

1. Iodine Monochloride

Three classes of solutes seem to exist[58,59]: (1) very soluble: KCl and NH_4Cl (ionized); (2) appreciably soluble: RbCl, CsCl, KBr, KI (ionized); $AlCl_3$, $AlBr_3$, PCl_5, $SbCl_5$ (ionized); $SiCl_4$, $TiCl_4$, $NbCl_5$, $SOCl_2$ (solvated); and pyridine, acetonitrile, and benzamide (ionized); and (3) slightly soluble: LiCl, NaCl, AgCl, and $BaCl_2$ (ionized). Iodine monobromide has comparable properties.

2. Bromine Trifluoride

According to Stein's review,[2] all the alkali metal fluorides except for lithium, as well as the silver, nitrosyl, ammonium, and barium fluorides (which are Lewis bases), and AuF_3, SnF_4, PtF_4, PdF_4, phosphorus, arsenic, antimony, bismuth, tantalum, ruthenium, and niobium fluorides (which are Lewis acids) are soluble in bromine trifluoride and are all ionized. XeF_2 and XeF_4[60] are dissolved but not ionized. Sheft[61] gave quantitative data about the solubility of a number of metal fluorides in BrF_3.

3. Bromine Pentafluoride

BrF_5 dissolves a limited number of fluorides like AsF_5, SbF_5, and UF_6, but no ions are formed. Bromine pentafluoride is not widely used as a solvent but in certain specific cases, such as in the reaction with the uranium hexafluoride, it can be very useful and selective. Xenon fluorides are also soluble and very recently Schrobilgen *et al.*[62] studied the Xe NMR spectra of solutions of XeF_2, $FXeSO_3F$, $FXeFMoOF_4$, $FXeFWOF_4$, $(FXe)_2SO_3F^+$, $(FXe)_2F^+$, and XeF_4 in bromine pentafluoride. The spectra obtained are particularly well resolved.

4. Iodine Pentafluoride

IF_5 appears to be a relatively good ionizing solvent, like bromine trifluoride. It undergoes self-ionization, although not to the same extent, and it dissolves the same compounds. It is not considered as useful as BrF_3 because in IF_5 during neutralization reactions there is a competition with several reactions, one of which is solvation, which makes pure complexes difficult to obtain. SbF_5, SO_3, BF_3, HF, and KIF_6 are dissolved, however, as are certain iodides.[2] Sladky and Bartlett showed[63] that XeF_2 and XeF_4 were soluble are not perceptibly ionized.[64] The addition of anhydrous hydrogen fluoride to IF_5 greatly increases its solvent properties.[2]

C. The Problems of Self-Ionization and Acid–Base Reactions

The basic characteristic of these solvents which explains their properties is their partial self-ionization. The fact that all of them can be halide ion donors (Lewis bases) or halide ion acceptors (Lewis acids)[64a] when solutes are added explains their noticeable conducting properties when the ions are sufficiently mobile (Table VIII). The interesting feature of these solvents is that the addition of a solute increases their electrical conductivity. They are ionizing solvents, except for BrF_5, and consequently conductometry is the method of choice for following reactions in these media.

1. Iodine Monochloride

For this solvent the following three schemes of ionization have been proposed[3]:

$$2\,ICl \rightleftharpoons I^+ + ICl_2^- \tag{31}$$

$$3\,ICl \rightleftharpoons I_2Cl^+ + ICl_2^- \tag{32}$$

$$4\,ICl \rightleftharpoons I_2Cl^+ + I_2Cl_3^- \tag{33}$$

Gillespie[65] has suggested that I_2Cl^+ cations disproportionate into I_3^+ and ICl_2^+. Further studies[66] on I_2Cl^+ with respect to I_3^+ and ICl_2^+, however, have shown the stability of I_2Cl^+ against disproportionation. In fact, I_3^+ and ICl_2^+ in a 1 : 1 molar ratio undergo a halogen redistribution reaction to form I_2Cl^+,[67] indicating the last ion's relative stability. Cornog and Karges[58,59] have extensively studied iodine monochloride as an ionizing solvent by conductometric methods. KCl and NH_4Cl greatly increase its

TABLE VIII

Electrical Conductivity of the Interhalogens Used as Solvents[a]

	ICl	BrF_3	BrF_5	IF_5
Specific conductivity at 25°C ($\Omega^{-1}\,cm^{-1}$)	4.4×10^{-3}	8.3×10^{-3}	9.9×10^{-8}	5.4×10^{-6}

[a] See Table VI.

conductivity, but experimental results are not always consistent because important parameters, such as viscosity, density, and dissociation coefficients vary with dilution.

Compounds like $AlCl_3$, $SnCl_4$, and $SbCl_5$ behave as Lewis acids in the solvent ICl, increasing the concentration of I_2Cl^+ ions, and forming adducts which can be described by the ionic formulas $I_2Cl^-AlCl_4^-$, $I_2Cl^+SbCl_6^-$, and $I_2Cl^+SnCl_5^-$.[68–70] On the other hand, PCl_5 and the alkali metal chlorides are Lewis bases promoting the formation of the ICl_2^- anion,[3] and conductometric titrations have been performed on various acid–base systems such as $SbCl_5$, RbCl; $NbCl_5$, KCl; $SnCl_4$, NH_4Cl. The amphoteric nature of PCl_5 is well illustrated by its acidity toward KCl and its basicity toward $SbCl_5$ and ICl.

Fialkov *et al.*[71–73] have used electrochemistry and cryoscopy to study the organic and inorganic chemistry of ICl and their work confirms the self-ionization of the liquid. The electrolysis of ICl in nitrobenzene presumably leads to the decomposition of ICl_2^- at the anode into iodine and chlorine. The solutions of $AlCl_3$, SO_2, SO_2Cl_2, nitrobenzene, diethylether, and acetic acid in ICl are conductive.[7,46,74]

2. Bromine Trifluoride

BrF_3 is the most extensively used interhalogen as reaction medium. Emeléus and his group at Cambridge University opened wide the way to the modern chemistry of the interhalogens.[75] By measuring the specific conductivities of ClF_3, BrF_3, and IF_5[76] they noticed the exceptionally high values obtained for BrF_3 ($8 \times 10^{-3}\ \Omega^{-1}\ cm^{-1}$ at 25°C). They postulated its self-ionization to be

$$2BrF_3 \rightleftharpoons BrF_2^+ + BrF_4^- \qquad (34)$$

They prepared Lewis bases by dissolving KF, AgF, and BaF_2 in BrF_3 and, when the solution was evaporated, the solid compounds $KBrF_4$, $AgBrF_4$, and $Ba(BrF_4)_2$ were isolated and characterized. SbF_5 and SnF_4 were found to be Lewis acids in the BrF_3 solvent. When they carried out conductometric titrations between these bases and acids they obtained diagrams which confirmed the assumption of the acid–base neutralization process in these solutions and, thus, of the self-ionization of the solvent.

A typical acid–base neutralization reaction in BrF_3 would be

$$BrF_2^+SbF_6^- + Ag^+BrF_4^- \rightarrow Ag^+SbF_6^- + 2BrF_3 \qquad (35)$$

Later, Edwards *et al.* deduced the structure of the adduct BrF_3SbF_5[77,78] and confirmed the proposed ionic formula $BrF_2^+SbF_6^-$. Brown *et al.*[79] found that the solid GeF_42BrF_3 was probably not ionic, and lately Edwards *et al.*[80] gave a structure which is consistent with an ionic formula although there are strong interactions between the ions through fluorine bridging; in solution in bromine trifluoride Emeléus' assumption seem to be confirmed.

Bouy[81] described reactions between the solvosystem "base" NOF and the "acid" SnF_4 in BrF_3 with the same scheme. Later on, the study of Chrétien and Martin,[82,83] in agreement with these principles, showed the acid–base reactions in BrF_3 between uranium tetrafluoride and the alkali metal fluorides by conductometry and cryoscopy. Martin later observed that xenon di- and tetrafluorides dissolved in an acid bromine trifluoride (which contained PF_6^-, AsF_6^-, or SbF_6^-), seem to form several complexes.[60,84] Richards and Woolfe[85] have fruitfully used BrF_3 as a reaction medium for calorimetry.

It is also worth noting that the acidic solutions that are obtained when $BrF_2^+SbF_6^-$ or $BrF_2^+AsF_6^-$ are added to BrF_3 can also result from the direct reaction of the oxides Sb_2O_3 and As_2O_3 or the chlorides[2,60] on BrF_3 at room temperature with a yield of 100%.

The behavior of BrF_3 in the chemistry of the interhalogens, apart from the others, is an interesting problem for the scientist. Are the high conductivity and the chemical confirmation of its self-ionization real good answers? Woolfe[75] estimated the ionic product $[BrF_2^+][BrF_4^-]$ to be near 4×10^{-4}. Surles, in his Ph.D. thesis,[86] calculated this product from the Raman-laser spectra and obtained an even higher value of 0.8. Calculations (unpublished results) from the conductometry data obtained by the Fuoss-Onsager method performed by Martin and Tantot give a result not far from unity. Bromine trifluoride is known to be an ionizing solvent, but the liquid also associates into polymers which are not yet well defined. The recently measured dielectric constant,[25] 106.8 at 25°C, shows that BrF_3 is one of the most polar liquids, and this is in full agreement with its extraordinary ionizing and complexing power. As Stein[87] observed, BrF_3 prepared by the reaction of fluorine with bromine is never pure. On the other hand, Quarterman *et al.*[29] showed that the usual impurities BrF, Br_2, and BrF_5, when added in known quantities, do not greatly affect its conductance. In addition their values for the specific conductance of BrF_3, which have been obtained on an ultrapure product just after distillation with a Kel-F apparatus, are very near Emeléus' values. Martin[25] assumes that the solvent, impossible to handle with real purity, ionizes any traces of impurities and that this influences the equilibrium of self-ionization. Perhaps the real self-ionization is not as considerable as has been previously measured and proposed. This would not be in contradiction with the observed negative coefficient of temperature of the conductance noted by Emeléus and by Woolfe,[76,88] which might be due to a slow evaporation of the impurities.

These considerations are a good example of the complexity of physical phenomena in liquids, where the data obtained by various techniques are not easy to correlate because the model used for the interpretation never fits closely enough to reality.

3. Bromine Pentafluoride

Bromine pentafluoride has not been extensively used as a solvent; its self-ionization is quite small, it does not have a high complexing power, and it does not dissolve many compounds. However, in certain precise cases it can be useful. For instance, it is a good solvent for uranium hexafluoride. Several groups have studied the acid–base properties of UF_6 in this solvent.

Meinert *et al.*[89] first showed that SbF_5 could be an acid when dissolved in BrF_5. Surles[90] and Christe[91] confirmed the existence of the ionic complex $BrF_4^+Sb_2F_{11}^-$. Surles, trying to prepare other acidic solutions, noted that AsF_5, BiF_5, NbF_5, and BF_3 do not react with BrF_5.[90] Bougon *et al.* observed[92] that complexes of the type $M^+BrF_6^-$ are obtained when an alkali metal fluoride such as KF, RbF, and CsF was added to BrF_5.

The following neutralization reaction has been observed in the molten complex BrF_5, $2SbF_5$[89,91]:

$$BrF_4^+Sb_2F_{11}^- + 2Cs^+BrF_6^- \rightarrow 3BrF_5 + 2Cs^+SbF_6^- \quad (36)$$

4. Iodine Pentafluoride

The specific conductance of IF_5 allows one to predict the self-ionization according to

$$2IF_5 \rightleftharpoons IF_4^+ + IF_6^- \quad (37)$$

Dissolving KF in IF_5 increases the conductivity of IF_5, and, after evaporation of the solvent, KIF_6 is obtained; thus, KF behaves as a base in IF_5. On the other hand, SbF_5, SO_3, HF, and KIO_3 are Lewis acids in this solvent and various acid–base reactions can be carried out, such as

$$K^+IF_6^- + IF_4^+SbF_6^- \rightarrow K^+SbF_6^- + 2IF_5 \quad (38)$$

Christe and Sawodny[93] observed recently that $IF_5 : SbF_5$ is predominantly ionic and compared the fundamental vibrations of IF_4^+ to those of ClF_4^+ and BrF_4^+. The complexes $KSbF_6$, KBF_4, $KPtF_6$, and $CsRbF_6$ have also been prepared in this manner, but the reactions are usually incomplete and solvation occurs. Generally, bromine trifluoride is preferred as a reaction medium for this type of synthesis.

VI. Spectroscopic and Structural Studies of Some Halogen Fluorides

A. General

Many structural studies have been performed on the halogen fluorides, especially in the liquid and the gas phase. A point group symmetry of

C_{2v}, distorded T, and C_{4v}, square based pyramid, have been determined for the halogen tri- and pentafluorides, respectively, by vibrational spectroscopy,[32,94–107] microwave spectra,[108–112] nuclear magnetic resonance spectroscopy,[113–115] and crystallography.[116–118] These geometrical structures are in very good agreement with Gillespie's VSEPR theory,[119] which predicts maximum repulsion between the electron pairs of the valence shell of the central atom, taking into account the lone pairs as well as the bonding pairs (Fig. 3). For the trifluorides as well as the pentafluorides, their geometry shows two types of halogen-fluorine bonds. One is an axial bond with a covalent character, and the others are "equatorial" bonds with a semi-ionic character. Chemical shift measurements, determination of bond lengths, and calculations of the stretching force constants confirm this interpretation with an exception for the ^{19}F chemical shifts in ClF_3 (see Table IX).

In the case of the IF_7 molecule the situation is more complex because several stable geometries exist for a seven-coordinate molecule.[119,120] The relative stability of each structure depends on interatomic forces between fluorine atoms. The most recent spectroscopic data[121–126] suggest an average D_{5h} pentagonal bipyramidal structure which is subjected to dynamic deformations and internal pseudorotation movements.

B. Vibrational Spectroscopic Data in the Liquid Phase

The Raman and infrared spectra of the halogen fluorides either in the gas or liquid phase are in good agreement with the assumptions of point group theory. For the trifluorides

$$\Gamma = 3A_1 + 2B_1 + B_2 \tag{39}$$

where all the modes are Raman and IR active.

For the pentafluorides

$$\Gamma = 3A_1 + 2B_1 + B_2 + 3E \tag{40}$$

All the modes are Raman active, but only the A_1 modes and the E modes

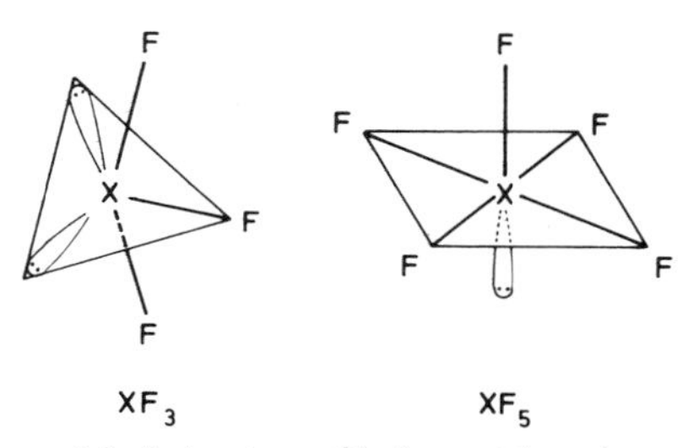

FIG. 3. Geometrical structure of halogen tri- and pentafluorides.

TABLE IX

HALOGEN–FLUORINE DISTANCES, FLUORINE–HALOGEN–FLUORINE ANGLES, STRETCHING FORCE CONSTANTS, CHEMICAL SHIFTS, AND SCALAR COUPLING CONSTANTS IN HALOGEN FLUORIDES

	$d(X–F_{ax})$[a] (Å)	$d(X–F_{eq})$[a] (Å)	$(F_{ax}–X–F_{eq})$[a]	$f_{R_{ax}}$[a] (mdyne/Å)	$f_{R_{eq}}$[a] (mdyne/Å)	σ_{ax}[a] (ppm/F_2)	σ_{eq}[a] (ppm/F_2)	$J_{F_{ax}} = F_{eq}$[a] (Hz)
ClF	1.628[b]	—	—	4.34[c]	—	877[d]	—	—
ClF_3	1.598[e]	1.698[e]	87° 29′[e]	4.19[f]	2.07[f]	425[g]	313[g]	441[d]
ClF_5	1.58[h]	1.67[h]	86°[h]	3.01[i]	2.57[i]	3[j]	170[j]	131[k]
BrF_3	1.721[l]	1.810[l]	86° 12′[l]	4.08[f]	3.01[f]	461[j,m]	461[j,m]	—
BrF_5	1.689[n]	1.774[n]	84° 48′[n]	4.07[i]	3.19[i]	153[j]	290[j]	76[k]
IF_5	1.844[n]	1.869[n]	81° 54′[n]	4.68[i]	3.64[i]	369[j]	418[j]	81[g]
IF_7	1.786[o]	1.858[o]	—	—	—	261[j,m]	261[j,m]	—

[a] The subscripts ax and eq are used for "axial" and "equatorial" fluorine atoms.
[b] D. A. Gilbert, A. Roberts, and P. A. Griswold, *Phys. Rev.* **76**, 1723 (1959).
[c] E. A. Jones, T. F. Parkinson, and T. F. Burke, *J. Chem. Phys.* **18**, 235 (1950).
[d] L. G. Alexacos and C. D. Cornwell, *J. Chem. Phys.* **41**, 2098 (1964).
[e] D. F. Smith, *J. Chem. Phys.* **21**, 609 (1953).
[f] R. Frey, R. L. Redington, and A. L. K. Aljibury, *J. Chem. Phys.* **54**, 344 (1971).
[g] E. L. Muetterties and W. D. Phillips, *J. Am. Chem. Soc.* **79**, 322 (1957).
[h] P. Goulet, R. Jurek, and J. Chanussot, *J. Phys.* **37**, 495 (1976).
[i] G. M. Begun, W. H. Fletcher, and D. F. Smith, *J. Chem. Phys.* **42**, 2236 (1964).
[j] H. S. Gutowsky and C. J. Hoffman, *J. Chem. Phys.* **19**, 1259 (1951).
[k] H. S. Gutowsky, D. M. McCall, and C. P. Slichter, *J. Chem. Phys.* **21**, 279 (1953).
[l] D. W. Magnuson, *J. Chem. Phys.* **27**, 223 (1957).
[m] Single line.
[n] A. G. Robiette, R. H. Bradley, and P. N. Brier, *Chem. Commun.* p. 1567 (1971).
[o] W. J. Adams, H. B. Thompson, and L. S. Bartell, *J. Chem. Phys.* **53**, 4040 (1970).

are IR active. The vibrational frequencies of the halogen fluorides shown in Table X, lead to the following observations.

For ClF_3 $\nu_4(b_1)$, a very strong infrared active band, is not observed in the Raman spectrum. In the liquid phase $\nu_3(a_1)$ and $\nu_6(b_2)$ are coincident at $320\ cm^{-1}$, but this accidental coincidence does not exist in the gas phase[99] nor in the solid phase as shown by an infrared study in a noble gas matrix[101] and by the Raman study of the solid.[105] Finally, $\nu_2(a_1)$ is particularly broad in the Raman spectrum of the liquid and splits into two lines when the temperature decreases sufficiently. This evolution can be interpreted[105] by the progressive appearance of associated species. This phenomenon disappears when ClF_3 is dissolved at low concentrations in hydrogen fluoride,[107] which confirms the previous assumptions about the existence of dimers in liquid ClF_3.

BrF_3 is known to be associated with its high Trouton constant. In fact, the Raman spectrum of the liquid is not well resolved. Moreover, the spectrum is perturbed by self-ionization, promoting the presence of bands due to the ionic species BrF_2^+ and BrF_4^-. A mathematical decomposition of the Raman spectrum into bands attributed to the monomer, polymers, and ionic species has been proposed.[86,103]

For the three halogen pentafluorides $\nu_7(e)$ and $\nu_5(b_1)$ do not appear in Raman spectra of the liquids.[96] Moreover, for ClF_5 a triple coincidence among $\nu_4(b_1)$, $\nu_3(a_1)$, and $\nu_8(e)$ occurs. This coincidence disappears in the matrix-isolated infrared spectrum[104] and in the solid phase.[106] No associated species seem to exist in liquid ClF_5. On the contrary, a great freedom in molecular reorientation is suggested by a study of shapes of the Raman bands.[107]

In BrF_5, dimers might exist at low temperature.[105] The presence of associated species in IF_5 is proved and justifies the existence of two polarized bands in the $\nu_1(a_1)$ region. One of these two bands disappears as the temperature increases,[98] or when IF_5 is dissolved in hydrogen fluoride.[106] Another band at $218\ cm^{-1}$ exhibits the same behavior.

C. NMR Data in the Liquid Phase

The ^{19}F-NMR spectra of all the stable halogen fluorides have been observed. If the anomalous chemical shift of ClF is excepted, a displacement of the ^{19}F line toward high field indicates an increase in the ionicity of the halogen–fluorine bond. The chemical shifts and the fluorine–fluorine scalar coupling constants were listed in Table IX. In contrast to the high resolution for the liquids, NMR spectra for the solids are sensitive to the variation of the chemical shift with the orientation of the applied magnetic field vis-à-vis their molecular axis. In ClF_5 a chemical shift anisotropy of 810 ± 70 ppm

TABLE X

VIBRATIONAL FREQUENCIES OF SOME HALOGEN FLUORIDES[a]

	ClF		ClF_3		BrF_3		ClF_5		BrF_5		IF_5	
	gas[b]	liquid[b]	IR[c] gas	Raman[d] liquid	IR[c] gas	Raman[e] liquid	IR[f] gas	Raman[f] liquid	IR[f] gas	Raman[d] liquid	IR[f] gas	Raman[h] liquid
$\nu_1(a_1)$	772	758	760	753	682	673	(712)[g]	709	683	689	710	705
			742		668							697
$\nu_2(a_1)$	—	—	538	511	557	531	541	538	587	572	(595)[g]	600
			522		547							
$\nu_3(a_1)$	—	—	328	324	241	236	486	480	363	370	318	316
$\nu_4(b_1)$	—	—	702	—	621	—	(488)[g]	480	(547)[g]	539	—	578
					614							
					604							
$\nu_5(b_1)$	—	—	442	430	359	341	—	—	—	—	—	—
					350							
					342							
$\nu_6(b_2)$	—	—	328	324	242	265	—	375	—	316	—	275
$\nu_7(e)$	—	—	—	—	—	—	732	—	644	—	640	—
$\nu_8(e)$	—	—	—	—	—	—	—	480	415	419	372	375
$\nu_9(e)$	—	—	—	—	—	—	302	296	—	240	—	218
												191

[a] Frequencies are given in cm^{-1}.
[b] E. A. Jones, T. F. Parkinson, and T. G. Burke, *J. Chem. Phys.* **18**, 235 (1950).
[c] H. Selig, H. H. Claassen, and J. H. Holloway, *J. Chem. Phys.* **52**, 3517 (1970).
[d] R. Rousson and M. Drifford, *J. Chem. Phys.* **62**, 1806 (1975).
[e] T. Surles, H. H. Hyman, L. A. Quarterman, and A. I. Popov, *Inorg. Chem.* **10**, 611 (1971).
[f] G. M. Begun, W. H. Fletcher, and N. H. Smith, *J. Chem. Phys.* **42**, 2236 (1964).
[g] These frequencies have been estimated from band combinations.
[h] H. Selig and H. Holzman, *Isr. J. Chem.* **7**, 417 (1969).

for axial fluorine and of 500 ± 200 ppm for equatorial fluorine was found from an analysis of the NMR absorption line at 56 MHz.[127] The chlorine–fluorine scalar coupling constants have also been measured in ClF_5 by studying the ^{19}F- and ^{35}Cl-NMR line width.[128] The $Cl–F_{axial}$ coupling (<20 Hz) is much smaller than the $Cl–F_{equatorial}$ coupling (190 Hz), indicating a strong difference between the two types of chemical bonds.

NMR spectroscopy also gives some information about association, chemical exchange, and molecular motions. The difference of chemical shifts between gaseous and liquid ClF_3 allows a description of the association of the molecules in liquid ClF_3.[129] In BrF_3, ClF_3, and IF_5 the disappearance of the line structure indicates a chemical exchange between nonequivalent fluorine atoms.[115] The exchange, occurring at a rate of about 5×10^3 sec^{-1} at 273°K for ClF_3 and 10^3 sec^{-1} at 322°K for IF_5, is very sensitive to impurities.[130] This is to be related to the difficulties in purification mentioned in Section II,E. In BrF_5 and ClF_5, this exchange, if it exists, is much slower.[115,130] The existence of two different longitudinal relaxation times for nonequivalent fluorine atoms shows that the exchange rates are lower than 1 sec^{-1} at 200°K.[131] In IF_7 only one ^{19}F line is observed.[113] This is presumably due to the intramolecular exchange inferred from other measurements.[123,132] The ^{35}Cl-NMR observed in ClF_3 and ClF_5 allows a measurement of the molecular rotational speed in the liquid phase.[131] The rotational correlation times are plotted in Fig. 4. The activation energy of the rotational motion is higher in ClF_3 (2.05 kcal/mole) than in ClF_5 (1.25 kcal/mole), as is expected since strong associations are known to occur in ClF_3 but not in ClF_5.

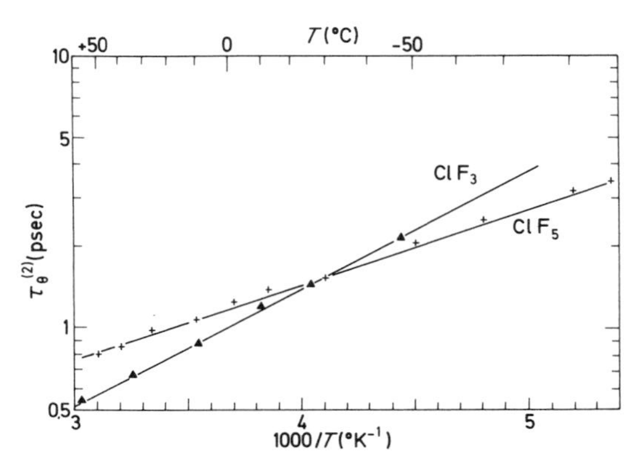

FIG. 4. Temperature dependence of the orientational correlation time $\tau_\theta^{(2)}$ in liquid ClF_3 and ClF_5 (from Ref. 131).

D. Recent Data on the Solid Phases of the Halogen Fluorides

Knowledge of the halogen fluorides in the solid phase has been recently completed in the structural and dynamical field by vibrational spectroscopy,[105,106] nuclear magnetic resonance,[127,133] and calorimetric measurements.[134] The crystallographic data are shown in Table XI. Two general features of the structure of these compounds in the solid phase are discussed: (1) the existence of phase transitions, and (2) the existence of fast molecular motions.

1. The Phase Transitions

For ClF_3, the transition suggested by Grisard,[135] based upon calorimetric measurements, have been characterized by Raman spectroscopy.[105] The values of the melting points and transition temperatures have been precisely determined by differential enthalpimetric analysis.[134] The Raman spectrum of the intermediate phase is consistent with the evolution of the band shapes in the liquid.[105] Thus, the preferential interactions in the solid phase are probably related to the associated states exhibited in the liquid. The structure of the low temperature phase has been known for quite some time.[116]

For ClF_5 three solid phases have been detected.[106] Solid I (181°–161°K) is a plastic solid phase where the reorientational motions have rates near those of the liquid. In fact, the Raman spectrum of solid I is practically identical to that of the liquid.

Solid II (161°–117°K) exhibits a vibrational spectrum comparable with that for solid BrF_5; the intermolecular interactions, however, are weaker than for bromine pentafluoride. Solid III ($T < 117°K$) does not seem to be very different from solid II from a structural point of view (this structure is not known). The parameters of the phases I and II have been determined by x-ray and neutron diffraction.[106]

For IF_7 the existence of a disordered cubic phase over 120°K was detected by Burbank.[117] This disorder is confirmed by the Raman spectrum which exhibits comparable and abnormally large band widths in the gas phase as well as in the liquid and the solid phases.[107] These band widths are correlated with the internal motion of "pseudorotation." The low temperature phase is said to be orthorhombic, but its structure is controversial.[136–139] The most recent measurements with NMR and NQR spectroscopy show a supplemental phase transition near 96°K.[140] For BrF_3 and BrF_5 no phase transition has been detected and the vibrational spectra[105,106] are in agreement with the crystallographical data.[117]

2. Dynamic Studies by Nuclear Magnetic Resonance Spectroscopy in the Solid State

In some halogen fluorides, measurements of the different relaxation times of the nuclear spins (longitudinal relaxation, relaxation in the rotating frame, and relaxation of the dipolar energy) have allowed the description of the molecular motions of characteristic frequency between 10^5 Hz and 10^{10} Hz. Low symmetry and strong intermolecular interactions prevent molecular reorientations in ClF_3 and BrF_3.

On the other hand, in ClF_5, several types of molecular motions have been found. Their characteristic times are shown in Fig. 5. The high temperature phase I is a plastic phase in which molecules are reorienting quasi-isotropically at a rate almost as fast as in the liquid. This easy rotation of the molecules in the solid phase shows the weakness of the intermolecular interactions. In phase I a fast diffusion of the molecules, probably by vacancy migration, also occurs.

In phases II and III the molecules still rotate, but only around their C_4 axes by 90° jumps. A discontinuity of the rotation rate at 117°K corresponds to the phase transition (S II/S III). In addition, the pure nuclear quadrupole relaxation time in phase II is due to another motion which is assumed to be infrequent jumps of the molecular axes between two close angular positions. An order of magnitude of the characteristics time, τ_{tilt}, of this tilting motion is given in Fig. 5.

BrF_5 and IF_5 have no plastic phase, but IF_7 has a very broad one. In the ordered phase of these three compounds, rotations of the molecules around their axes exist as in ClF_5. The existence of rotations in IF_5, which is highly associated in the liquid state, shows that molecular shape (C_4 axis) plays an important part in the possibilities of motion.

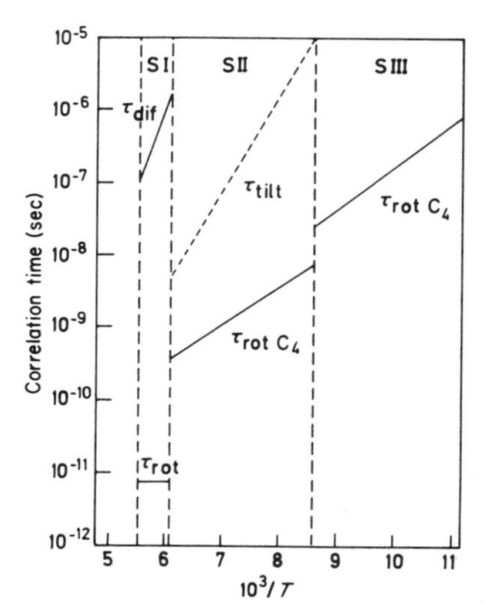

FIG. 5. Temperature dependence of the characteristic times describing the molecular motions in solid ClF_5: rotational (τ_{rot}) and diffusional (τ_{dif}) motion in the plastic phase; rotational motion (τ_{C_4}) around the C_4 axis and tilting motion (τ_{tilt}) in the ordered phases.

TABLE XI

SOME STRUCTURAL DATA ON HALOGEN FLUORIDES IN THE SOLID STATE

Compound	Transition	Phase	Transition	Phase	Transition	Phase
ClF_3	$T = 197.5°K$ $S = 9.01$ e.u.	S I	$T = 192.7°K$ $S = 1.85$ e.u.[b]	S II D_{2h}^{16} ($a = 8.825, b = 6.09, c = 4.52$)[a]		
BrF_3	$T = 282°K$[c]	C_{2v}^{12} ($a = 5.34, b = 7.35, c = 6.61$)[d]				
ClF_5	$T = 181°K$ $S = 2.12$ e.u.[b]	S I plastic phase cubic $a = 5.7$[e]	$T = 161°K$ $S = 5.20$ e.u.	S II orthorhombic ($a = 6.17, b = 7.22, c = 7.66$)	$T = 117°K$ $S = 0.18$ e.u.	S III

BrF_5

C_{2v}^{12} ($a = 6.442$, $b = 7.245$, $c = 7.846$)[d]

$T = 217.5$°K
$S = 11.7$ e.u.[b]

IF_5

C_{2h}^{6} ($a = 15.07$, $b = 6.836$, $c = 18.24$, $b = 92.96$)[f]

$T = 275.5$°K
$S = 11.7$ e.u.[b]

IF_7

	S I		S II
	plastic phase cubic $a = 6.28$[d]		orthorhombic $a = 8.74$, $b = 8.87$, $c = 6.14$)[i]
$T = 278$°K[g,h]		$T = 153$°K	$T = 96$°K[j]

[a] R. D. Burbank and F. N. Bensey, *J. Chem. Phys.* **21,** 602 (1953).
[b] P. Barberi, *Eur. Symp. Fluorine Chem., 5th, 1974,* p. 15.
[c] G. D. Oliver and J. W. Grisard, *J. Am. Chem. Soc.* **74,** 2705 (1952).
[d] R. D. Burbank and F. N. Bensey, *J. Chem. Phys.* **27,** 382 (1957).
[e] R. Rousson and J. M. Weulersse, to be published.
[f] R. D. Burbank and G. R. Jones, *Inorg. Chem.* **13,** 1071 (1974).
[g] L. Stein, *in* "Halogen Chemistry" (V. Gutmann, ed.), Vol. 1, p. 133. Academic Press, New York, 1967.
[h] Sublimation temperature.
[i] R. D. Burbank, *Acta Cryst.* **15**, 1207 (1962).
[j] J. M. Weulersse, to be published.

The data obtained by these different physical methods in the solid phases of halogen fluorides are consistent. The compounds known to be associated in the liquid phase, such as BrF_3, ClF_3, and IF_5, exhibit little or no molecular motion in the solid phase. On the other hand, ClF_5 and IF_7 have weak intermolecular interactions and a great reorientation freedom which allow the existence of plastic phases with a weak entropy variation at the melting point. BrF_5 is an intermediate case between the two preceding ones.

In conclusion, although the intermolecular forces are different, the symmetries of the intermolecular interactions in the condensed phases seem to be the same (particularly in liquid ClF_3, solid I in ClF_3, liquid and solid BrF_5, and solid BrF_3).

The symmetry of these interactions can be interpreted as association by fluorine bridges, which may be formed by the approach of one "equatorial" fluorine of one molecule to the central atom of another molecule.[101,105,129]

Acknowledgments

We are grateful indeed to Dr. W. W. Wilson for his helpful advice in the wording of the final draft.

References

1. A. G. Sharpe, *in* "Non-Aqueous Solvent Systems" (T. C. Waddington, ed.), p. 285. Academic Press, New York, 1965.
2. L. Stein, *in* "Halogen Chemistry" (V. Gutmann, ed.), p. 133. Academic Press, New York, 1967.
3. A. I. Popov, *in* "Halogen Chemistry" (V. Gutmann, ed.), Vol. 1, p. 225. Academic Press, New York, 1967; *Inorg. Chem., Ser. One* **3**, 53 (1972).
4. A. J. Downs and C. J. Adams, *Compr. Inorg. Chem.* **2**, 1476 (1973).
5. T. A. O'Donnell, *Compr. Inorg. Chem.* **2**, 1054 (1973).
6. M. Schmeisser, P. Sartori, and D. Naumann, *Chem. Ber.* **103**, 590 and 880 (1970).
7. N. N. Greenwood, "Mellor's Comprehensive Treatise on Inorganic and Theoretical Chemistry," Vol. II, Suppl. I. Longmans, Green, New York, 1956.
   Part I. Longmans, Green, New York, 1956.
8. N. N. Greenwood, *Rev. Pure Appl. Chem.* **1**, 84 (1951).
9. H. Gutmann, H. Lewin, and B. Perlmutter-Hayman, *J. Phys. Chem.* **72**, 3671 (1968).
10. G. Brauer, ed., "Handbook of Preparative Inorganic Chemistry," 2nd ed., Vol. 1. Academic Press, New York, 1963.
11. R. E. Dodd and P. L. Robinson, *in* "Experimental Inorganic Chemistry," p. 223. Elsevier, Amsterdam, 1954.
12. "Inorganic Syntheses," Vol. 1, H. S. Booth, ed. Vol. 3, L. F. Audrieth, ed. Vol. 9, S. Y. Tyree, ed. McGraw-Hill, New York, 1939–1967.
13. "Gmelin's Handbuch der anorganischen Chemie," 8th ed., Syst. No. 7. (R. J. Meyer, ed.) Verlag Chemie, Weinheim, 1931; "Iod," Syst. No. 8. Verlag Chemie, Weinheim, 1933.
14. H. Meinert, *Z. Chem.* **7**, 41 (1967).
15. M. Schmeisser and E. Scharf, *Angew. Chem.* **72**, 324 (1960).
16. M. Schmeisser, W. Ludovici, D. Naumann, P. Sartori, and E. Scharf, *Chem. Ber.* **101**, 4214 (1968).
17. "Gmelin's Handbuch der anorganischen Chemie," 8th ed., Syst. No. 6, Part B, No. 2. (R. J. Meyer, ed.) Verlag Chemie, Weinheim, 1969.

18. R. Gatti, R. L. Krieger, J. E. Sicre, and H. J. Schumacher, *J. Inorg. Nucl. Chem.* **28**, 655 (1966); R. L. Krieger, R. Gatti, and H. J. Schumacher, *Z. Phys. Chem.* **51**, 240 (1966).
19. H. H. Rogers, S. Evans, and J. H. Johnson, *J. Electrochem. Soc.* **116**, 601 (1969); H. H. Rogers, R. Keller, and J. H. Johnson, *ibid.* p. 604.
20. D. Pilipovich, W. Maya, E. A. Lawton, H. F. Bauer, D. F. Sheehan, N. N. Ogimachi, R. D. Wilson, F. C. Gunderloy, Jr., and V. E. Bedwell, *Inorg. Chem.* **6**, 1918 (1967).
21. G. A. Hyde and M. M. Boudakian, *Inorg. Chem.* **7**, 2648 (1968).
22. H. Selig, C. W. Williams, and G. J. Moody, *J. Phys. Chem.* **71**, 2739 (1967); H. H. Claassen, E. L. Gasner, and H. Selig, *J. Chem. Phys.* **49**, 1803 (1968).
23. N. Bartlett and L. E. Levchuk, *Proc. Chem. Soc., London* p. 342 (1963).
24. H. Schmidt and H. Meinert, *Angew. Chem.* **72**, 109 (1960).
25. D. Martin and G. Tantot, *J. Fluorine Chem.* **6**, 477 (1975).
26. H. C. Hodge and F. A. Smith, *in* "Fluorine Chemistry" (J. H. Simons, ed.), Vol. IV. Academic Press, New York and London, 1965.
27. L. M. Vincent and J. Gillardeau, *Commis. Energ. At.* [*Fr.*], *Rapp.* **CEA-R-2360** (1963).
28. D. W. Osborne, F. Schreiner, and H. Selig, *J. Chem. Phys.* **54**, 3790 (1971).
29. L. A. Quarterman, H. H. Hyman, and J. J. Katz, *J. Phys. Chem.* **61**, 912 (1957).
30. O. Ruff and A. Braida, *Z. Anorg. Chem.* **214**, 81 (1933).
31. D. A. Johnson, *in* "Some Thermodynamic Aspects of Inorganic Chemistry," p. 53. Cambridge Univ. Press, London and New York, 1968.
32. R. Bougon, *Commis. Energ. At.* [*Fr.*], *Rapp.* **CEA-R-3924** (1970).
33. J. A. Blauer, H. G. McMath, and F. C. Jaye, *J. Phys. Chem.* **73**, 2683 (1969); G. Mamantov, E. J. Vasini, M. C. Moulton, D. G. Vickroy, and T. Maekawa, *J. Chem. Phys.* **54**, 3419 (1971).
34. F. W. Bennett and A. G. Sharpe, *J. Chem. Soc.* p. 1383 (1950).
35. V. Gutmann, *Z. Anorg. Allg. Chem.* **264**, 169 (1951); *Monatsh. Chem.* **82**, 280 (1951).
36. F. Nyman and H. L. Roberts, *J. Chem. Soc.* p. 3180 (1962).
37. J. J. Pitts and A. W. Jache, *Inorg. Chem.* **7**, 1661 (1968).
38. R. Veyre, M. Quenault, and C. Eyraud, *C. R. Hebd. Seances Acad. Sci., Ser. C* **268**, 1480 (1969); C. J. Schack and W. Maya, *J. Am. Chem. Soc.* **91**, 2902 (1969); D. E. Gould, L. R. Anderson, D. E. Young, and W. B. Fox, *ibid.* p. 1310; C. J. Schack, R. D. Wilson, J. S. Muirhead, and S. N. Cohz, *ibid.* p. 2907; D. E. Young, L. R. Anderson, and W. B. Fox, *Inorg. Chem.* **9**, 2602 (1970).
39. L. Stein, *J. Am. Chem. Soc.* **91**, 5396 (1969); *Yale Sci.* **44**, 2 (1970); *Science* **168**, 362 (1970).
40. N. Bartlett, unpublished results.
41. C. J. Schack and R. D. Wilson, *Inorg. Chem.* **9**, 311 (1970).
42. L. D. Hall and J. F. Manville, *Can. J. Chem.* **47**, 361 and 379 (1969).
43. W. K. R. Musgrave, *Adv. Fluorine Chem.* **1**, 1 (1960).
44. R. D. Chambers, W. K. R. Musgrave, and J. Savory, *Proc. Chem. Soc., London* p. 113 (1961); *J. Chem. Soc.* p. 3779 (1961); P. Sartori and A. J. Lehnen, *Chem. Ber.* **104**, 2813 (1971).
45. G. Tantot and R. Bougon, *C. R. Hebd. Seances Acad. Sci., Ser. C* **281**, 271 (1975).
45a. R. Bougon, Tue Bui Huy, P. Chaipin, and G. Tantot, *C. R. Hebd. Séances Acad. Sc. Ser. C* **283**, 71 (1976).
46. A. G. Sharpe, *Q. Rev., Chem. Soc.* **4**, 115 (1950).
47. D. Pilipovich and C. J. Schack, *Inorg. Chem.* **7**, 386 (1968).
48. F. Q. Roberto, *Inorg. Nucl. Chem. Lett.* **8**, 737 (1972).
48a. K. O. Christe, *Inorg. Nucl. Chem. Lett.* **8**, 741 (1972).
49. R. J. Gillespie and G. J. Schrobilgen, *J. Chem. Soc., Chem. Commun.* p. 90 (1974).
50. R. J. Gillespie and G. J. Schrobilgen, *Inorg. Chem.* **13**, 1230 (1974).
51. K. O. Christe and R. D. Wilson, *Inorg. Chem.* **14**, 694 (1975).
52. P. Delvalle, *Bull. Soc. Chim. Fr.* p. 1611 (1963).

53. H. H. Hyman, T. Surles, L. A. Quarterman, and A. I. Popov, *Appl. Spectrosc.* **24**, 464 (1970).
54. R. Didchenko and R. H. Toenis Koetter, NASA Accession No. N64-22126, Rep. No. AD433910 (1964).
55. H. Meinert and U. Gross, *Z. Chem.* **12**, 150 (1972).
56. D. Martin and J. Clement, *Rev. Chim. Miner.* **10**, 621 (1973).
57. A. M. Bond, T. A. O'Donnell, and A. B. Waugh, *J. Electroanal. Chem. Interfacial Electrochem.* **39**, 137 (1972).
58. J. Cornog and R. A. Karges, *J. Am. Chem. Soc.* **54**, 1882 (1932).
59. J. Cornog, R. A. Karges, and H. W. Horrabin, *Proc. Iowa Acad. Sci.* **39**, 159 (1932).
60. D. Martin, *C. R. Hebd. Seances Acad. Sci., Ser. C* **265**, 919 (1967).
61. J. Sheft, H. H. Hyman, and J. J. Katz, *J. Am. Chem. Soc.* **75**, 5221 (1953).
62. G. J. Schrobilgen, J. H. Holloway, P. Granger, and C. Brévard, *C. R. Hebd. Seances Acad. Sci. Ser. C* **282**, 519 (1976).
63. F. O. Sladky and N. Bartlett, *J. Chem. Soc. A* p. 2188 (1969).
64. G. R. Jones, R. D. Burbank, and N. Bartlett, *Inorg. Chem.* **9**, 2264 (1970).
64a. H. Meinert and U. Gross, *J. Fluorine Chem.* **2**, 381 (1972–1973).
65. R. J. Gillespie and M. J. Morton, *Q. Rev., Chem. Soc.* **25**, 553 (1971).
66. W. W. Wilson and F. Aubke, *Inorg. Chem.* **13**, 326 (1974).
67. W. W. Wilson, J. R. Dalziel, and F. Aubke, *J. Inorg. Nucl. Chem.* **37**, 665 (1975).
68. J. Shamir and M. Lustig, *Inorg. Chem.* **12**, 1108 (1973).
69. D. J. Merryman and J. D. Corbett, *Inorg. Chem.* **13**, 1258 (1974).
70. D. J. Merryman, J. D. Corbett, and P. A. Edwards, *Inorg. Chem.* **14**, 428 (1975).
71. Ya. A. Fialkov and K. Ya. Kaganskaya, *J. Gen. Chem. USSR (Engl. Transl.)* **18**, 289 (1948).
72. Ya. A. Fialkov and I. D. Muzyka, *J. Gen. Chem. USSR (Engl. Transl.)* **18**, 802 (1948); **19**, 1416 (1949); **20**, 385 (1950).
73. Ya. A. Fialkov and O. I. Shor, *J. Gen. Chem. USSR (Engl. Transl.)* **19**, 1787 (1949).
74. N. N. Greenwood, *Rev. Pure Appl. Chem.* **1**, 84 (1951).
75. A. A. Woolfe and H. J. Emeléus, *J. Chem. Soc.* p. 2865 (1949).
76. A. A. Banks, H. J. Emeléus, and A. A. Woolfe, *J. Chem. Soc.* p. 2861 (1949).
77. A. J. Edwards and G. R. Jones, *J. Chem. Soc. A* p. 1936 (1969).
78. A. J. Edwards and G. R. Jones, *Chem. Commun.* p. 1304 (1967); *J. Chem. Soc. A* p. 1467 (1969).
79. D. H. Brown, K. R. Dixon, and D. W. A. Sharp, *Chem. Commun.* p. 654 (1966).
80. A. J. Edwards and K. O. Christe, *J. Chem. Soc., Dalton Trans.* p. 175 (1976).
81. P. Bouy, Doctorate Thesis No. 4, p. 183. University of Paris (1959).
82. A. Chrétien and D. Martin, *C. R. Hebd. Seances Acad. Sci., Ser. C* **263**, 235 (1966).
83. D. Martin, *Rev. Chim. Miner.* **4**, 367 (1967).
84. D. Martin, *C. R. Hebd. Seances Acad. Sci., Ser. C* **268**, 1145 (1969).
85. G. W. Richards and A. A. Woolfe, *J. Chem. Soc. A* p. 1072 (1969).
86. T. Surles, Ph.D. thesis, Michigan State University, East Lansing (1970).
87. L. Stein, *J. Am. Chem. Soc.* **81**, 1269 (1959).
88. A. A. Woolfe, *Adv. Inorg. Chem. Radiochem.* **9**, 267–275 (1966).
89. H. Meinert, U. Gross, and A. R. Grimmer, *Z. Chem.* **10**, 226 (1970); U. Gross and H. Meinert, *ibid.* p. 441.
90. T. Surles, A. Perkins, L. A. Quarterman, H. H. Hyman, and A. I. Popov, *J. Inorg. Nucl. Chem.* **34**, 3561 (1972).
91. M. D. Lind and K. O. Christe, *Inorg. Chem.* **11**, 608 (1972).
92. R. Bougon, P. Charpin, and J. Soriano, *C. R. Hebd. Seances Acad. Sci., Ser. C* **272**, 565 (1971).
93. K. O. Christe and W. Sawodny, *Inorg. Chem.* **12**, 2879 (1973).

94. R. C. Lord, M. A. Lynch, W. C. Schumb, and E. J. Slowinski, *J. Am. Chem. Soc.* **72**, 522 (1950).
95. C. V. Stephenson and E. A. Jones, *J. Chem. Phys.* **20**, 1830 (1952).
96. G. M. Begun, W. H. Fletcher, and N. H. Smith, *J. Chem. Phys.* **42**, 2236 (1964).
97. R. J. Gillespie and H. J. Clase, *J. Chem. Phys.* **47**, 1071 (1967).
98. H. Selig and H. Holzman, *Isr. J. Chem.* **7**, 417 (1969).
99. H. Selig, H. H. Claassen, and J. H. Holloway, *J. Chem. Phys.* **52**, 3517 (1970).
100. M. Drifford, D. Martin, and R. Bougon, *Rev. Chim. Miner.* **7**, 1069 (1970).
101. R. Frey, R. L. Redington, and A. L. K. Aljibury, *J. Chem. Phys.* **54**, 344 (1971).
102. K. O. Christe, E. C. Curtis, and D. Pilipovich, *Spectrochim. Acta. Part A* **27**, 931 (1971).
103. T. Surles, H. H. Hyman, L. A. Quarterman, and A. I. Popov, *Inorg. Chem.* **10**, 611 (1971).
104. K. O. Christe, *Spectrochim. Acta, Part A* **27**, 631 (1971).
105. R. Rousson and M. Drifford, *J. Chem. Phys.* **62**, 1806 (1975).
106. R. Rousson and J. M. Weulersse, to be published.
107. R. Rousson and M. Drifford, unpublished results.
108. D. F. Smith, *J. Chem. Phys.* **21**, 609 (1953).
109. R. H. Bradley and M. J. Whittle, *Chem. Phys. Lett.* **11**, 192 (1971).
110. A. G. Robiette, R. H. Bradley, and P. N. Brier, *Chem. Commun.* No. 23, p. 1567 (1971).
111. R. H. Bradley and P. N. Brier, *J. Mol. Spectrosc.* **44**, 536 (1972).
112. R. Jurek, P. Suzeau, J. Chanussot, and J. P. Champion, *J. Phys.* (*Paris*) **35**, 533 (1974).
113. H. S. Gutowsky and C. J. Hoffman, *J. Chem. Phys.* **19**, 1959 (1951).
114. H. S. Gutowsky, D. M. McCall, and C. P. Slichter, *J. Chem. Phys.* **21**, 279 (1953).
115. E. L. Muetterties and W. D. Phillips, *J. Am. Chem. Soc.* **79**, 322 (1957).
116. R. D. Burbank and F. N. Bensey, *J. Chem. Phys.* **21**, 602 (1953).
117. R. D. Burbank and F. N. Bensey, *J. Chem. Phys.* **27**, 982 (1957).
118. R. D. Burbank and G. R. Jones, *Inorg. Chem.* **13**, 1071 (1974).
119. R. J. Gillespie, *Can. J. Chem.* **38**, 818 (1960).
120. H. Bradford Thompson and C. J. Bartell, *Inorg. Chem.* **7**, 488 (1968).
121. W. E. Falconer, A. Büchler, J. L. Stauffer, and W. Klemperer, *J. Chem. Phys.* **48**, 312 (1968).
122. H. H. Claassen, E. L. Gasner, and H. Selig, *J. Chem. Phys.* **49**, 1803 (1968).
123. W. J. Adams, H. B. Thompson, and L. S. Bartell, *J. Chem. Phys.* **53**, 4040 (1970).
124. E. W. Kaiser, J. S. Muenter, W. Klemperer, and W. E. Falconer, *J. Chem. Phys.* **53**, 53 (1970).
125. M. Brownstein and H. Selig, *Inorg. Chem.* **11**, 656 (1972).
126. H. H. Eysel and K. Seppelt, *J. Chem. Phys.* **56**, 5081 (1972).
127. J. M. Weulersse, P. Rigny, and J. Virlet, *J. Chem. Phys.* **63**, 5190 (1975).
128. M. Alexandre and P. Rigny, *Can. J. Chem.* **52**, 3676 (1974).
129. L. G. Alexacos and C. D. Cornwell, *J. Chem. Phys.* **41**, 2098 (1964).
130. A. N. Hamer, *J. Inorg. Nucl. Chem.* **9**, 98 (1959).
131. M. Alexandre, Ph.D. Thesis, University of Orsay (1974); *Commis. Energ. At.* [*Fr.*], *Rep.* **CEA-R-4557** (1975).
132. R. D. Burbank and N. Bartlett, *Chem. Commun.* **11**, 645 (1962).
133. J. M. Weulersse, J. Virlet, and L. Guibe, *J. Chem. Phys.* **63**, 5201 (1975).
134. P. Barberi, *Eur. Symp. Fluorine Chem., 5th, 1974*.
135. J. W. Grisard, H. A. Bernhardt, and G. D. Oliver, *J. Am. Chem. Soc.* **72**, 5725 (1951).
136. J. Donohue, *J. Chem. Phys.* **30**, 1618 (1959).
137. R. D. Burbank, *J. Chem. Phys.* **30**, 1619 (1959).
138. R. D. Burbank, *Acta Crystallogr.* **15**, 1207 (1962).
139. J. Donohue, *Acta Crystallogr.* **18**, 1018 (1965).
140. J. M. Weulersse, P. Rigny, and J. Virlet, to be published.

~ 4 ~

Inorganic Halides and Oxyhalides as Solvents

RAM CHAND PAUL AND GURDEV SINGH

Department of Chemistry, Panjab University
Chandigarh-160014 (India)

I. Introduction

A large number of inorganic halides and oxyhalides have been examined as nonaqueous solvents. Notable among these are the antimony(III), arsenic(III), nitrosyl, phosphoryl, thiophosphoryl, and thionyl halides; sulfuryl and seleninyl chlorides; disulfur dichloride; and tin(IV) chloride. Most of these solvents are highly susceptible to hydrolysis and must be handled under strictly anhydrous conditions. The following account of the solvent properties of the inorganic halides and oxyhalides is confined to their possible autoionization; a description of purification methods; the qualitative or quantitative solubilities of inorganic and organic compounds; solvolytic reactions; conductivity measurements; conductometric, potentiometric, spectrophotometric, and visual titrations; and the isolation of solid complexes in these solvents.

Many reactions in halide and oxyhalide solvents have conveniently been described on the basis of the solvent system concept. However, it has been increasingly recognized that the solvent system concept, based on the mode of self-ionization of the solvent, has been rather over emphasized and an alternative coordination model has been suggested for the explanation of solute–solvent interactions. Both models have been used to give a full account of the properties of solutions in these solvents.

II. Antimony Halides

Numerous investigators[1–13] have employed fused antimony(III) chloride as a medium for conductometric, potentiometric, cryoscopic, and preparative work. It has also been suggested as a convenient solvent for the physicochemical study of interionic attractions[14–16] and as a dispersion medium for the study of infrared and nuclear magnetic resonance spectra.[17–22] Similar studies have been undertaken in fused antimony(III) bromide.[23–25] However, antimony(V) chloride has not proved to be a good solvent for inorganic compounds.[26,27]

A. Antimony(III) Chloride

Pure antimony(III) chloride is available as a commercial product and can be further purified by repeated distillation over antimony metal in a current of dry carbon dioxide or oxygen-free nitrogen.[11,14] Some of its important physical constants are listed in Table I.

Antimony(III) chloride has been shown to be a good solvent for a large number of inorganic and organic compounds.[11] Thus, potassium, rubidium, cesium, ammonium, thallium, tetraethylammonium, triphenylmethyl,

TABLE I

PHYSICAL CONSTANTS OF ANTIMONY(III) CHLORIDE[a]

Melting point	73.17°C[b]
Boiling point	219°–223°C[c]
Density	3.14 g/cm^3 at 20°C[d]
	2.681 g/cm^3 at 73.7°C[c,e]
	2.621 g/cm^3 at 100.3°C[c,e]
Viscosity	0.0416 dyn · sec · cm^{-2} at 75°C[c]
	0.0201 dyn · sec · cm^{-2} at 100°C[c]
Cryoscopic constant	15.6 ± 0.2°K · kg · $mole^{-1}$[b]
	14.7 ± 0.4°K · kg · $mole^{-1}$[f]
Dielectric constant	33.2 at 75°C[g]
Dipole moment	3.12–4.11 Debye[c]
Specific electrical conductivity	0.85 × 10^{-6} ohm^{-1} cm^{-1} at 95°[h]

[a] G. Jander and K. H. Swart, *Z. Anorg. Chem.* **299**, 252 (1959).
[b] G. B. Porter and E. C. Baughan, *J. Chem. Soc.* p. 744 (1958).
[c] Gmelin, "Handbuch der anorganischen Chemie," 8th ed., Vol. 18B. Verlag Chemie, Weinheim, 1949.
[d] E. Cohn and T. Strengers, *Z. Phys. Chem.* (*Leipzig*) **52**, 164 (1905).
[e] D. I. Zuravlev, *J. Phys. Chem.* **13**, 684 (1939).
[f] J. R. Atkinson, E. C. Baughan, and B. Dacre, *J. Chem. Soc. A* p. 1377 (1970).
[g] H. Schlundt, *J. Phys. Chem.* **5**, 512 (1901).
[h] Z. Klemensiewicz, *Bull. Int. Acad. Pol. Sci. Lett., Cl. Sci. Math. Nat.* p. 487 (1908).

mercury(II), aluminum(III), tellurium(IV), and selenium(IV) chlorides; potassium, mercury(II), and antimony(III) acetates; and tetramethylammonium chloride, sulfate, and chlorate have appreciable solubilities in antimony(III) chloride. Antimony(III) oxide and sulfide, arsenic(III) sulfide, and the nonmetals iodine and sulfur are also readily soluble. Lithium, sodium, tin(II), iron(III), and bismuth(III) chlorides are only slightly soluble. Oxides, sulfates, nitrates, carbonates, and chlorides of other metals are either insoluble or dissolve with decomposition.

Antimony(III) chloride forms adducts with a variety of compounds. A large number of addition compounds with aromatic hydrocarbons and other organic compounds containing oxygen, sulfur, and halogen (usually with an $SbCl_3$: organic compound ratio of 1 : 1 and 2 : 1) have been reported.[28–32] Nuclear quadrupole resonance[29,33] and infrared and Raman spectral studies[30,31,34] on some of these compounds have been undertaken. Antimony(III) chloride also forms adducts with amines and heterocyclic bases.[35–37] Among the adducts with inorganic compounds may be mentioned $SbCl_3 \cdot POCl_3$,[38] $SbCl_3 \cdot GaCl_3$,[39] $SbCl_3 \cdot SbOCl$,[11] $2PCl_5 \cdot 4SbCl_3$,[40] and numerous double and complex salts of alkali metal chlorides having the

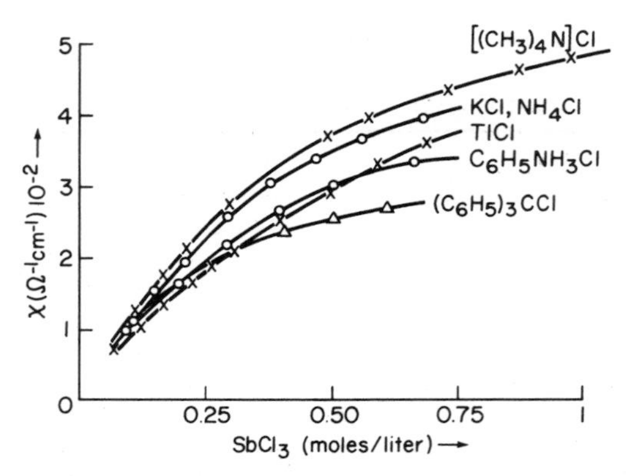

FIG. 1. Specific conductivity of bases in molten $SbCl_3$ at 99°C.

formulas R^ISbCl_4, $R_2{}^ISbCl_5$, $R^ISb_2Cl_7$, and $R_3{}^ISb_2Cl_9$ (where R^I = Li, Na, K, Rb, Cs, or NH_4); and $R^{II}SbCl_5$ and $R^{II}(SbCl_4)_2$ (where R^{II} = Be, Mg, Ca, Sr, or Ba).[28]

Specific conductance of antimony(III) chloride (0.85×10^{-6} ohm^{-1} cm^{-1} at 95°C) has been attributed to its dissociation[11] as,

$$2SbCl_3 \rightleftharpoons SbCl_2{}^+ + SbCl_4{}^- \tag{1}$$

Its conductance is considerably increased upon addition of chloride ion donors (base analogs) such as $(CH_3)_4NCl$, KCl, NH_4Cl, $(C_6H_5)_3CCl$, $C_6H_5NH_3Cl$, and TlCl (Fig. 1), and chloride ion acceptors (acid analogs) such as $AlCl_3$, $SeCl_4$, and $SbCl_2 \cdot ClO_4$.[11] The last is presumably present in solutions prepared by dissolving $AgClO_4$ in $SbCl_3$. The increase in conductance on the addition of $TeCl_4$, $SbCl_5$, $Sb(OOCCH_3)_3$, and $FeCl_3$ is comparatively much less. The change in conductance is presumably due to the formation of appropriate chloroanions in solution

$$AlCl_3 + SbCl_3 \rightleftharpoons [SbCl_2 \cdot AlCl_4] \rightleftharpoons SbCl_2{}^+ + AlCl_4{}^- \tag{2}$$

Cryoscopic data[11] over the concentration range 0.01–0.30 *M* show that $(C_6H_5)_3CCl$, $(CH_3)_4NCl$, $C_6H_5NH_3Cl$, TlCl, KCl, and SbOCl are strong electrolytes in fused antimony(III) chloride with the Van't Hoff factor *i* approaching 2 with increasing dilution. The *i* factor for tellurium(IV) and selenium(IV) chlorides is approximately 1 against an expected value of 3 if the chloride ion transfer and dissociation as

$$TeCl_4 + 2SbCl_3 \rightleftharpoons (SbCl_2)_2TeCl_6 \rightleftharpoons 2SbCl_2{}^+ + TeCl_6{}^{2-} \tag{3}$$

is presumed. It would, therefore, be necessary to assume considerable association. With dilution, the *i* factor for Sb_2O_3 increases from less than 1 to more than 3. It is suggested that Sb_2O_3 is solvolyzed to give three molecules of SbOCl, which undergo association/dissociation processes in concentrated and dilute solutions, respectively.

$$Sb_2O_3 + 4SbCl_3 \rightleftharpoons 3(SbOClSbCl_3) \qquad (4)$$

Neutralization reactions in fused antimony(III) chloride solvents have been studied by conductometric and potentiometric titration and also by preparative methods.[12] These titrations have been carried out at 80°–100°C. Antimony(III) sulfate, aluminum(III), iron(III), and antimony(V) chlorides and $[SbCl_2]^+ClO_4^-$ (from $AgClO_4$ solution in $SbCl_3$) act as acids in this solvent, whereas tetramethylammonium chloride, triphenylmethyl chloride, potassium chloride, lead chloride, and antimony oxychloride act as bases. Tellurium(IV) and selenium(IV) chlorides are amphoteric and can be titrated against bases such as tetramethylammonium and potassium chlorides as well as against acids such as aluminum(III) chloride. A typical plot of a conductometric titration between tetramethylammonium chloride and antimony(III) sulfate (Fig. 2) shows two breaks in the base to acid molar ratio of 6 : 1 and 6 : 4. The first break has been attributed to the reaction

$$6[(CH_3)_4N]SbCl_4 + Sb_2(SO_4)_3 \rightleftharpoons 3[(CH_3)_4N]_2SO_4 + 8SbCl_3 \qquad (5)$$

Tetramethylammonium sulfate then reacts further with antimony(III) sulfate

$$3[(CH_3)_4N]_2SO_4 + 3Sb_2(SO_4)_3 \rightleftharpoons 6[(CH_3)_4N][Sb(SO_4)_2] \qquad (6)$$

giving an overall reaction

$$6[(CH_3)_4N]SbCl_4 + 4Sb_2(SO_4)_3 \rightleftharpoons 6[(CH_3)_4N][Sb(SO_4)_2] + 8SbCl_3 \qquad (7)$$

corresponding to the second break. In another conductometric titration between selenium(IV) chloride and aluminum(III) chloride, the curve shows breaks at the $AlCl_3 : SeCl_4$ molar ratio of 2 : 1 and 1 : 1 which appear to correspond to the reactions

$$2[SbCl_2]AlCl_4 + SeCl_4 \rightarrow [SeCl_2][AlCl_4]_2 + 2SbCl_3 \qquad (8)$$

$$[SeCl_2][AlCl_4]_2 + SeCl_4 \rightarrow 2[SeCl_3][AlCl_4] \qquad (9)$$

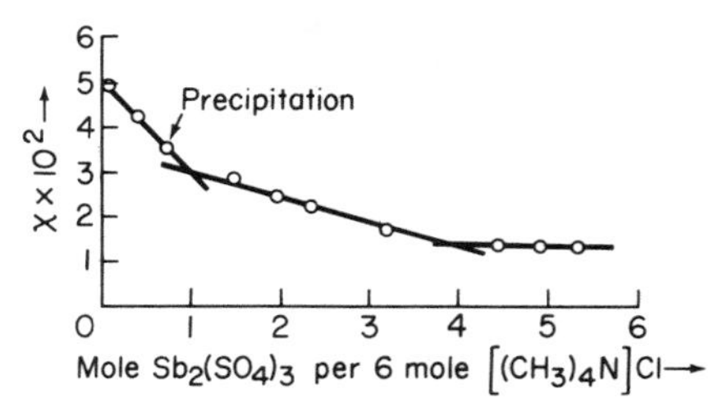

FIG. 2. Conductometric titration of 1.97g $(CH_3)_4NCl$ in 14.15 ml $SbCl_3$ with $Sb_2(SO_4)_3$ at 100°C.

Potentiometric titrations of aluminum(III) chloride with tetramethylammonium chloride and triphenylmethyl chloride using a gold reference electrode confirm the results obtained from conductometric titrations.[12] The compounds indicated by conductometric titration curves have been isolated by mixing appropriate amounts of acidic and basic compounds in molten $SbCl_3$ at 80°–100°C. The hot mixture was allowed to stand, and excess $SbCl_3$ was extracted with carbon disulfide.[12] Compounds such as $2[(CH_3)_4N]_2SO_4 \cdot 3SbCl_3$, $[(CH_3)_4N][Sb(SO_4)_2] \cdot SbCl_3$, $[(CH_3)_4N]ClO_4$, $KAlCl_4$, $KFeCl_4$, $KSbCl_6$, K_2SeCl_6, $K_2SeCl_6 \cdot SeCl_4$, K_2TeCl_6, $SbOAlCl_4$, $C_6H_5NH_3[AlCl_4]$, and $Pb[AlCl_4]_2$ have been prepared in this manner.

Solvolytic reactions in fused antimony(III) chloride have also been studied.[13] A large number of metal carbonates, acetates, sulfides, and oxides are converted to the corresponding metal chlorides

$$3M_2^{I}CO_3 + 3SbCl_3 \rightarrow 6M^{I}Cl + 3SbOCl + 3CO_2 \quad (10)$$

$$3M^{I}(CH_3COO) + SbCl_3 \rightarrow 3M^{I}Cl + Sb(CH_3COO)_3 \quad (11)$$

Cryoscopic measurements in antimony(III) chloride show that anthracene, fluorene, benzophenone, dibenzyl, water, and SbOCl are normal solutes in this solvent.[14,41] The cryoscopic constant has been found to have a value of 14.7 ± 0.4°K kg/mole. The Van't Hoff *i* factors for tetramethylammonium, triphenylmethyl, cesium, and potassium chlorides when plotted against molality (Fig. 3) show that tetramethylammonium and triphenylmethyl chlorides are strong 1 : 1 electrolytes in antimony(III) chloride while cesium and potassium chlorides are completely dissociated only at infinite dilution.[14] The general form of the curve in Fig. 3 agrees with the usual extension to the Debye–Hückel theory. Tetramethylammonium bromide has to be excluded as it has a limiting *i* factor of 2.2 which may be the result of partial halogen exchange

$$(CH_3)_4NBr + SbCl_3 \rightleftharpoons (CH_3)_4N^+ + Cl^- + SbCl_2Br. \quad (12)$$

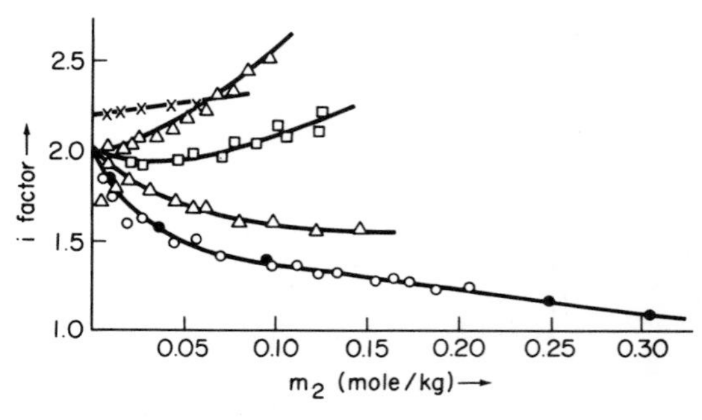

FIG. 3. i-Factors for electrolytes. × = Me_4NBr, △ = Ph_3CCl, □ = Me_4NCl, ▽ = CsCl, ○ = KCl, ● = KCl (see Ref. 2).

The osmotic coefficients for these electrolytes can be interpreted in terms of solvation and ion-pair formation by comparison with the Debye–Hückel equation. Linear Kohlrausch plots, which can be compared with theoretical Debye–Hückel–Onsagar slopes,[14] are obtained from the conductivity data of Klemensiewicz and co-workers[6-10] for the chlorides and bromides of thallium(I), potassium and ammonium, and rubidium chloride in molten antimony(III) chloride. The limiting conductivities of all these halides are almost identical and suggest that both chloride and bromide ions have abnormally high mobilities. The conductance data for tetramethylammonium and triphenylmethyl chloride solutions in molten antimony(III) chloride[15] also show that in both these solutions the Cl^- ion exists as the singly charged species (presumably $SbCl_4^-$) which is highly mobile [the transport number of $(C_6H_5)_3CCl$ at 75°C is 0.85–0.90]. The conductivities of the solutions of bornyl, *n*-decyl, 1-ethylcyclohexyl, cyclohexyl, diphenylmethyl, cinnamyl, and benzyl chlorides in molten antimony(III) chloride at 75°C indicate that these halides ionize principally as

$$2RCl \rightleftharpoons R_2Cl^+ + Cl^- \tag{13}$$

although in dilute solutions the simple ionization

$$RCl \rightleftharpoons R^+ + Cl^- \tag{14}$$

can also be detected.

Baughan *et al.*[16] have shown fused antimony(III) chloride to be a good solvent for investigating the formation of positive ions of the aromatic hydrocarbons.* When perylene, naphthacene, and pentacene are dissolved in freshly sublimed antimony(III) chloride at 75°C, highly colored solutions are formed which give well-resolved electron spin resonance spectra. Later work by Baughan and co-workers[42,43] shows that such cations are not formed if the solvent is completely free from oxygen or antimony(V) chlorides. In pure liquid $SbCl_3$ the aromatic hydrocarbon perylene behaves as a 1 : 1 weak electrolyte. This is attributed to the reaction

$$Pn + 2SbCl_3 \rightleftharpoons [Pn \rightarrow SbCl_2]^+ + SbCl_4^- \tag{15}$$

with the hydrocarbon acting as a Lewis base to enhance the self-ionization of the solvent. The solution shows no ESR signals and is greenish when dilute and yellow when more concentrated. On addition of oxygen or antimony(V) chloride, one $SbCl_4^-$ anion and one free radical cation are produced per molecule of perylene.

$$O_2 + 4Pn + 6SbCl_3 \rightarrow 4\mathring{P}\mathring{n}^+ + 4SbCl_4^- + 2SbOCl \tag{16}$$

$$2Pn + SbCl_5 + SbCl_3 \rightarrow 2\mathring{P}\mathring{n}^+ + 2SbCl_4^- \tag{17}$$

* See Chapter 5 of Volume IV of this treatise for a complete description.

Dry hydrogen chloride does not give conducting solutions in pure $SbCl_3$. Upon addition of perylene to this solution no ESR signals are obtained, the solutions are green and, therefore, do not contain free-radical cations. This is interpreted as

$$Pn + HCl + SbCl_3 \rightarrow PnH^+ + SbCl_4^- \quad (18)$$

However, when oxygen is admitted into the solution, the color immediately changes to purple and ESR signals become marked. Addition of water in equivalent proportions to perylene in pure antimony(III) chloride has no effect. The oxidation of perylene in molten $SbCl_3$ has also been studied by magnetic susceptibility measurements.[44] The results show that radical cation formation is complete at low concentrations but dimerization, polymerization, and formation of a chloroantimonate complex occurs at high concentrations.

Molten antimony(III) chloride has been employed for electrochemical studies by Texier.[45,46] Current–voltage studies indicate that the domain of electrochemical activity of the vitreous carbon electrode is dependent upon chloride ion concentration.[45] This system is a good indicator of chloride ion concentration except at very low concentration ($P_{Cl} > 7$). Gibbs free energies ΔG° have been determined experimentally for the processes of the type

$$3MCl_n(\text{dissolved}) + n\text{Sb(solid)} \rightleftharpoons n\text{SbCl}_3(\text{solvent}) + 3M(\text{solid}) \quad (19)$$

in fused antimony(III) chloride at 99°C.[47,48] From a comparison of these values with theoretical free energies, the solvation energies have been estimated for various metal chlorides in antimony(III) chloride. Such free energy assignments have been used to predict general trends of redox equilibria in antimony(III) chloride.

Molten antimony(III) chloride has been proved to be a good solvent for observing the fundamental stretching frequencies of the O–H, N–H, and C–H groups in the infrared region and as such has been used extensively for the study of a large number of biochemical compounds.[17–20] Sharp infrared and nuclear magnetic resonance spectra of polyester and polyamides dissolved in arsenic(III) chloride and antimony(III) chloride have also been obtained.[22] A new sampling technique, which utilizes solid antimony(III) chloride as the matrix in which the sample is dispersed or dissolved, has been described by Szymanski and co-workers.[21]

B. Antimony(III) Bromide

The solvent properties of antimony(III) bromide (mp 97°C, bp 280°C, dielectric constant 20.9 at 100°C) have been investigated by Jander and Weis[23,24] and Puente.[25] The type of investigations and the results obtained in this case are comparable to those obtained by Jander and Swart[11–13] for antimony(III) chloride. Several compounds that are readily soluble and also those that are only moderately soluble in molten antimony(III) bromide have been listed.[23] Crystalline solvates such as $3KBr \cdot 2SbBr_3$, $3NH_4Br \cdot 2SbBr_3$, $3RbBr \cdot 2SbBr_3$, $TlBr \cdot SbBr_3$, $(CH_3)_4NBr \cdot SbBr_3$, $KF \cdot SbBr_3$, and $[(CH_3)_4N]_2SO_4 \cdot 2SbBr_3$ have been isolated.[23] Metal oxides, carbonates, and acetates are solvolyzed in the solvent to give the corresponding bromides.

The specific conductivity of pure antimony(III) bromide (0.9–1.0×10^{-5} ohm^{-1} cm^{-1} at 100°C) increases when various acidic and basic substances are dissolved in it. The solvent system has been examined by preparative experiments, conductance measurements, and conductometric titrations. The Van't Hoff *i* factors of several salts have been determined.[23]

The hydrobromide of 2,4,6-collidine acts as a strong base in antimony(III) bromide, its solution in $SbBr_3$ reacts with boron(III) bromide, and the conductance of the solution falls to a constant value upon addition of excess of BBr_3. The formation of 2,4,6-collidinium bromoborate has been postulated.[25]

Determination of the viscosities, conductivities, and densities of the mixtures of antimony(III) bromide and aluminum bromide have shown the presence of maxima in the curves corresponding to the composition $SbBr_3 \cdot AlBr_3$.[49]

C. Antimony(III) Iodide

Few references are available on the use of molten antimony(III) iodide (mp 167°C, bp 401°C, dielectric constant 13.9)[50] as solvent. Metallic antimony dissolves in antimony(III) iodide to the extent of 3.5 mole % at 300°C and 5.8 mole % at 400°C.[51] The oxidation–reduction equilibrium in liquid SbI_3 containing "dissolved" antimony metal has been investigated in concentration cells of the type, $C \mid Sb_{(sat)}, SbI_{3(l)} \vdots Sb_{(soln)}, SbI_{3(l)} \mid C$. An electrode reaction involving two equivalent of reduction per mole of solute was observed.[52] This was found to be consistent with the formation of the catenated Sb_2I_4, analogous to P_2I_4 and As_2I_4. The assumption that ions derived from the solvent, presumably SbI_2^+ and SbI_4^-, carry the current within the cell is supported by the fact that addition of potassium iodide, even to the extent of 2%, to molten SbI_3 results in a hundredfold increase in the conductance of the solvent. Vapor pressure studies of solutions of antimony metal in liquid SbI_3 confirm the formation of Sb_2I_4 in this system.[53]

Alkylammonium salts of the anions SbI_4^-, SbI_5^{2-}, SbI_6^{3-}, and $Sb_2I_9^{3-}$ have been prepared by the reaction of aminehydroiodide salts with antimony(III) iodide.[54] The existence of the sulfur adduct $SbI_3 \cdot 3S_8$ has been confirmed by crystal structure determination.[55]

III. Arsenic Halides

The solvent properties of arsenic(III) halides, which were first studied by Walden[56,57] have been reinvestigated by Gutmann,[58-66] Lindqvist,[67-70] and Jander.[71-73] Arsenic(III) chloride has been more thoroughly investigated than any of the other arsenic halides.

A. Arsenic(III) Chloride

Arsenic(III) chloride is readily available and can be easily purified by distillation after allowing it to stand over sodium metal.[67] Some of its important physical constants, along with those of other arsenic(III) halides, are listed in Table II. It is an ampholytic solvent and can accept or donate chloride ions

$$AsCl_3 + Cl^- \rightleftharpoons AsCl_4^- \text{ or } AsCl_3 \rightleftharpoons AsCl_2^+ + Cl^- \quad (20)$$

Arsenic(III) chloride, therefore, has a characteristic autoionization as represented by,

$$2AsCl_3 \rightleftharpoons AsCl_2^+ + AsCl_4^- \quad (21)$$

A specific conductivity of 1.0×10^{-7} ohm^{-1} cm^{-1} at 19°C,[69] however, suggests that the extent of this self-ionization is very small. The existence of $AsCl_4^-$ ions in arsenic(III) chloride solutions has only been proved indirectly by conductometric studies, transport measurements,[65] and radiochlorine exchange studies,[74] while evidence for the existence of $AsCl_2^+$ ion is based on the interpretation of the data on conductometric titrations.[58,60,62,64]

With its wide liquid range (−16.2°–130.2°C) and a dielectric constant of 12.8 at 20°C, arsenic(III) chloride has proved to be a good solvent for a variety of substances.[58] Aluminum(III), titanium(IV), tin(IV), antimony(III and V), iron(III), vanadium(IV), and phosphorus(V) chlorides; mercury(II), cobalt(II), arsenic(III), potassium, and rubidium iodides, and the complex salts $(CH_3)_4N \cdot AsCl_4$, $(CH_3)_4N \cdot VCl_5$, and $(CH_3)_4N \cdot SbCl_6$ are readily soluble, whereas alkali metal, ammonium, niobium(V), and tantalum(V) chlorides and the complex salts $[(CH_3)_4N]_2SnCl_6$ and $(CH_3)_4N \cdot TiCl_5$ are only moderately soluble in arsenic(III) chloride. Metals are mostly insoluble while nonmetals such as phosphorus, sulfur, iodine, and bromine have a high solubility. Oxy salts such as permanganates, sulfates, nitrates, and carbonates are only sparingly soluble. Oxides and salts of some metals react with arsenic(III) chloride to give solvolyzed products. Gutmann[58] has listed the reaction products of some such solvolytic reactions with metal oxides and other salts.

Arsenic(III) chloride is a polar, pyramidal molecule which forms several

TABLE II

PHYSICAL CONSTANTS OF ANHYDROUS ARSENIC(III) CHLORIDE, BROMIDE, AND FLUORIDE[a]

	$AsCl_3$	$AsBr_3$	AsF_3
Melting point (°C)	−16.2°[b]	31.2[i]	−5.95[k]
Boiling point (°C)	130.2°[c]	221[j]	62.8[k]
Specific conductance (ohm^{-1} cm^{-1})	1×10^{-7} (19°C)[d]	1.6×10^{-7} (35°C)[j]	2.4×10^{-5} (25°C)[k]
Dielectric constant	12.8 (20°C)[e]	9.3 (35°C)[j]	5.7 (−6°C)[f]
Density (g/cm^3)	2.16 (20°C)[f]	3.33 (50°C)[f]	2.67 (0°C)[f]
Dipole moment	2.15 (benzene)[g] 3.11 (dioxane)[h]	1.6 (CS_2)[l]	

[a] For additional data, please see V. Gutmann, *Z. Anorg. Allg. Chem.* **266**, 331 (1951); G. Jander and K. Gunther, *ibid.* **297**, 81 (1958); J. C. Bailar, H. J. Emeléus, R. Nyholm, and A. F. Trotman-Dickenson, eds., "Comprehensive Inorganic Chemistry," Vol. 2, p. 589. Pergamon, Oxford, 1973.

[b] A. A. Woolf and N. N. Greenwood, *J. Chem. Soc.* p. 2200 (1950).

[c] W. Blitz and E. Meinecke, *Z. Anorg. Allg. Chem.* **131**, 1 (1933).

[d] I. Lindqvist and L. H. Andersson, *Acta Chem. Scand.* **9**, 79 (1955).

[e] P. Walden, *Z. Anorg. Chem.* **25**, 214 (1900).

[f] T. C. Waddington, ed., "Non-Aqueous Solvent Systems," p. 303. Academic Press, New York, 1965.

[g] J. W. Smith, *Proc. R. Soc. London, Ser. A* **136**, 256 (1932).

[h] P. A. McCusker and B. C. Curran, *J. Am. Chem. Soc.* **64**, 614 (1942).

[i] G. Brauer, ed., "Handbook of Preparative Inorganic Chemistry," 2nd ed., Vol. 1, pp. 596, 608, and 621. Academic Press, New York, 1963.

[j] Z. E. Jollès, "Bromine and its Compounds," p. 222ff. E. Benn, London, 1966.

[k] R. D. W. Kemmitt and D. W. A. Sharp, *Adv. Fluorine Chem.* **4**, 142 (1965).

[l] M. G. Malone and A. L. Ferguson, *J. Chem. Phys.* **2**, 99 (1934).

solid, crystalline solvates (Table III). These solvates dissolve in arsenic(III) chloride to give conducting solutions. For example, antimony(V) and tin(IV) chlorides form the solid solvates $SbCl_5 \cdot AsCl_3$ and $SnCl_4 \cdot 2AsCl_3$, respectively, which dissolve in $AsCl_3$ to give ions of the type MCl_6^-, MCl_6^{2-}, or MCl_5^- by accepting chloride ions. The dissociation constant (0.936×10^{-3}) of the solvoacid $SbCl_5 \cdot AsCl_3$, which ionizes to give $AsCl_2^+$ and $SbCl_6^-$ ions, shows that it is a stronger acid in arsenic(III) chloride than HF is in water.[64]

Tetraalkylammonium chlorides give highly conducting solutions in arsenic(III) chloride. Gutmann[58] obtained a compound of the composition $(CH_3)_4N \cdot AsCl_4$ by the evaporation of excess solvent from a solution of tetramethylammonium chloride in arsenic(III) chloride followed by warming to constant weight in a vacuum at 40°C. Through conductometric measurements he was able to show that the compound contained $AsCl_4^-$ ions. Lindqvist and Andersson,[67] however, obtained the solvate $(CH_3)_4N \cdot AsCl_4 \cdot 2AsCl_3$ under milder evaporation. The loss in weight after heating the solvate to 100°C at 9 mm mercury shows that two of the three $AsCl_3$ molecules are loosely bound and that $(CH_3)_4N \cdot AsCl_4$ is quite stable. In the corresponding tetraethylammonium chloride–arsenic(III) chloride system the existence of $(C_2H_5)_4NCl \cdot 2AsCl_3$ and $3(C_2H_5)_4NCl \cdot 5AsCl_3$ has been recognized.[70] Alkali metal chlorides which are only slightly soluble in arsenic(III) chloride give rise to compounds of the type $KAsCl_4$,[58] $Rb_3As_2Cl_9$, and $Cs_3As_2Cl_9$.[75,76]

Pyridine dissolves in arsenic(III) chloride to give conducting solutions and two solvates of the composition $2C_5H_5N \cdot AsCl_3$[77,78] and $C_5H_5N \cdot AsCl_3$[79,80] have been isolated. Conductance of both the solutions has been attributed to the production of Cl^- ions

$$C_5H_5N + AsCl_3 \rightleftharpoons C_5H_5N \cdot AsCl_3 \rightleftharpoons (C_5H_5N \cdot AsCl_2^+) + Cl^- \quad (22)$$

Dissociation constant of the solvobase $C_5H_5N \cdot AsCl_3$ (4.99×10^{-5}) makes it about three times as strong as aqueous ammonia.[64]

Acid–base neutralization reactions between tetramethylammonium chloride and antimony(V), tin(IV), titanium(IV), and vanadium(IV) chlorides have been studied conductometrically in arsenic(III) chloride.[58] Breaks in the titration curves correspond to the formation of compounds with a $(CH_3)_4NCl$: Lewis acid molar ratio of 1 : 1 with $SbCl_5$ and VCl_4 and ratios of 1 : 1 and 2 : 1 with $SnCl_4$ and $TiCl_4$. A plot of the conductometric titrations of tin(IV) chloride against tetramethylammonium chloride is given in Fig. 4. When the tetramethylammonium chloride solution in $AsCl_3$ is added to that of tin(IV) chloride in the same solvent, the conductance of the solution begins to rise due to the formation of a highly ionized 1 : 1 complex. At concentrations higher than the 1 : 1 molar ratio, a precipitate of

TABLE III

FORMATION OF SOLVATES IN ARSENIC(III) CHLORIDE

Substance	Solvate	Melting point (°C)
$(CH_3)_4NCl$	$(CH_3)_4NCl \cdot AsCl_3$[a]	—
$(CH_3)_4NCl$	$(CH_3)_4NCl \cdot 3AsCl_3$[b]	—
$(C_2H_5)_4NCl$	$3(C_2H_5)_4NCl \cdot 5AsCl_3$[c]	—
RbCl	$3RbCl \cdot 2AsCl_3$[d]	—
CsCl	$3CsCl \cdot 2AsCl_3$[d]	—
KCl	$KCl \cdot AsCl_3$[a]	—
$SbCl_5$	$SbCl_5 \cdot AsCl_3$[a]	—
$SiCl_4$	$SiCl_4 \cdot AsCl_3$[e]	−29.5
PCl_5	$2PCl_5 \cdot 5AsCl_3$[f]	40
NOCl	$2NOCl \cdot AsCl_3$[g]	—
$SnCl_4$	$SnCl_4 \cdot 2AsCl_3$[c]	—
$POCl_3$	$POCl_3 \cdot AsCl_3$[c]	−23.5
Dioxane	$(C_4H_8O_2)_3 \cdot 2AsCl_3$[h]	62
Benzene	$C_6H_6 \cdot 2AsCl_3$[i]	—
Pyridine	$C_5H_5N \cdot AsCl_3$[j]	145.5
	$2C_5H_5N \cdot AsCl_3$[k]	64
Diethylether	$(C_2H_5)_2O \cdot AsCl_3$[l]	—
Nitrobenzene	$2C_6H_5NO_2 \cdot AsCl_3$[m]	—
Acetone	$(CH_3)_2CO \cdot AsCl_3$[n]	—

[a] V. Gutmann, *Z. Anorg. Allg. Chem.* **266**, 331 (1951).

[b] I. Lindqvist and L. H. Andersson, *Acta Chem. Scand.* **8**, 128 (1954).

[c] M. Agerman, L. H. Andersson, I. Lindqvist, and M. Zackrisson, *Acta Chem. Scand.* **12**, 477 (1958).

[d] H. L. Wheeler, *Z. Anorg. Chem.* **4**, 452 (1893).

[e] H. H. Sisler, B. Pfahler, and W. J. Wilson, *J. Am. Chem. Soc.* **70**, 3825 (1948).

[f] L. Kolditz, *Z. Anorg. Allg. Chem.* **289**, 118 (1957).

[g] J. Lewis and D. B. Sowerby, *Recl. Trav. Chim. Pays-Bas* **75**, 615 (1956).

[h] G. O. Doak, *J. Am. Pharm. Assoc.* **23**, 541 (1934).

[i] M. P. Shulgina, *J. Gen. Chem. USSR* (*Engl. Transl.*) **4**, 225 (1934).

[j] B. P. Kondratenko, *J. Gen. Chem. USSR* (*Engl. Transl.*) **4**, 246 (1934); O. Dafert and Z. A. Melinski, *Ber. Dtsch. Chem. Ges. B* **59**, 788 (1926); C. S. Gibson, J. D. A. Johnson, and D. C. Vining, *J. Chem. Soc.* p. 1710 (1930).

[k] W. B. Shirey, *J. Am. Chem. Soc.* **52**, 1720 (1930).

[l] F. I. Terpugov, *J. Gen. Chem. USSR* (*Engl. Transl.*) **2**, 868 (1932).

[m] B. P. Kondratenko, *J. Gen. Chem. USSR* (*Engl. Transl.*) **4**, 244 (1934).

[n] M. Zackrisson and K. I. Alden, *Acta Chem. Scand.* **14**, 994 (1960).

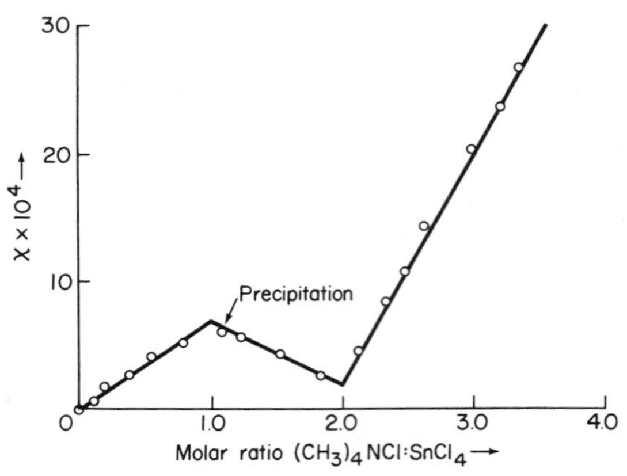

FIG. 4. Conductometric titration of a solution of $SnCl_4$ in $AsCl_3$ with $(CH_3)_4NCl$ at 20°C.

$[(CH_3)_4N]_2SnCl_6$ begins to separate and the conductance of the solution begins to fall. At the molar ratio 2 : 1, formation of the normal salt $[(CH_3)_4N]_2SnCl_6$ is complete and the increase in conductance thereafter is due to the addition of the quaternary ammonium salt solution. The breaks in the titration curve at 1 : 1 and 2 : 1 molar ratios have been interpreted in terms of the following equilibria:

$$(CH_3)_4N \cdot AsCl_4 + (AsCl_2)_2SnCl_6 \rightleftharpoons (CH_3)_4N \cdot AsCl_2 \cdot SnCl_6 + 2AsCl_3 \quad (23)$$

$$(CH_3)_4N \cdot AsCl_2 \cdot SnCl_6 + (CH_3)_4N \cdot AsCl_4 \rightleftharpoons [(CH_3)_4N]_2SnCl_6 + 2AsCl_3 \quad (24)$$

The acid salt $(CH_3)_4N \cdot AsCl_2 \cdot SnCl_6$ may be desolvated to give $(CH_3)_4N \cdot SnCl_5$. Both the complexes, $(CH_3)_4N \cdot SnCl_5$ and $[(CH_3)_4N]_2SnCl_6$, have been isolated.

When a concentrated solution of phosphorus(V) chloride in arsenic(III) chloride is evaporated, colorless, hygroscopic, prismatic crystals of $2PCl_5 \cdot 5AsCl_3$ (mp 40°C) are formed,[40] which give a conducting solution in arsenic(III) chloride. Cryoscopic and conductometric data indicate that in a 0.01 molal solution of the solvate in arsenic(III) chloride, the equilibrium

$$2AsCl_3 + PCl_4^+ + PCl_6^- \rightleftharpoons 2PCl_4^+ + 2AsCl_4^- \quad (25)$$

lies 70% to the left. The solvate should, therefore, be better formulated as a hexachlorophosphate(V) rather than a tetrachloroarsenate(III).

A number of tetrachlorophosphonium(V) salts can be conveniently prepared in arsenic(III) chloride. For example, $[PCl_4][PF_6]$, a white hygroscopic salt, can be prepared quantitatively by the addition of AsF_3 to a solution of P_2Cl_{10} in $AsCl_3$.[81] Thermal decomposition of $[PCl_4][PF_6]$ in a vacuum yields liquid homopolar PCl_4F (mp −63°C, bp 67°C), whereas heating it in an arsenic(III) chloride suspension gives a solution from which an ionic solid, PCl_4F (mp 177°C under pressure, sublimes at 175°C) is crystallized.[82] Reaction of phosphorus(III) chloride and bromine in arsenic(III) chloride gives $[PCl_4][PCl_5Br]$ as the product.[83] When chlorine is passed through a suspension of $AlCl_3$, $GaCl_3$, $TaCl_5$, $AuCl_3$, $FeCl_3$, PCl_5, or $SbCl_5$ in arsenic(III) chloride, an increase in conductivity of all these solutions is noticed.[84] This has been attributed to the formation of compounds such as $[AsCl_4]_n[MCl_{x+n}]$. Two compounds, $[AsCl_4][AlCl_4]$ and $[AsCl_4][GaCl_4]$, were isolated in solid state from their concentrated solutions.[84] Raman and infrared studies by Birkmann and Gerding[85] indicate that $PCl_5 \cdot AsCl_5$ is stable only under high chlorine pressure and exists in the form of $PCl_4^+AsCl_6^-$ rather than in the form $AsCl_4^+PCl_6^-$, as was suggested by Kolditz and Schmidt.[84]

Although tellurium(IV) chloride does not form any solid solvate with arsenic(III) chloride, it dissolves in the latter to give a conducting solution which can be conductometrically titrated against $(CH_3)_4NCl$ solution in $AsCl_3$.[60] Both the acid salt $[(CH_3)_4N][AsCl_2]TeCl_6$ and the normal salt $[(CH_3)_4N]_2TeCl_6$ have been isolated. Tellurium(IV) chloride reacts with phosphorus(V) chloride in arsenic(III) chloride solutions and a conductometric titration curve shows breaks corresponding to the formation of the compounds $2TeCl_4 \cdot PCl_5$, $TeCl_4 \cdot PCl_5$, and $TeCl_4 \cdot 2PCl_5$, all of which are soluble in $AsCl_3$ and can be isolated by evaporating the solvent.[62]

Tellurium(IV) chloride also reacts with tin(IV) chloride, antimony(V) chloride, and vanadium(IV) chloride in arsenic(III) chloride,[62] and conductometric titration curves show inflections corresponding to the formation of $2SnCl_4 \cdot TeCl_4$, $SnCl_4 \cdot TeCl_4$, $SnCl_4 \cdot 2TeCl_4$, $SbCl_5 \cdot TeCl_4$, $SbCl_5 \cdot 2TeCl_4$, $2VCl_4 \cdot TeCl_4$, $VCl_4 \cdot TeCl_4$, and $VCl_4 \cdot 2TeCl_4$. In all these cases tellurium(IV) chloride appears to function as a solvobase and reacts with solvoacids to form such neutralization complexes as

$$AsCl_2^+ \cdot SbCl_6^- + TeCl_3^+ \cdot AsCl_4^- \rightarrow TeCl_3^+ \cdot SbCl_6^- + 2AsCl_3 \tag{26}$$

Solutions of pyridine in arsenic(III) chloride can be titrated against solvoacids, vanadium(IV) chloride, and tin(IV) chloride.[64] The curves show inflections at acid to base molar ratio of 1 : 2 and solid compounds containing the cation $C_5H_5N \cdot AsCl_2^+$ and the anions VCl_6^{2-} and $SnCl_6^{2-}$ have been obtained.

The silver–silver chloride electrode behaves as a pCl electrode in arsenic(III) chloride[69] and changes in the chloride ion concentration during the titration of acids and bases in arsenic(III) chloride can be measured by using a simple silver–silver chloride concentration cell

$$\text{Ag, AgCl} \quad | \quad Cl^-(C_1) \quad \| \quad Cl^-(C_2) \quad | \quad \text{AgCl, Ag}$$

Using such an electrode, reactions of the acids, iron(III) chloride, and antimony(V) chloride, and the bases, pyridine, diethylamine, and tetramethylammonium chloride have been followed potentiometrically in arsenic(III) chloride. The reaction between $(CH_3)_4NCl$ and $SbCl_5$ (Fig. 5) can be represented as

$$Cl^- + SbCl_5 \rightleftharpoons SbCl_6^- \tag{27}$$

If it is assumed that the liquid junction potential can be neglected and that both antimony(V) chloride and tetramethylammonium chloride are strong electrolytes in arsenic(III) chloride, a maximum value for the ionic product $K = C_{AsCl_2^+} \cdot C_{AsCl_4^-}$ can be obtained from Fig. 5.

$$\tfrac{1}{2}\mathrm{p}K = \frac{460 - 215}{60} - \log C_{SbCl_5} \tag{28}$$

which gives p$K > 15$ or $[AsCl_2^+][AsCl_4^-] < 10^{-15}$.

Acid–base titrations using crystal violet[86] and benzanthrone[87] as internal indicators have been carried out by Paul *et al.* in arsenic(III) chloride solvents. Only the results given by basic titrants are subject to quantitative interpretation because the end point with acidic titrants cannot be precisely detected. Quinoline, pyridine, α-picoline, and dimethylaniline have been used as titrants against the acids tin(IV) chloride and titanium(IV) chloride. Crystal violet gives a dirty green color with the pure solvent $AsCl_3$, light orange with solutions of $SnCl_4$ and $TiCl_4$, violet with solutions of quinoline and α-picoline, and blue with solutions of dimethylaniline. The color at the end point is generally dirty green. These results are in good agreement with the calculated values. Benzanthrone gives an orange color with the solvent $AsCl_3$, pink-violet with solutions of $TiCl_4$, red with solutions of $SnCl_4$, and

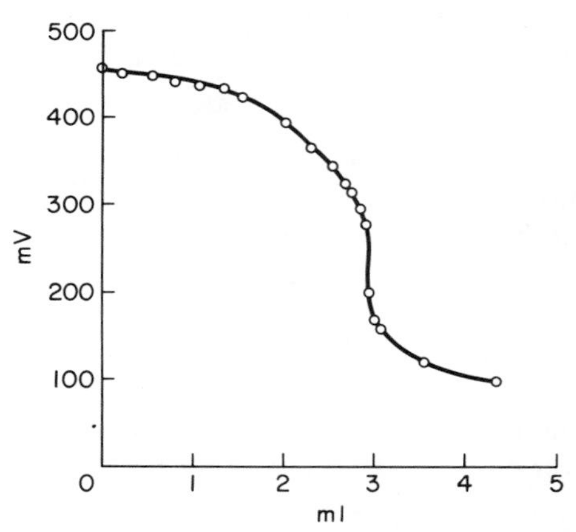

FIG. 5. Potentiometric titration curve of $SbCl_5$ with $(CH_3)_4NCl$.

light yellow with solutions of quinoline, α-picoline, and dimethylaniline. At the end point the color of the indicator is yellow in the case of $SnCl_4$ and red in the case of $TiCl_4$. Arsenic(III) chloride has also been suggested as a solvent for NMR spectroscopy.[88]

B. Arsenic(III) Bromide

The pioneer work on the use of fused arsenic(III) bromide as a solvent has been done by Jander and Gunther.[71–73] It can be easily prepared by the action of bromine on metallic arsenic and can be purified by distilling twice at atmospheric pressure, followed by distillation under vacuum.[71] Some of its important physical constants were listed in Table II.

Fused arsenic(III) bromide dissolves a considerable number of inorganic and organic compounds, especially the predominantly covalent compounds and salts with large cations. The qualitative solubilities of many organic and inorganic compounds have been reported by Jander[71] and other workers.[89–93]

Arsenic(III) bromide has a specific conductivity[71] of 1.6×10^{-7} ohm^{-1} cm^{-1} at 35°C. By analogy with other halide systems, the self-dissociation of the solvent has been proposed as

$$2AsBr_3 \rightleftharpoons AsBr_2^+ + AsBr_4^- \qquad (29)$$

Predominantly covalent bromides of aluminum(III), gallium(III), indium(III), boron(III), gold(III), tin(IV), mercury(II), bismuth(III), and tellurium(IV) act as electrolytes in fused $AsBr_3$ and behave as acids due to their ability to form complex bromoanions.[71] Solutions of substituted ammonium bromides, heterocyclic nitrogen bases, and amines in arsenic(III) bromide show high conductivity (Fig. 6) and thus act as bases in this solvosystem. Quaternary ammonium bromides and hydrobromides of heterocyclic nitrogen bases generally form solvates of the type $R_4NBr \cdot AsBr_3$ and $B \cdot HBr \cdot AsBr_3$. Pyridine forms two solvates, $C_5H_5N \cdot AsBr_3$ and $2C_5H_5N \cdot AsBr_3$, while 2,4-lutidine, dimethylaniline, and tri-*n*-butylamine give solvates of the type $2B \cdot 3AsBr_3$.[71,79] Jander[71] has listed a series of such solvates with quaternary ammonium bromides and other organic bases. Cryoscopic and conductance data show that both association and electrolytic dissociation occur both in acidic and basic solutions. However, most of the solutes are weak electrolytes. The electrolytic dissociation generally increases with dilution.

Neutralization reactions with the acids $AlBr_3$, $GaBr_3$, $InBr_3$, $BiBr_3$, $SnBr_4$, and $HgBr_2$, and the bases $(C_2H_5)_4NBr$, $(C_2H_5)_3NHBr$, $(C_2H_5)_2NH_2Br$, KBr, CsBr, CuBr, AgBr, TlBr, $(CH_3)_2NH$, $C_6H_5N(CH_3)_2$, and 2,4-lutidine have been followed both conductometrically and potentiometrically.[72] Some of these titration curves are presented in Fig. 7.

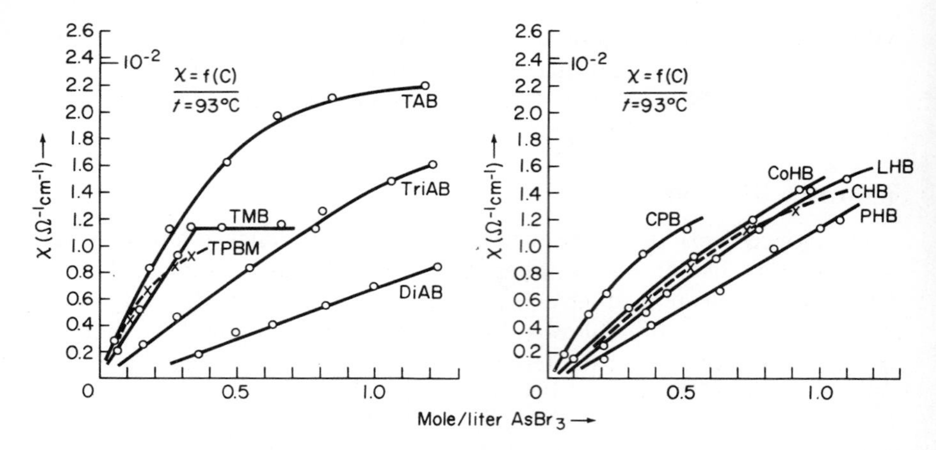

FIG. 6. Specific conductivity of bases in $AsBr_3$. TMB = $(CH_3)_4NBr$, TAB = Et_4NBr, TriAB = Et_3NHBr, DiAB = Et_2NH_2Br, CHB = quinaldine-hydrobromide, CPB = cetylpyridiniumbromide, LHB = lutidinehydrobromide, CoHB = collidinehydrobromide, PHB = pyridinehydrobromide, and TPBM = triphenylbromomethane.

Lead(II), tellurium(IV), and possibly zinc(II) bromides are amphoteric in fused $AsBr_3$ and react with both acids and bases. As a result of these neutralization reactions, compounds of the type $[(C_2H_5)_4N]AlBr_4$, $[(C_2H_5)_3NH]AlBr_4$, $Cs[GaBr_4]$, $[C_6H_5NH_3]InBr_4$, $Pb[AlBr_4]_2$, and $[TeBr_3]InBr_4$ have been isolated.[72]

Several anhydrous metal bromides can be prepared through solvolytic reactions in this solvent.[73] For example, metal oxides, and carbonates sol-

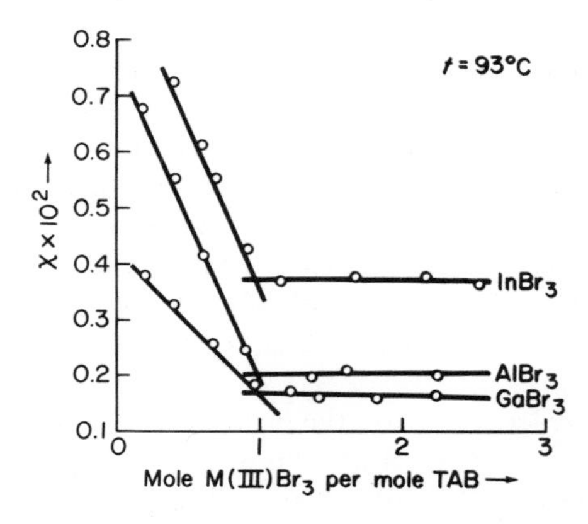

FIG. 7. Conductometric titration of Et_4NBr with $AlBr_3$, $GaBr_3$, and $InBr_3$ in $AsBr_3$ at 93°C (TAB = Et_4NBr).

volyze in $AsBr_3$ to give anhydrous metal bromides

$$3PbO + 2AsBr_3 \rightarrow 3PbBr_2 + As_2O_3 \tag{30}$$

$$3CdCO_3 + 2AsBr_3 \rightarrow 3CdBr_2 + As_2O_3 + 3CO_2 \tag{31}$$

Arsenic(III) oxide, sulfide and selenide react with $AsBr_3$ to give oxy-, thio-, and selenobromides

$$As_2S_3 + AsBr_3 \rightarrow 3AsSBr \tag{32}$$

C. Arsenic(III) Fluoride

As compared to other arsenic(III) halides, no systematic work has been done to study the solvent behavior of arsenic(III) fluoride. It can be prepared by the action of concentrated sulfuric acid on a well-dried, intimate mixture of arsenic(III) oxide and calcium fluoride. The product is purified by refluxing over sodium fluoride and fractionating.[94] Arsenic(III) fluoride is very rapidly hydrolyzed and is toxic, but can be easily and safely manipulated in a vacuum line. The pure solvent has a specific conductivity of 2.4×10^{-5} ohm^{-1} cm^{-1} at 25°C, which is largely increased by the addition of small amounts of potassium and antimony(V) fluorides.[94] From these solutions the solid compounds $KF \cdot AsF_3$ and $SbF_5 \cdot AsF_3$ have been isolated. Equimolar quantities of these two compounds react in arsenic(III) fluoride to give $KSbF_6$ which can be isolated by the removal of solvent under vacuum. The ionization of these two solvates and their neutralization in AsF_3 can be represented as

$$KF \cdot AsF_3 \rightleftharpoons K^+ + AsF_4^- \tag{33}$$

$$AsF_3 \cdot SbF_5 \rightleftharpoons AsF_2^+ + SbF_6^- \tag{34}$$

$$K^+ + AsF_4^- + AsF_2^- + SbF_6^- \rightleftharpoons KSbF_6 + 2AsF_3 \tag{35}$$

Addition of boron(III) fluoride to arsenic(III) fluoride increases the conductance of the latter but no solid compound has been isolated. However, when BF_3 is passed through a solution of potassium fluoride in AsF_3, potassium tetrafluoroborate is produced. In view of the above neutralization reactions, the self-ionization of arsenic(III) fluoride into AsF_2^+ and AsF_4^- has been postulated.[94]

The fluorides of rubidium, calcium, and thallium react exothermally with arsenic(III) fluoride to yield solid fluoroarsenites of the composition $RbAsF_4$, $CsAsF_4$, and $TlAsF_4$ which are extremely hygroscopic.[95] The ^{19}F magnetic resonance spectra of solutions of these fluoroarsenites consist of a single resonance, indicating a fluorine exchange process in this system. It was, therefore, not possible to establish whether F^- or AsF_4^- is the major conducting species in arsenic(III) fluoride system.[95] Although Woolf and Greenwood[94] have observed a slight increase in the conductance of arsenic fluoride solutions on the addition of boron(III) fluoride, yet NMR studies do not show any evidence for the interaction of these two compounds. The ^{19}F magnetic resonance spectrum of the arsenic(III) fluoride–antimony(V) fluoride system consists of a single resonance peak against two or more peaks expected in any compound formed between antimony and arsenic fluorides.[95] This can be attributed to a rapid fluorine exchange, possibly through a structure of the type

$$F_2As \begin{smallmatrix} F \\ \\ F \end{smallmatrix} SbF_4 \rightleftharpoons F_2As \begin{smallmatrix} F \\ \\ F \end{smallmatrix} SbF_4 \quad (36)$$

Refluxing antimony(V) chloride with arsenic(III) fluoride at 100°–110° for an hour results in partial fluorination of antimony(V) chloride and colorless ionic crystals of $SbCl_4F$ (mp 83°C with decomposition) are obtained after distilling off $AsCl_3$ and excess AsF_3.[96]

$$3SbCl_5 + AsF_3 \rightleftharpoons 3SbCl_4F + AsCl_3 \quad (37)$$

Cryoscopic and conductance measurements indicate that in arsenic(III) fluoride, $SbCl_4F$ dissociates into $SbCl_4^+$ and F^- ions. $SbCl_4F$ vigorously fluorinates PCl_3 giving PF_3 and $[PCl_4][SbCl_6]$.[96]

Crystalline compound $[AsCl_4][AsF_6]$ is slowly formed when a stream of chlorine is rapidly passed through AsF_3 at 0°C.[97] Dess *et al.*,[98] however, have reported that the reaction is dependent on the water content of the system and that tetrachloroarsenic(V) hexafluoroarsenate is obtained only if a small amount of water is present. In a completely anhydrous system no reaction occurs. Oxidation of arsenic(III) fluoride with nitrogen(III) chloride also yields the same compound.[98]

$$6AsF_3 + 4NCl_3 \xrightarrow{H_2O} 3[AsCl_4][AsF_6] + 2N_2 \quad (38)$$

Oxidation with bromine and iodine under similar conditions did not give the corresponding tetrabromo- and tetraiodoarsenic(V) complexes.[99] Kolditz and Schaefer[100] have reported several methods for the preparation of hexafluoroarsenates of the type $[SCl_3][AsF_6]$, $[SeCl_3][AsF_6]$, and $[TeCl_3][AsF_6]$.

IV. Nitrosyl and Nitryl Halides

A. Nitrosyl Chloride

Liquid nitrosyl chloride (dielectric constant 19.7 at −10°C, specific conductance 2.88×10^{-6} ohm^{-1} cm^{-1} at −20°C) has been found to be a fairly effective ionizing solvent in several instances. A number of methods for the preparation of nitrosyl chloride have been listed by Beckham *et al.*[101]

However, methods employing reactions between potassium chloride[102] or hydrogen chloride[103] with nitrosonium hydrogen sulfate and nitrogen dioxide with moist potassium chloride[104] have commonly been used by a number of workers. Complete instructions for the preparation of nitrosyl chloride from hydrogen chloride and nitrosonium hydrogen sulfate are available in "Inorganic Syntheses."[105] For precise electrical measurements, nitrosyl chloride prepared by this method was further purified, first by a rough fractional condensation and then by distillation through a helix-packed fractionating column, attached to a high-vacuum line, in a fairly dark room.[106]

Nitrosyl chloride has a nitrogen–oxygen bond distance of 1.14 ± 0.02 Å, a nitrogen–chlorine bond distance of 1.95 ± 0.01 Å and a bond angle of 116 ± 2°.[107,108] The exceptionally large nitrogen–chlorine bond length has been explained on the basis of resonance between the homopolar structure O=N—Cl and an ionic structure $[N{\equiv}O]^+Cl^-$. Some of the important physical constants of the solvent are given in Table IV.

The small specific conductivity of pure nitrosyl chloride has been attributed to its ionization as

$$NOCl \rightleftharpoons NO^+ + Cl^- \tag{39}$$

in which both the cation and anion are solvated. The nitrosyl ion should be solvated very strongly because of the resonance structures I, II, and III,

$$[:\ddot{O}{=}\ddot{N}{-}\ddot{Cl}{-}\ddot{N}{=}\ddot{O}:]^+, \quad [:\ddot{O}{=}\ddot{N}{-}\ddot{Cl}:]:N^+{\equiv}O:, \quad :O{\equiv}N:^+[:\ddot{Cl}{-}\ddot{N}{=}\ddot{O}:]$$

(I) **(II)** **(III)**

whereas the chloride ion should not be appreciably solvated because the

TABLE IV

PHYSICAL CONSTANTS OF NITROSYL CHLORIDE

Melting point (°C)	−61.5[a]
Boiling point (°C)	−5.8[a]
Density (g/ml)	1.373 at −10°C[b]
Dipole moment (Debye)	1.83[c]
Dielectric constant	19.7 at −10°C[d]
	21.4 at −19.5°C[d]
	22.5 at −27°C[d]
Specific conductance (ohm^{-1} cm^{-1})	2.73×10^{-6} at −10°C[e]
	2.88×10^{-6} at −20°C[f]

[a] M. Trautz and W. Gerwig, *Z. Anorg. Allg. Chem.* **134**, 409 (1924).
[b] E. Briner and Z. Pylkoff, *J. Chim. Phys.* **10**, 640 (1912).
[c] J. A. Ketelaar, *Recl. Trav. Chim. Pays-Bas* **62**, 289 (1943).
[d] A. B. Burg and D. E. McKenzie, *J. Am. Chem. Soc.* **74**, 3143 (1952).
[e] C. C. Addison and J. Lewis, *J. Chem. Soc.* p. 2843 (1951).
[f] A. B. Burg and G. W. Campbell, *J. Am. Chem. Soc.* **70**, 1964 (1948).

electronic structure of NOCl will not be affected by an electron donor as weak as a chloride ion.[109] In agreement with this position, potassium chloride is insoluble in nitrosyl chloride, whereas a number of nitrosonium salts such as $NOAlCl_4$, $NOFeCl_4$, and $NOSbCl_6$ are readily soluble because of the solvation of nitrosonium ions, and are strong electrolytes in nitrosyl chloride[109] (Table V). A study of the pressure composition isotherms shows the existence of further solvated compounds, $NOAlCl_4 \cdot NOCl$ and $NOFeCl_4 \cdot NOCl$, having dissociation pressures of 180 mm and 224 mm at 0°C, respectively.[106] The freezing-point diagram of the NOCl–$AlCl_3$ system also indicates the formation of $AlCl_3 \cdot 2NOCl$.[110] The magnetic susceptibility measurements on the iron compound[106] are in favor of its formulation as a solvate of the type $[NO \cdot NOCl]^+[FeCl_4]^-$. The solvated nitrosonium ion may contain either a chlorine bridge or more probably an oxygen bridge.[111] Compounds such as $(NO)_2SnCl_6$, $(NO)_2TiCl_6$, $NOHSO_4$, and $NOBF_4$, which appear to be quite insoluble and nonconducting,[109] have been shown to be unsolvated,[106] probably because the nitrosonium group is strongly attached to the metal resulting in a high stability of these compounds.

The conductance of solutions of $NOFeCl_4$ in nitrosyl chloride has been studied by Burg and McKenzie.[106] Equivalent conductance values at −10°C for different concentrations of $NOFeCl_4$ when fitted to the Shedlovsky equation gave a limiting equivalent conductance value λ_0 as 401.2 and ionization constant value K_{ion} as 3.73×10^{-3}. Assuming that the Debye–Hückel–Onsagar theory is applicable, values for the degree of dissociation for the complex $NO^+[FeCl_4]^-$ have been calculated (Table VI). The ions involved in it have been identified by carrying out electrolysis of solutions of $NOFeCl_4$ in nitrosylchloride, nitric oxide being evolved at the cathode and chlorine at the anode.[106] Tentative values of 0.12 and 0.88 have

TABLE V

CONDUCTANCE IN LIQUID NITROSYL CHLORIDE[a]

Solute	Molarity	Temperature (°C)	Specific conductance (mhos)	Molar conductance
$NOAlCl_4$	0.098	−20	1.17×10^{-2}	119
$NOFeCl_4$	0.0099	−20	1.34×10^{-3}	136
$NOFeCl_4$	0.0094	−21	1.26×10^{-3}	134
$NOFeCl_4$	0.0094	−44	1.00×10^{-3}	106
$NOSbCl_6$	0.140	−20	2.35×10^{-2}	168
$NOSbCl_6$	0.140	−44	2.20×10^{-2}	157
NOCl	pure	−20	2.88×10^{-6}	—

[a] A. B. Burg and G. W. Campbell, *J. Am. Chem. Soc.* **70**, 1964 (1948).

TABLE VI

DISSOCIATION OF $NOFeCl_4$ IN NOCl AT $-10°C$[a]

Concentration (mmole/liter)	0.328	0.481	0.704	1.53	2.19	4.65	6.13
Degree of dissociation	0.937	0.916	0.885	0.820	0.792	0.717	0.693

[a] A. B. Burg and D. E. McKenzie, *J. Am. Chem. Soc.* **74**, 3143 (1952).

been assigned for the transport numbers of the $FeCl_4^-$ and NO^+ ions respectively. Should this evidence of high mobility of NO^+ be confirmed, it could imply that some chain mechanism analogous to the transport of H^+ ions in water is operative in NOCl for the transport of NO^+.[106]

Reactions of nitrosyl chloride with inorganic compounds are generally of two types. In one it may react with metals, nonmetals, metal oxides, and metal salts to form the corresponding chlorides, and the remaining portions of the reacting molecules form one or more separate products, which really corresponds to solvolysis. Thus nitrosyl chloride reacts with potassium, silver, and cadmium to give the corresponding metal chlorides and nitric oxide.[101,112] Sulfur, selenium, and iodine react to give S_2Cl_2, Se_2Cl_2, and ICl along with nitric oxide,[112,113] and silver perchlorate, iodide, and sulfite react to give silver chloride and a variety of other products.[114,115]

$$AgClO_4 + NOCl \rightleftharpoons NOClO_4 + AgCl \quad (40)$$

$$2AgI + 2NOCl \rightleftharpoons 2NO + I_2 + 2AgCl \quad (41)$$

$$3Ag_2SO_3 + 6NOCl \rightleftharpoons 6AgCl + 4NO + SO_2 + S_2N_2O_9 \quad (42)$$

Silver salts of various phosphoric acids react with nitrosyl chloride to yield silver chloride and nitrosyl salts of condensed phosphoric acids,[116]

$$4Ag_3PO_4 + 12NOCl \rightleftharpoons (NO)_2P_4O_{11} + 5N_2O_3 + 12AgCl \quad (43)$$

$$3Ag_4P_2O_7 + 12NOCl \rightleftharpoons (NO)_4P_6O_{17} + 4N_2O_3 + 12AgCl \quad (44)$$

The second mode of reaction involves the direct formation of addition compounds of the type $MCl_n \cdot xNOCl$, between metal chloride and nitrosyl chloride. A large number of such addition compounds (Table VII) have been prepared by reacting either the metal or the metal chloride with nitrosyl chloride.

The existence of nitrosonium ions is well established by infrared studies[117–119] and most of the adducts listed above may indeed be complex chloroanion salts of the nitrosonium ion. A comparison of the Raman spectra of the compounds $NaCl \cdot AlCl_3$ and $NOCl \cdot AlCl_3$ indicates the presence of NO^+ and $AlCl_4^-$ ion in the latter adduct.[120,121] The force

TABLE VII

STABLE SOLVATES OF METAL HALIDES WITH NITROSYL CHLORIDE

Metal halides	Solvation number[a]	Metal halides	Solvation number[a]
$MgCl_2$[b]	1	$MnCl_2$[f,k]	1
BF_3[c,d]	1	—	—
BCl_3[e,f]	1	—	—
$AlCl_3$[b,g,h,i,j]	1, 2	$FeCl_3$[i,m,v]	1, 2
$GaCl_3$[f]	1	$PdCl_2$[f]	2
$InCl_3$[f]	1	—	—
$TlCl_3$[f]	1	$PtCl_4$[f,i]	2
$SnCl_4$[f,h,i,k,l,m]	2	$CuCl$[k,l,v]	1
SnF_4[n]	2	—	—
$PbCl_4$[i,k]	2	$AuCl_3$[f,l]	1
$AsCl_3$[c,o]	2	$ZnCl_2$[f,k,l]	1
$SbCl_3$[c]	1	$HgCl_2$[f,i]	1
$SbCl_5$[h,i,p,q]	1	—	—
SbF_5[c]	1	$ThCl_4$[s]	2
$BiCl_3$[d,e,f,i]	1	UO_2Cl_2[w]	1
$TiCl_4$[f,i]	2	VCl_4[x]	2
$TiCl_3(R*COO)$[r]	1	$TaCl_5$[u]	1
$ZrCl_4$[s,t]	2	$NbOCl_3$[u]	1
$NbCl_5$[u]	1	$TaOCl_3$[u]	1

* R = butyl, isobutyl, propionyl, etc.

[a] Nitrosyl chloride molecules per metal halide molecule.

[b] H. Gall and H. Mengdehl, *Ber. Dtsch. Chem. Ges.* **60**, 86 (1927).

[c] T. C. Waddington and F. Klanberg, *Z. Anorg. Allg. Chem.* **304**, 185 (1960).

[d] G. A. Olah and W. S. Tolgyesi, *J. Org. Chem.* **26**, 2319 (1961).

[e] F. Hewitt and A. K. Holliday, *J. Chem. Soc.* p. 532 (1953).

[f] J. R. Partington and A. L. Whynes, *J. Chem. Soc.* p. 1952 (1948); p. 3135 (1949).

[g] H. Gerding and H. Houtgraaf, *Recl. Trav. Chim. Pays-Bas* **72**, 21 (1953).

[h] A. B. Burg and G. W. Campbell, *J. Am. Chem. Soc.* **70**, 1964 (1948).

[i] H. Rheinbolt and R. Wasserfuhr, *Ber. Dtsch. Chem. Ges.* **60**, 732 (1927).

[j] H. Houtgraaf and A. M. de Roos, *Recl. Trav. Chim. Pays-Bas* **72**, 963 (1953).

[k] R. W. Asmussen, *Z. Anorg. Allg. Chem.* **243**, 127 (1939).

[l] J. J. Sudborough, *J. Chem. Soc.* **59**, 655 (1891).

[m] W. J. Van Heteren, *Z. Anorg. Chem.* **22**, 277 (1899).

[n] A. A. Woolf, *J. Inorg. Nucl. Chem.* **3**, 285 (1956).

[o] J. Lewis and D. B. Sowerby, *J. Chem. Soc.* p. 1617 (1957).

[p] F. Seel, *Z. Anorg. Chem.* **252**, 24 (1943).

[q] F. Seel and H. Bauer, *Z. Naturforsch., Teil B* **12**, 397 (1947).

[r] J. Amaudruct and C. Devin, *C. R. Hebd. Seances Acad. Sci.* **264**, 2156 (1967).
[s] R. Perrot and C. Devin, *C. R. Hebd. Seances Acad. Sci.* **246**, 772 (1958).
[t] V. Gutmann and R. Himml, *Z. Anorg. Allg. Chem.* **287**, 199 (1956).
[u] J. MacCordick and R. Rohmer, *C. R. Hebd. Seances Acad. Sci.* **263**, 1369 (1966).
[v] A. B. Burg and D. E. McKenzie, *J. Am. Chem. Soc.* **74**, 3143 (1952).
[w] C. C. Addison and N. Hodge, *J. Chem. Soc.* p. 2490 (1961).
[x] J. MacCordick, *Experientia* **26**, 1289 (1970).

constant for the NO group, however, suggests that the compound is intermediate between an ionic and molecular compound

$$NO^+[AlCl_4]^- \rightleftharpoons AlCl_3 \cdots ClNO \qquad (45)$$

with the ionic form making a larger contribution. The existence of a similar equilibrium has been suggested by Seel[122,123] while interpreting the conductivity of $NOCl \cdot SbCl_5$ in liquid sulfur dioxide.

The compounds $SnCl_4 \cdot 2NOCl$ and $PtCl_4 \cdot 2NOCl$ have been shown by crystallographic methods to contain the nitrosonium ion and to be isomorphous with $(NH_4)_2SnCl_6$ and $(NH_4)_2PtCl_6$.[124] On the basis of the measurement of the magnetic susceptibility of some of the nitrosyl addition compounds, Asmussen[125] has classified addition compounds such as $ZnCl_2 \cdot 2NOCl$ and $HgCl_2 \cdot NOCl$, which readily lose nitrosyl chloride, as molecular addition compounds and compounds such as $SbCl_5 \cdot NOCl$, $SnCl_4 \cdot 2NOCl$, and $PtCl_4 \cdot 2NOCl$, which are more stable to heat, as nitrosonium salts. Other compounds, such as $CuCl \cdot NOCl$ and $MnCl_2 \cdot NOCl$, have been formulated as $[CuNO]Cl_2$ and $[MnNO]Cl_3$. Treatment of pure liquid sulfurtrioxide with nitrosyl chloride in dry nitrogen atmosphere yielded crystalline $NOCl \cdot SO_3$ and $NOCl \cdot 2SO_3$.[126-128] Infrared studies show that both these compounds contain the NO^+ ion.[128]

Investigations involving isotope exchange reactions have greatly helped in understanding the chemistry of nitrosyl chloride solutions and the nature of the solute–solvent interaction in them. Lewis and Wilkins[129] have observed complete and rapid exchange of the Cl ion between tetramethyl or tetraethylammonium chloride enriched with ^{36}Cl when dissolved in liquid nitrosyl chloride. The results have been interpreted as favoring the self-ionization of nitrosyl chloride rather than an ion transfer process. Complete and rapid exchange of radiochlorine has also been observed between aluminum, gallium, indium, thallium(III), iron(III), and antimony(V) chlorides and liquid nitrosyl chloride as solvent.[74,130-132] This is considered as favoring the formulation of the addition compounds of the above chlorides with nitrosyl chloride as nitrosonium salts. The exchange of radiochlorine between zinc, mercury, and cadmium chlorides, which are almost insoluble, and liquid nitrosyl chloride, has also been studied by Lewis and Sowerby.[133] Rapid exchange was observed between the adsorbed nitrosyl chloride and the metal chloride, followed by a slow heterogenous exchange with excess of the solvent. These results have been interpreted as favoring the formation of unstable nitrosonium salts which decompose to give rise to heterogenous exchange. This agrees with the fact that no exchange was noticed when a stable nitrosonium salt, nitrosonium chlorostannate, or chlorides which do not form addition complexes with nitrosyl chloride, for example, sodium or potassium chlorides, were employed.

Nitrosyl chloride forms a white solid addition compound $(CH_3)_3NO \cdot NOCl$ with trimethylamine oxide.[106] Dimethylaniline gives $C_6H_5N(CH_3)_2 \cdot NOCl$, which has been shown by measurements of electrical conductivity and absorption spectra to be the nitrosonium salt $[C_6H_5N(CH_3)_2 \cdot NO]^+Cl^-$.[134] Paul and Chadha[135] have reported 1 : 1 addition compounds of pyridine, β and γ-picolines, quinoline, and isoquinoline with nitrosyl chloride. A shift in the $\nu N{=}O$ stretching frequency of nitrosyl chloride to a higher spectral region on complex formation, has been considered as suggesting the presence of $(N{\equiv}O)^+$ in the complexes. Measurements of the molar conductivities of these addition compounds in nitrobenzene show them to be uniunivalent electrolytes

$$C_5H_5N \cdot NOCl \rightleftharpoons [C_5H_5N \cdot NO]^+ + Cl^- \tag{46}$$

Neutralization reactions between acids and bases specific to this solvent system have been studied conductometrically. A conductometric titration[106] between $NOFeCl_4$ and $(CH_3)_4NCl$ in nitrosyl chloride shows a break in the conductance versus composition curve at an acid/base molar ratio of 1 : 1 (Fig. 8).

$$NOFeCl_4 + (CH_3)_4NCl \rightleftharpoons (CH_3)_4NFeCl_4 + NOCl \tag{47}$$

The increase in the resistance before the stoichiometric point has been attributed to the replacement of very mobile NO^+ ion by the far less mobile $(CH_3)_4N^+$ ion, and the decrease in the resistance beyond this point is due to the addition of $(CH_3)_4NCl$ until the solution becomes saturated. Similar titrations with $NOBF_4$ and $NOClO_4$ as acids show corresponding changes in conductance but the reactions do not go to completion, probably because the low solubility of the acids favors the reverse reaction.[106] A conductometric titration between $AsCl_3$ and tetraethylammonium chloride in liquid nitrosyl chloride at $-20°C$ also gives a break in the conductivity curve at a 1 : 1 molar ratio.[130]

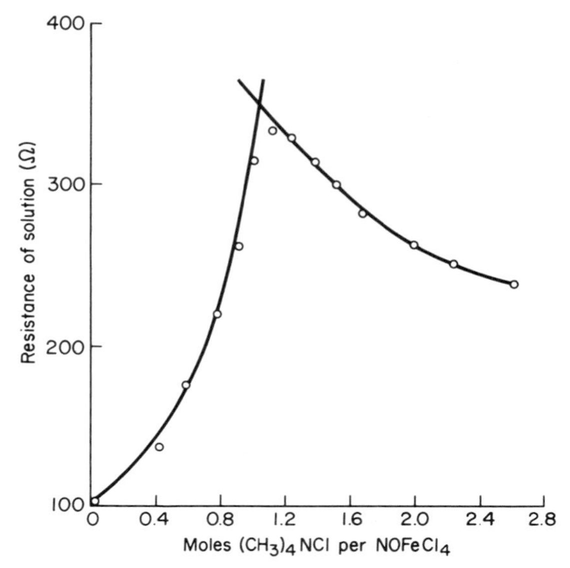

FIG. 8. Conductometric titration of $NOFeCl_4$ with $(CH_3)_4NCl$ in NOCl.

B. Nitrosyl Fluoride, Nitryl Fluoride, and Nitryl Chloride

Because of the very low boiling points, nitrosyl fluoride, nitryl fluoride, and nitryl chloride have not been studied for their solvent properties. Investigations carried out thus far include the preparation of fluoroanion salts of nitrosonium or nitronium ions.

Nitrosyl fluoride (mp −132.5°C, bp −59.9°C, dipole moment 1.81D) can be prepared by allowing fluorine to mix by diffusion with a stoichiometric excess of nitric oxide at 25°C and a total pressure of approximately 1 atmosphere.[136] The chief method for the preparation of fluoroanion salts of the nitrosonium ion involves the reaction of nitrosyl fluoride with the element itself or its oxide or halide. Some of the fluoroanion complexes which have already been reported are $(NO)_2TiF_6$,[137] $NOVF_6$,[138] $NOAuF_4$,[137] $NOBF_4$,[124,137,139] $(NO)_2GeF_6$,[137] $(NO)_2SnF_6$,[137] $NOPF_6$,[140] $NOAsF_6$,[137] $NOSbF_6$,[137] $NOSO_3F$,[137] $(NO)_2WF_8$,[141,142] $(NO)_2ReF_8$,[141,142] $(NO)_2OsF_8$,[142] $(NO)_2PtF_6$,[143] $NOTcF_6$,[141] $NONbF_6$,[144] $NOTaF_6$,[144] $NOMoF_6$,[136] $NOUF_6$,[136] and $NOFeF_4$.[145] The infrared spectra of some of these complexes have been studied. The sharp absorption bands found between 2310 and 2330 cm^{-1} in the spectra of all these complexes has been ascribed to the presence of NO^+ stretching mode.[141,143,146] Moody and Selig[147] have recently reported the synthesis of $2NOF \cdot XeF_6$ and

$NOF \cdot XeOF_4$. Raman and infrared spectroscopic studies indicate that the structure $2NOF \cdot XeF_6$ may be in equilibrium with the structure represented by NO^+ and XeF_8^{2-} ions. Addition of BF_3, PF_5, and AsF_5 to nitrosyl fluoride increases the conductance of the solvent.[148] These fluorides function as Lewis acids by accepting fluoride ions to become BF_4^-, PF_6^-, and AsF_6^-. Standard free energies have been calculated for cells with over all reaction

$$M + 2NOF \rightleftharpoons 2NO + MF_2 \qquad (48)$$

where M is lead, tin, or cadmium.

Nitryl fluoride (mp $-166°C$, bp $-72°C$, dipole moment 0.466D) can be prepared by allowing a slight excess of fluorine to mix by diffusion with dinitrogen tetroxide,[146] by direct fluorination of sodium nitrite[149] or by slow backward and forward distillation of pure and dry dinitrogen tetraoxide across a bed of silver difluoride at 120°C.[150] Many nonmetals react with nitrylfluoride to give nitronium salts.[149] Compounds such as NO_2BF_4, $(NO_2)_2SiF_6$, $(NO_2)_2GeF_6$, NO_2AsF_6, NO_2PF_6, NO_2SbF_6, NO_2SeF_5, and NO_2IF_6 have also been prepared by this method. Metal and nonmetal fluorides also react directly with nitrylfluoride to yield nitronium salts. The compounds prepared by this method include NO_2PtF_6,[143] NO_2TcF_7,[141] NO_2VF_6,[144] NO_2WF_7, NO_2MoF_7, NO_2UF_7,[146] NO_2BF_4, NO_2PF_6, NO_2AsF_6, NO_2SbF_6, $(NO_2)_2SiF_6$, and $(NO_2)_2SnF_6$.[151] Nitrylfluoride also reacts with oxides such as SO_3, SeO_2, TeO_2, and V_2O_5 to yield the nitronium salts NO_2SO_3F, NO_2SeF_5, NO_2VOF_4, and NO_2VF_6.[146] The infrared spectra of some of these nitronium salts have been studied and a strong absorption band in the region 2360–2380 cm^{-1} has been assigned to NO_2^+.[141,143,146]

Nitryl chloride (mp $-145°C$, bp $-15°C$, specific conductance 10^{-7} ohm^{-1} cm^{-1} at $-40°C$, dipole moment 0.53D) can be prepared by the action of chlorosulfuric acid on anhydrous nitric acid and then purified by passing it through three or four traps containing sulfuric acid.[152,153] Cryoscopic and conductometric studies in disulfuric acid show that nitryl chloride is a source of nitronium ions.[153] The Raman and infrared spectra of a concentrated solution of nitryl chloride in disulfuric acid show sharp absorption bands at 1400 cm^{-1} and 2350 cm^{-1}, respectively, indicating the presence of nitronium ions.[153]

Treatment of pure liquid SO_3 with nitryl chloride gives $NO_2Cl \cdot 2SO_3$.[126] Solutions of antimony(V) chloride and boron(III) chloride in liquid sulfur dioxide at $-40°C$ react with excess of nitryl chloride to yield the adducts $NO_2Cl \cdot SbCl_5$ and $NO_2Cl \cdot BCl_3$.[153] The molar conductance of the solutions of these adducts in nitromethane is comparable with those of uniunivalent electrolytes.[153] The infrared spectra of $NO_2Cl \cdot SbCl_5$ and

$NO_2Cl \cdot BCl_3$ show strong absorption bands, characteristic of nitronium ion, at 2355 and 2360 cm^{-1}, respectively.

Adducts of nitryl chloride with organic nitrogen and oxygen bases such as pyridine (1 : 2 and 2 : 1), α-picoline (1 : 1), quinoline, piperidine, diethylamine, triethylamine, morpholine, ethylenediamine, dimethylsulfoxide, and urea (all 1 : 2) have been prepared in acetonitrile.[153] Molar conductivities of these complexes in various solvents indicate that all these complexes are ionic. The infrared spectra of these complexes show a lowering in the NO_2^+ stretching frequency to about 2150 cm^{-1} indicating that the bases combine with nitryl chloride to form ionic complexes.

V. Phosphoryl and Thiophosphoryl Halides

A. Phosphoryl Chloride

The solvent system phosphoryl chloride has been a subject of extensive study by a large number of workers. By far the largest contribution has come from Gutmann and his co-workers.[154-187] The relevant physical properties of phosphoryl chloride are given in Table VIII. Its liquid range (+1.3 to 107.7°C) and dielectric constant (13.9 at 20°C) make it a useful solvent for a number of chemical reactions. The chief impurities present in phosphoryl chloride are polyphosphoryl chlorides and hydrogen chloride, which are formed by hydrolysis of the solvent. They can be easily removed by repeated fractional distillations. Another difficulty is the removal of the last traces of water from the solvent and protection against moisture during the experimental manipulations. It has been shown that in phosphoryl chloride small amounts of water form $H_3O^+Cl^-$ which is dissociated in the solvent. Even carefully purified phosphoryl chloride contains about 10^{-4} moles of water per liter. Its presence can be neglected in concentrated solution but must be taken into consideration in the case of extremely dilute solutions.[164] Experimental measurements are always made in special apparatus, designed to minimize its contact with moisture.

The specific conductivity of pure phosphoryl chloride (2×10^{-8} ohm^{-1} cm^{-1}) has been ascribed[164] to the presence of ions produced by its self-dissociation

$$2POCl_3 \rightleftharpoons POCl_2^+ + POCl_4^- \quad (49)$$

The low specific conductivity, however, indicates that the extent of self-dissociation cannot be very high. The ionic product of phosphoryl chloride has been evaluated to be less than 9×10^{-14}.[164] Electrolysis of a 0.14 *M* solution of triethylammonium chloride in phosphoryl chloride produces chlorine at the anode and a polymeric solid of composition $(PO)_x$ is preci-

TABLE VIII

PHYSICAL CONSTANTS OF PHOSPHORYL HALIDES AND THIOHALIDES[a,b]

Solvent	Melting point (°C)	Boiling point (°C)	Dielectric constant (ε)	Specific conductance (ohm^{-1} cm^{-1})	Density (g/ml)	Dipole moment (Debye)
$POCl_3$	1.3	107.7	13.9 (20°)	2×10^{-8} (10°C)	1.645 (25°C)	2.4
$POBr_3$	56	193	—	$<4 \times 10^{-8}$ (56°C)	2.82 (45°C)	—
$PSCl_3$	−35	125	5.8 (21.5°)	1.3×10^{-9} (25°C)	1.636 (22°C)	1.41
$PSBr_3$	37.8	212 dec	3.7 (solid) 6.2 (liquid)	7.8×10^{-8} (50°C)	2.85 (17°C)	1.58

[a] G. Brauer, ed., "Handbook of Preparative Inorganic Chemistry," 2nd ed., Vol. 1, pp. 534–536. Academic Press, New York, 1963.

[b] J. C. Bailar, H. J. Emeléus, R. Nyholm, and A. F. Trotman-Dickenson, eds., "Comprehensive Inorganic Chemistry," Vol. 2, pp. 432 and 439. Pergamon, Oxford, 1973.

pitated out at the cathode.[188] The Cl : PO ratio is 3 : 1. The reactions occurring at the electrodes have been represented as

$$POCl_3 \rightleftharpoons POCl_2^+ + Cl^-$$

$$Cl^- \rightarrow \tfrac{1}{2}Cl_2 + e \text{ (anode)}$$

$$3POCl_2^+ + 3e \rightarrow PO + 2POCl_3 \text{ (cathode)} \quad (50)$$

The results have been interpreted as providing strong evidence in favor of the autoionization of the solvent. However, an alternative explanation[189] could be that the triethylammonium ion was discharged at the cathode and reacted with the solvent to give

$$Et_3NH^+ + e \rightarrow Et_3N + [H]$$

$$3[H] + POCl_3 \rightarrow PO + 3HCl \quad (51)$$

overall reaction could then be represented as

$$3ET_3NH^+ + 3e + POCl_3 \rightarrow PO + 3Et_3NH^+ + 3Cl^- \quad (52)$$

Anhydrous covalent chlorides such as BCl_3, $AlCl_3$, $GaCl_3$, $InCl_3$, PCl_3, $AsCl_3$, $SbCl_3$, CCl_4, $SiCl_4$, $GeCl_4$, $SnCl_4$, $TiCl_4$, $ZrCl_4$, $TeCl_4$, $SeCl_4$, PCl_5, $SbCl_5$, $NbCl_5$, $TaCl_5$, $MoCl_5$, WCl_6, and $FeCl_3$ have a fairly good solubility in phosphoryl chloride.[190 192] Many of them form conducting solutions and yield solid solvates. On the other hand, the solubilities of ionic chlorides are usually small. For example, alkali metal chlorides have extremely low solubilities in it, although the solubilities tend to increase with increasing size of the cation[155] (Table IX). Quarternary ammonium chlorides such as tetraethyl and tetrabutylammonium chlorides are, on the other hand, readily soluble.[165] Solubilities are sometimes considerably enhanced upon addition of complex-forming agents. Thus barium chloride, which is insoluble in phosphoryl chloride, readily dissolves in the presence of iron(III) chloride forming $Ba(FeCl_4)_2$. Similarly, alkali metal chlorides show higher solubilities in the presence of antimony(V) chloride because of the formation of soluble hexachloroantimonates. Several bromides, iodides, chlorates, permanganates, and dichromates are also soluble but the color changes accompanying dissolution suggest that some reaction occurs during the process.[177,192] Tetraalkylammonium permanganate gives a blue solution and tetraethylammonium chlorate gives a yellow solution in phosphoryl chloride.

Quarternary ammonium chlorides and alkali metal chlorides, up to the limit of their solubilities, are ionized in phosphoryl chloride

$$R_4NCl \rightleftharpoons R_4N^+ + Cl^- \quad (53)$$

The dissociation is incomplete because of the moderate value of the dielectric constant, which favors association to ion pairs even at higher

TABLE IX

SOLUBILITY[a] AND SPECIFIC CONDUCTIVITY OF SATURATED SOLUTIONS OF ALKALI METAL SALTS

Salt	Solubility (g/liter)	Specific conductivity of saturated solution[b] ($ohm^{-1}\ cm^{-1}$)	Equivalent conductance at $V = 1000$ liters/mole
LiCl	0.05	6.6×10^{-6}	4.0
NaCl	0.31	3.0×10^{-5}	6.4
KCl	0.60	3.4×10^{-5}	6.7
NH_4Cl	0.46	3.6×10^{-5}	6.9
RbCl	0.87	8.3×10^{-5}	14.6
CsCl	1.26	1.1×10^{-4}	16.0
KF	0.40	2.6×10^{-5}	6.4
KBr	0.51	4.3×10^{-5}	14.5
KI	1.71	1.2×10^{-4}	23.1
KCN	0.73	3.3×10^{-5}	7.2
KCNO	0.80	3.1×10^{-5}	9.0
KCNS	0.76	2.9×10^{-5}	6.6

[a] V. Gutmann, *Monatsh. Chem.* **83**, 279 (1952).
[b] Measured at 20°C.

dilutions.[165] The molar conductance of Et_4NCl in $POCl_3$ has been determined at 20°C over the concentration range 5.97×10^{-7} to 5.43×10^{-2} M. Within this range Et_4NCl behaves as a typical polar compound, dissolved in a solvent of low dielectric constant. Its dissociation constant has been evaluated to be as 7.1×10^{-4}.[164] The results of similar measurements on Bu_4NCl, Pr_4NCl, Et_4NClO_4, Et_4NBr, and Et_4NI over the concentration range 10^{-5}–10^{-3} M resemble those of Et_4NCl.[165] The ions show normal mobilities (Table X) and the Walden rule is obeyed. For R_4N^+, ClO_4^-, and Cl^- the Bjerrum parameters are proportional to the Stokes radii. The iodides and bromides behave differently and this has been attributed to solvolysis.

The molar conductivity of triethylamine in phosphoryl chloride solution has been determined over the concentration range 4.1×10^{-5} to 1.8×10^{-3} M.[167] The limiting conductance value is 48.5 ± 1. Calculations of ionic mobilities and Stokes radii show that $Et_3NPOCl_2^+$ is the cation. The overall equilibrium is represented as

$$Et_3N + POCl_3 \rightleftharpoons [Et_3N \cdot POCl_2]^+Cl^- \rightleftharpoons Et_3N \cdot POCl_2^+ + Cl^- \quad (54)$$

In a dilute solution of triethylammonium chloride in $POCl_3$, the ions Et_3NH^+ and $Et_3N \cdot POCl_2^+$ occur in approximately equal quantities.[170]

TABLE X

IONIC MOBILITIES (m), STOKE RADII (R), AND SOLVATION NUMBER (n) OF TETRAALKYLAMMONIUM COMPOUNDS IN PHOSPHORUS OXYCHLORIDE[a]

Ion	m	m in H_2O (25°C)	R	n
$[Et_4N]^+$	26	32.7	2.98	0
$[Pr_4N]^+$	19	23.4	4.13	0
$[Bu_4N]^+$	16	19.5	5.10	0
$[Cl]^-$	27	—	2.46	0.5–1.0
$[ClO_4]^-$	28	—	2.43	0

[a] M. Baaz and V. Gutmann, *Monatsh. Chem.* **90**, 256 (1959).

Thus, in phosphoryl chloride, the coordination affinities of H^+ and $POCl_2^+$ towards triethylamine are similar. The equilibria involved are

$$\begin{array}{lll} Et_3NH^+ + POCl_3 & \rightleftharpoons & Et_3N \cdot POCl_2^+ + HCl \\ -Cl^- \upharpoonleft\downharpoonright \; +Cl^- & & -Cl^- \upharpoonleft\downharpoonright \; +Cl^- \\ Et_3N \cdot HCl & & Et_3N + POCl_3 \end{array} \quad (55)$$

A large number of solid solvates of Lewis acid and bases with phosphoryl chloride have been isolated (Table XI). The formation of others has been shown from the phase diagram studies carried out by thermal and x-ray analysis. The structure of a few solid solvates has been investigated by x-ray diffraction. Structural studies on $SbCl_5 \cdot POCl_3$,[193,194] $TaCl_5 \cdot POCl_3$,[195] $NbCl_5 \cdot POCl_3$,[194] $TiCl_4 \cdot POCl_3$,[196] $TiCl_4 \cdot 2POCl_3$,[195] and $SnCl_4 \cdot 2POCl_3$[197] have established that in all these solvates coordination to the metal halide is through the oxygen atom of the phosphoryl chloride. Infrared and Raman spectra of several compounds have been studied which also favor coordination through oxygen.[198–202]

On dissolution in phosphoryl chloride, most of the solid solvates give highly conducting solutions which have been employed in potentiometric, conductometric, and spectrophotometric titrations.[158,161,162,168,169,174,175,181,182,184] Ionization in such a solution can take place (1) by halide ion transfer, (2) by ionization of the solvates, and (3) with the formation of autocomplex ions. Conflicting views have been expressed regarding the relation between the species present in solution and the solid from which the solution is prepared. Whereas the results of potentiometric and conductometric titrations can be interpreted on the basis of equilibria involving chloride-ion transfer (solvent

TABLE XI

PHOSPHORYL CHLORIDE SOLVATES OF METAL HALIDES AND OTHER SUBSTANCES

Solute : $POCl_3$ ratio	Solute
1 : 1	$TiCl_4$, $SbCl_5$, $SbCl_3$, $TaCl_5$, $SnCl_4$, $AlCl_3$, $AlBr_3$, $FeCl_3$, $HfCl_4$, $ZrCl_4$, UCl_4, BF_3, BCl_3, BBr_3, $AsCl_3$, $GaCl_3$, $TeCl_4$, $MoCl_5$, $MoOCl_3$, $MoOCl_4$, $WOCl_4$, $ReOCl_4$
1 : 2	$TiCl_4$, $SnCl_4$, $VOCl_3$, VCl_4, $AlCl_3$, $HfCl_4$, $ZrCl_4$, $BiCl_3$, $GaCl_3$, $MgBr_2$, $SnOCl_2$, $TiOBr_2$, $TiOCl_2$, CaO, MgO, C_5H_5N
2 : 1	$MgCl_2$
1 : 3	$MgBr_2$, $BeCl_2$, CaO, MgO, MnO, ZnO, C_5H_5N, Me_3N, Et_3N, quinoline, isoquinoline, piperidine, β-picoline, γ-picoline, lutidine
2 : 3	$AlCl_3$, $FeCl_3$
1 : 4	UCl_4, Zr_2OCl_6
1 : 6	$AlCl_3$
Ternary systems	$TiCl_4 \cdot SbCl_5 \cdot 3POCl_3$, $TiCl_4 \cdot NbCl_5 \cdot 3POCl_3$, $TiCl_4 \cdot TaCl_5 \cdot 3POCl_3$, $AlCl_3 \cdot SbCl_5 \cdot 3POCl_3$, $AlCl_3 \cdot NbCl_5 \cdot 3POCl_3$, $AlCl_3 \cdot TaCl_5 \cdot 3POCl_3$, $GaCl_3 \cdot TiCl_4 \cdot 3POCl_3$, $GaCl_3 \cdot ZrCl_4 \cdot 3POCl_3$, $ZrCl_4 \cdot SbCl_5 \cdot 2POCl_3$, $2AlCl_3 \cdot SnCl_4 \cdot 2POCl_3$

system concept),[158,168,169,175,181,182] investigations by Meek and Drago[203] and Drago and Purcel[204] indicate that the chloride-ion transfer mechanism has probably been overemphasized for oxychloride solvents. They suggest an alternative coordination model, which does not require the ionization of the solvent as a prerequisite to provide chloride ions, to explain the equilibria present in the solutions of chloride ion acceptors in phosphoryl chloride.*

Antimony(V) chloride forms a solid solvate, $SbCl_5 \cdot POCl_3$, with phosphoryl chloride. X-ray structure determination shows that the coordination around antimony is octahedral with an O atom from $POCl_3$ in the sixth corner.[193,194] The solid dissolves in phosphoryl chloride and conductivity measurements show the presence of binary dissociation equilibria involving ions with normal mobilities.[168] This has been interpreted in terms of the equilibrium

$$SbCl_5 \cdot POCl_3 \rightleftharpoons POCl_2{}^+ + SbCl_6{}^- \quad (K = 4 \times 10^{-6}) \tag{56}$$

Thus the formation of $SbCl_6{}^-$ is regarded as a ligand-exchange reaction involving a change from O coordination in the undissociated species,

* See Chapter 1 of Volume I of the Treatise for a detailed discussion of this interpretation.

$Cl_5SbOPCl_3$, to Cl coordination followed by dissociation.[168] The reaction may be represented as

$$SbCl_6^- \xrightleftharpoons{\pm POCl_3} Cl_5SbOPCl_3 + Cl^- \quad (K \leq 10^{-9}) \tag{57}$$

Although coordination of chlorine at antimony provides a higher stability to the species than oxygen coordination, due to the low chloride-ion-donating tendency of the solvent, as compared to the capacity of the oxygen atom to donate, a considerable quantity of the O-coordinated form is present in solution, which is almost quantitatively transformed into coordinated chlorocomplex by the addition of chloride ions. It has also been observed that in concentrated solutions of antimony(V) chloride, colloidal aggregates of indefinite composition, $(SbCl_5)_n(POCl_3)_m$ are formed on standing. These are presumed to be oxygen coordinated.[168] Alternative ionization processes which may be suggested in this system are

$$2Cl_5SbOPCl_3 \rightleftharpoons [(POCl_3)_2SbCl_4]^+ + SbCl_6^- \tag{58}$$

$$Cl_5SbOPCl_3 \rightleftharpoons [Cl_4SbOPCl_3]^+ + Cl^- \tag{59}$$

$$Cl_5SbOPCl_3 \rightleftharpoons [Cl_5SbOPCl_2]^+ + Cl^- \tag{60}$$

The last two equilibria are to be excluded since such equilibria should be suppressed by the presence of tetraethylammonium chloride which definitely produces chloride ions in solution. This is not shown by experimental results. Further, potentiometric and conductometric titrations[154,157,161,162] between antimony(V) chloride and tetraethylammonium chloride show sharp breaks in the curves at a 1 : 1 molar ratio, corresponding to the formation of $(C_2H_5)_4NSbCl_6$ which could be isolated from the solutions, a fact which supports the presence of antimony containing anion in solution.

Leman and Tridot[205] have reported, from thermal analysis of the $POCl_3$–$SbCl_5$ system, the formation of a compound $POCl_3 \cdot 2SbCl_5$ which melts incongruently and transforms into the 1 : 1 compound at 88.5°C. During similar studies, however, Voitovich[38,206] could not find any indication of the existence of this compound.

Niobium(V) and tantalum(V) chlorides form the monosolvates $NbCl_5$–$POCl_3$ and $TaCl_5 \cdot POCl_3$.[207] X-ray crystal structure determinations show that these compounds are isostructural with $SbCl_5 \cdot POCl_3$.[194,195] Potentiometric titrations[162] show that both the chlorides take up one chloride ion to give the hexacoordinated chloro complex. Molybdenum(V) chloride also forms a monosolvate with phosphoryl chloride.[208] Photometric and potentiometric studies[178] show that the addition of chloride ion donors such as Et_4NCl, KCl, and CsCl to a red solution of tungsten(VI) chloride in phosphoryl chloride causes the formation of the yellow solvated WCl_7^- complex, possibly, Cl_7WOPCl_3. The complex is stable only in the presence of the solvent. No evidence was found for the formation of WCl_8^{2-}. The hexachloride of tungsten does not act as a chloride ion donor toward antimony(V) chloride.

Two solvates of titanium(IV) chloride, namely, $TiCl_4 \cdot POCl_3$ and $TiCl_4 \cdot 2POCl_3$, have been reported.[209] From a study of the infrared spectra of these compounds, Sheldon and Tyree[202] have concluded that the oxygen atom in the $POCl_3$ functions as the donor atom. The crystal structures of both the adducts have been determined from three-dimensional x-ray data.[195,196] The structure of the monosolvate is build up of dimeric molecules, $(TiCl_4 \cdot POCl_3)_2$, with double chlorine bridges between the two titanium atoms. The coordination around titanium atom is octahedral and the oxygen atom in $POCl_3$ functions as the donor atom (Fig. 9). The disolvate $TiCl_4 \cdot 2POCl_3$ is isostructural with $SnCl_4 \cdot 2POCl_3$. The structure is comprised of discrete molecules of $TiCl_4 \cdot 2POCl_3$. The titanium atom is octahedrally coordinated with two $POCl_3$ groups lying in the *cis*-position.

Solutions of titanium(IV) chloride in phosphoryl chloride are yellow and their spectra are independent of the concentration of the solute. Conductometric and potentiometric titrations with titanium(IV) chloride show that, in addition to taking up one or two chloride ions from chloride ion donors such as tetraethylammonium chloride to give $[Cl_5Ti \cdot OPCl_3]^-$ and $TiCl_6^{2-}$ ions, it can also give up one chloride ion on reacting with strong chloride-ion acceptors like $SbCl_5$ and $FeCl_3$ to produce ions of the type $[Cl_3Ti(OPCl_3)_3]^+$.[157,174,210] The behavior of titanium(IV) chloride thus depends wholly on the relative donor or acceptor strength of the other chloride. Different equilibria involved can be represented as

$$Cl_4Ti(OPCl_3)_2 + Cl^- \rightleftharpoons [Cl_5TiOPCl_3]^- + POCl_3 \tag{61}$$

$$[Cl_5TiOPCl_3] + Cl^- \rightleftharpoons [TiCl_6]^{2-} + POCl_3 \tag{62}$$

$$Cl_4Ti(OPCl_3)_2 + POCl_3 \rightleftharpoons [Cl_3Ti(OPCl_3)_3]^+ + Cl^- \tag{63}$$

Donor or acceptor properties of $TiCl_4$ appear to be concentration dependent, the former dominating in dilute solutions while the latter in concentrated solutions. Upon addition of titanium(IV) chloride to a suspension of potassium chloride in phosphoryl chloride, a solid having the composition $KCl \cdot TiCl_4 \cdot POCl_3$ is formed. Ebullioscopic studies show that this compound acts as a binary electrolyte, presumably giving the $[Cl_5TiOPCl_3]^-$ ion.[179]

A study of the phase diagram of the $TiCl_4$–$SbCl_5$–$POCl_3$ system, obtained by thermal analysis and x-ray powder diffraction techniques shows the formation of a ternary compound, $TiCl_4 \cdot SbCl_5 \cdot 3POCl_3$.[206,210,211] On the basis of conductance measurements,[210] this compound has been formulated as $[Cl_3Ti(OPCl_3)_3]^+[SbCl_6]^-$. Similar studies[206,212] on the systems $TiCl_4$–$TaCl_5$–$POCl_3$ and $TiCl_4$–$NbCl_5$–$POCl_3$ also show the formation of

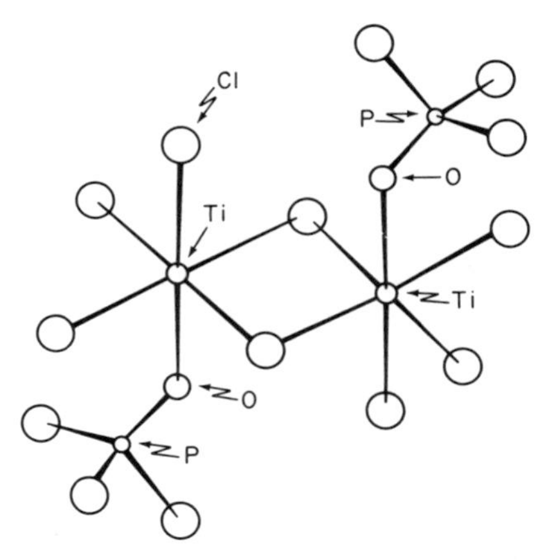

FIG. 9. The molecule of the addition compound $(TiCl_4 \cdot POCl_3)_2$.

ternary compounds $TiCl_4 \cdot TaCl_5 \cdot 3POCl_3$ and $TiCl_4 \cdot NbCl_5 \cdot 3POCl_3$. No such compound could be isolated in the systems $SnCl_4$–$SbCl_5$–$POCl_3$ or $AsCl_3$–$SbCl_5$–$POCl_3$.[210]

Concentrated solutions of ferric chloride in phosphoryl chloride ($\sim 10^{-2}$ *M*) are dark red.[166] However, on dilution the color changes and at a concentration of about 10^{-4} *M* the solution is yellow.[111,166] Upon removal of the solvent, the red solution gives a brown amorphous solid which is a highly condensed species rich in phosphoryl chloride. If the solvent is removed by warming the red-colored solution to 35°C under reduced pressure, red crystals of the composition $2FeCl_3 \cdot 3POCl_3$ are left. The latter, when subjected to further evacuation at 50°C under 2–3 mm pressure, yields a brown solid, $FeCl_3 \cdot POCl_3$.[213] Isolation of another solvate, $2FeCl_3 \cdot POCl_3$,[214] has not been confirmed.[213] The ultraviolet spectra of these solutions show that in the concentrated solutions, tetrachloroferrate(III) species, $FeCl_4^-$, are not present, but on dilution ($\sim 10^{-4}$ *M*), the color change to yellow corresponds to the presence of $FeCl_4^-$ as one of the principal absorbing species.[172] The color transformation from red to yellow is completely reversible upon dilution or concentration. The addition of chloride ion donors (Et_4NCl, KCl, etc.) to the red ferric chloride solution also changes the color of the solution. The yellow solution produced has the spectrum characteristic of the $FeCl_4^-$ ion and there is a corresponding change in conductivity.[172] On addition of chloride-ion acceptors such as antimony(V) chloride to dilute solutions of $FeCl_3$ in $POCl_3$, or to the solutions of $FeCl_3$ containing chloride ion donors, the color change yellow to red is affected.[172] Analysis of photometric titration curves shows that Et_4NCl, KCl, $ZnCl_2$, PCl_5, $HgCl_2$, BCl_3, $TiCl_4$, $SnCl_4$, and $SbCl_6^-$ donate one chloride ion each to $FeCl_3$, while $AlCl_3$ donates two chloride ions readily. Basic strength decreases in the order $Et_4NCl \sim KCl \sim ZnCl_2 > TiCl_4 > PCl_5 \sim AlCl_3 \gg SbCl_6^- \sim HgCl_2 > BCl_3 \sim SnCl_4$.[176] The formation of the tetrachloroferrate(III) ion can be followed conductometrically,[169,174,215] potentiometrically,[169,174,215] spectrophotometrically,[166,169,172,173,177] and ebullioscopically.[171,179]

Ebullioscopic investigations on $FeCl_3$ and $AlCl_3$ solutions in phosphoryl chloride show the existence of polymeric species.[171,179] The dissociation constant for $KFeCl_4$ in boiling $POCl_3$ is 3×10^{-4} as against a value of 9×10^{-4} for $KSbCl_6$, 5×10^{-4} for $RbSbCl_6$, 3×10^{-4} for $K[Cl_5TiOPCl_3]$, and 9×10^{-5} for $[Al(OPCl_3)_6][FeCl_4]_3$.

Photometric studies on 10^{-2} *M* phosphoryl chloride solutions of iron(III) chloride in the presence of a Lewis base such as pyridine or triethylamine show that chloride-ion transfer from $POCl_3$ to $FeCl_3$ is facilitated by the Lewis bases.[177] For example, pyridine assists in the transfer of chloride ion by the formation of the $C_5H_5N \cdot POCl_2^+$ ion,

$$C_5H_5N + POCl_3 + FeCl_3 \rightarrow C_5H_5N \cdot POCl_2^+ + FeCl_4^- \quad (64)$$

Tetraethylammonium permanganate gives an unstable blue solution in $POCl_3$, which converts four equivalents of $FeCl_3$ to $FeCl_4^-$. The reactions may be represented as follows:

$$MnO_4^- \rightarrow [(MnCl_8)^-] \rightarrow MnCl_3 + Cl^- + 2Cl_2$$

$$Cl^- + FeCl_3 \rightleftharpoons [FeCl_4]^-$$

$$MnCl_3 + 3FeCl_3 \rightarrow Mn^{3+} + 3[FeCl_4]^- \quad (65)$$

Tetraethylammonium chlorate gives a yellow solution in $POCl_3$ which reacts with one equivalent of ferric chloride.[177]

The conductivity of phosphoryl chloride solutions of iron(III) chloride can be explained on the basis of the formation of tetrachloroferrate(III) ion. The essential equilibria as represented by the solvent system concept are

$$FeCl_3 + POCl_3 \rightleftharpoons FeCl_3 \cdot POCl_3 \rightleftharpoons FeCl_4^- + POCl_2^+ \quad (66)$$

However, no conclusive evidence is available to establish the presence of $POCl_2^+$ in the solution. An alternative interpretation for the formation of $FeCl_4^-$ has been put forward by Meek and Drago.[203] The behavior of anhydrous iron(III) chloride in triethylphosphate has been shown to be similar to its behavior in $POCl_3$. Spectral investigations demonstrate that the tetrachloroferrate(III) ion is the principal absorbing species in dilute as well as concentrated solutions of $FeCl_3$ in $PO(OEt)_3$ (Fig. 10). Use of triethylphosphate as a solvent excludes the possibility of $FeCl_4^-$ being formed from a chloride ion produced as a result of the self-ionization of the

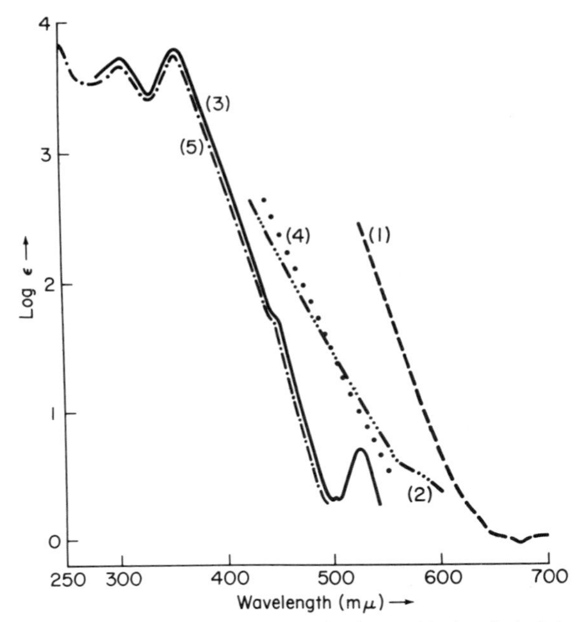

FIG. 10. Visible and ultraviolet spectra of iron(III) chloride in triethylphosphate and phosphoryl chloride. (1) $FeCl_3$ in $POCl_3 \cdot C \sim 1 \times 10^{-1}$ *M*; (2) $FeCl_3$ in $POCl_3 \cdot C \sim 3.3 \times 10^{-3}$ *M*; (3) $FeCl_3$ in $POCl_3 \cdot C \sim 1 \times 10^{-4}$ *M*; (4) $FeCl_3$ in $PO(OEt)_3 \cdot C \sim 4 \times 10^{-3}$ *M*; (5) $FeCl_3$ in $PO(OEt)_3 \cdot C \sim 1 \times 10^{-4}$ *M*.

solvent. Meek and Drago suggest the following equilibria to explain the formation of $FeCl_4$ in both $PO(OEt)_3$ and $POCl_3$:

$$FeCl_3 + Y_3PO \rightarrow FeCl_3 \cdot OPY_3 \rightleftharpoons [FeCl_{3-x}(OPY_3)_{1+x}]^+ + xFeCl_4^- \rightleftharpoons$$
$$[Fe(OPY_3)_6]^{3+} + 3FeCl_4^- \quad (67)$$

Boron(III) chloride dissolves in phosphoryl chloride to give conducting solutions. Conductometric, potentiometric, and spectrophotometric measurements show that, as a chloride ion acceptor, boron(III) chloride is weaker than the antimony(V) or iron(III) chlorides but stronger than tin(IV), titanium(IV), or aluminum(III) chlorides.[175] Absence of a reaction with a strong chloride ion acceptor in phosphoryl chloride excludes the presence of $[Cl_3POBCl_2]^+$ in this solvent. On the other hand, chloride ion donors give tetrachloroborate ions. The results of conductometric and potentiometric titrations can be readily interpreted in terms of the existence, in solution, of the equilibrium

$$BCl_3 \cdot POCl_3 \rightleftharpoons POCl_2^+ + BCl_4^- \quad (68)$$

However, conflicting views have been expressed regarding the structure of the solid solvate, $BCl_3 \cdot POCl_3$. From infrared spectral studies Waddington and Klanberg,[200] Peach and Waddington,[201] and Wartenberg and Goubeau,[198] have reported the coordinated structure $Cl_3PO \rightarrow BCl_3$ for the solvate whereas the ionic structure $[POCl_2]^+[BCl_4]^-$ has been postulated by Gerrard *et al.*[216,217]

There is a rapid exchange of radioactive chlorine between boron(III) chloride and phosphoryl chloride in solutions rich in phosphoryl chloride, but no exchange is observed when boron(III) chloride is in excess.[218] The absence of chlorine exchange between BCl_3 and $POCl_3$ in boron(III)-chloride-rich systems indicates that the bonding in the addition compound cannot be of the type $[POCl_2]^+[BCl_4]^-$. The higher dielectric constant of phosphoryl chloride (13.9 at 25°C) compared to that of boron(III) chloride (1.0 at 0°C) suggests that ionic processes may occur more readily in phosphoryl-chloride-rich solutions than in boron(III)-chloride-rich solutions. The observed rapid halogen exchange in $POCl_3$ rich systems can thus occur by the mechanism

$$POCl_3 \rightleftharpoons POCl_2^+ + Cl^-$$

$$Cl^- + \overset{*}{Cl}_3BOPCl_3 \rightleftharpoons [\overset{*}{Cl}_4BOPCl_3^-] \rightleftharpoons \overset{*}{Cl}^- + Cl_3BPOCl_3 \qquad (69)$$

Although it is difficult to see how, if solvated BCl_4^- ion is formed in a phosphoryl-chloride-rich system, it is also not formed in a boron-trichloride-rich system at least in an amount which could promote a rapid chlorine exchange.

Aluminum halides have fairly good solubilities in phosphoryl chloride and give highly conducting solutions. Four solvates of aluminum(III) chloride, namely, $AlCl_3 \cdot POCl_3$,[156,219] $AlCl_3 \cdot 2POCl_3$,[220,221] $AlCl_3 \cdot 6POCl_3$,[220] and $2AlCl_3 \cdot 3POCl_3$,[219,222] have been reported from the examination of phase diagrams. Of these only the first two have actually been isolated. Aluminum bromide and aluminum iodide also form solvates of the composition $AlBr_3 \cdot POCl_3$ and $AlI_3 \cdot 2POCl_3$.[156]

Conductometric titrations of zinc(II), tin(IV), and antimony(V) chlorides with aluminum(III) chloride in phosphoryl chloride give breaks in the titration curves at a molar ratio of 1 : 1 in the case of zinc(II) and antimony(V) chloride and 1 : 2 in the case of tin(IV) chloride, with simultaneous precipitation of the compounds such as $AlCl_3 \cdot SbCl_5 \cdot 3POCl_3$ and $2AlCl_3 \cdot SnCl_4 \cdot 2POCl_3$.[157] Later work by Baaz *et al.*[215] however, has shown that, in the case of a titration between $AlCl_3$ and $SbCl_5$, the nature of the curve depends on the direction of the titration. When antimony(V) chloride is titrated with aluminum(III) chloride, breaks are obtained at $AlCl_3 : SbCl_5$ molar ratio of 1 : 1, 2 : 1, and possibly also at 3 : 1. On the other hand, in the titration of $AlCl_3$ with $SbCl_5$, only one break occurs at a $AlCl_3 : SbCl_5$ molar ratio of 1 : 2. Colloidal precipitates are formed during the titration of antimony(V) chloride with aluminum(III) chloride and the establishment of the equilibrium between the solid and the solution is very slow. Ebullioscopic measurements on aluminum chloride solutions in phosphoryl chloride also show the existence of polymeric species which decompose slowly.[179]

Conductometric, potentiometric, spectrophotometric, and preparative investigations have shown that aluminum chloride acts both as a chloride ion donor and acceptor in phosphoryl chloride.[215] A titration of aluminum chloride and tetraethylammonium chloride gives a break at a molar ratio of 1 : 1 corresponding to the formation of $[Et_4N]AlCl_4$. The $AlCl_4^-$ ion appears to be the only anionic species and there is no evidence for the formation of $AlCl_5^{2-}$ and $AlCl_6^{3-}$ ions. Strong chloride ion acceptors such as $FeCl_3$ and $SbCl_5$, on the other hand, give rise to the formation of $AlCl^{2+}$ ions which are further solvated through O-coordination by the solvent. The third chloride ion of $AlCl_3$ ionizes only if insoluble complex compounds such as $AlCl_3(FeCl_3)_3(POCl_3)_6$ and $AlCl_3(SbCl_5)_3(POCl_3)_6$ are formed. These compounds have been formulated as $[Al(OPCl_3)_6][FeCl_4]_3$ and $[Al(OPCl_3)_6][SbCl_6]_3$ but no structural evidence is available. The above results can be explained on the basis of the formation of "autocomplex" ions.[215] Chloride ions are released by the ionization of aluminum chloride due to the preferred O-coordination by the solvent. These chloride ions may react with other solvate molecules to give the chlorocomplex

$$AlCl_3 + nPOCl_3 \rightleftharpoons Cl_3Al(OPCl_3)_n$$

$$Cl_3Al(OPCl_3)_n \rightleftharpoons [Cl_2Al(OPCl_3)_n]^+ + Cl^-$$

$$\rightleftharpoons [ClAl(OPCl_3)_n]^{2+} + 2Cl^-$$

$$Cl_3Al(OPCl_3)_n + Cl^- \rightleftharpoons [AlCl_4]^- + nPOCl_3$$

$$3Cl_3Al(OPCl_3)_n \rightleftharpoons [ClAl(OPCl_3)_n]^{2+} + 2[AlCl_4]^- \quad (70)$$

The presence of such ions accounts for the conductivities of the solutions and for the abnormal transference number of aluminum[160] since aluminum is transported to both the cathode and the anode.

Gallium chloride behaves in an analogous manner and gives the solvates $GaCl_3 \cdot POCl_3$ and $GaCl_3 \cdot 2POCl_3$.[223 225] The specific electrical conductivity, viscosity, density, and surface tension of the molten monosolvate have been determined over a range of temperatures. It is suggested that the complex is dissociated to the extent of about 0.5% into $POCl_2^+$ and $GaCl_4^-$.[223] The Raman spectrum of the molten compound was interpreted in favor of the structure $POCl_2^+GaCl_4^-$.[225,226] Ten lines were observed, four of which corresponded in position and polarization to the spectrum of $GaCl_4^-$ and six of which were assigned to frequencies due to $POCl_2^+$. Later Raman and infrared spectral studies, however, indicate that in the liquid complex the coordination occurs via the oxygen atom and that the Raman lines of the ionic species which may be present to the extent of 0.5% could not be detected in the presence of 99.5% of the covalent adduct $Cl_3PO \rightarrow GaCl_3$.[199] A small equilibrium concentration of ions such as $POCl_2^+$ and $GaCl_4^-$ in the melt could possibly be formed by a halide ion-transfer process or, more likely, the oxygen-coordinated adduct could be considered as undergoing autoionization of the type

$$2[Cl_3PO \rightarrow GaCl_3] \rightleftharpoons [GaCl_2(POCl_3)_2]^+ + [GaCl_4]^- \qquad (71)$$

Thermal analysis of the ternary systems $GaCl_3$–$TiCl_4$–$POCl_3$ and $GaCl_3$–$ZrCl_4$–$POCl_3$ indicates the formation of the solvates $GaCl_3 \cdot TiCl_4 \cdot 3POCl_3$ and $GaCl_3 \cdot ZrCl_4 \cdot 3POCl_3$.[227]

Tertiary nitrogen bases such as pyridine,[228–230] quinoline, isoquinoline, β and γ-picolines, piperidine,[230] triethylamine, and trimethylamine[228,229] form solid solvates in the $POCl_3$: base molar ratio of 1 : 3. The solvates are nonconducting in nitrobenzene and the infrared spectra show that the pyridine molecules coordinate to the phosphorus atom of phosphoryl chloride.[230] Substituted carboxylic amides[231] are also reported to form adducts in $POCl_3$: base ratio of 1 : 1, 1 : 2, and 1 : 3.

Various color indicators have been employed to visually detect the end point in acid/base titrations. Use of sulfonaphthalein indicators such as Cresol red, xylenol blue, bromophenol blue, and bromothymol blue has been investigated by following the titrations visually and by absorption spectroscopy.[232] The spectra of the solutions change at a certain pCl value and are essentially independent of both the nature of the solvent and the nature of the chloride ion donor or acceptor. Basic solutions are slightly yellow while acidic solutions are red. It has been concluded that the solvent cations replace phenolic hydrogen atoms in the indicator molecules on dissolution in $POCl_3$. The color changes have been explained as due to the chloride ion transfer reactions involving the equilibrium

$$[In(POCl_2)_2]^{2+} + Cl^- \rightleftharpoons [In(POCl_2)]^+ + POCl_3 \qquad (72)$$

Benzanthrone[87] and crystal violet[86,186] act as reversible internal indicators in phosphoryl chloride. The results obtained in visual titrations are fairly good. Benzanthrone gives a yellow color in pure phosphoryl chloride and in solutions of acids such as tin(IV) and titanium(IV) chlorides, while in the presence of bases quinoline, α-picoline, and dimethylaniline it gives a red color. The violet color of a solution of crystal violet in phosphoryl chloride changes first to green and then to yellow on addition of a base. The spectra of green and yellow solutions correspond to those of malachite green and $[Ph_3C]^+$, respectively. It is suggested that the weak indicator, crystal violet, coordinates stepwise the chloride ion acceptors to its dimethylamino-nitrogen atom. Addition of chloride ions results in the formation of chloro-complexes which are stronger than the coordination compounds between the acceptor halides and the indicator. This leads to a reversal of the color change.

B. Phosphoryl Bromide

Pure phosphoryl bromide (mp 56°C, bp 193°C, specific conductance $< 4 \times 10^{-8}$ ohm^{-1} cm^{-1} at 56°C) is a colorless crystalline solid which is very hygroscopic and is easily hydrolyzed. Its physical constants were given in Table VIII. Of the various methods available for the preparation of phosphoryl bromide, the method of digestion of phosphorus(V) oxide and bromide[233,234] has been found to be the simplest. A calculated amount of bromine is added to phosphorus(III) bromide and the phosphorus(V) bromide so formed is heated with a slight excess of phosphorus(V) oxide for several days at 70°C. The product is distilled in an all-glass apparatus at atmospheric pressure and purified by crystallization from dry carbon tetrachloride. Final purification can be affected by repeated fractional freezing of the molten compound in an evacuated vessel until cryoscopic homogeneity is attained.[235]

Paul and Vasisht[236] have qualitatively determined the solubility of a number of bromides in molten phosphoryl bromide at 100°C. The results show that, in general, the ionic bromides are either insoluble or only slightly soluble, whereas covalent bromides are mostly soluble. Thus tin(IV), phosphorus(III), arsenic(III), and antimony(III) bromides are freely miscible. Phosphorus(V), antimony(V), tellurium(IV), tetramethylammonium, tetraethylammonium, dimethylphenylbenzylammonium, diethylphenylbenzylammonium, and lithium bromides have a solubility of less than 10%. Magnesium, cadmium, and mercury(II) bromides have a solubility of less than 0.5% and sodium, potassium, ammonium, barium, and lead(II) bromides are insoluble.

Phosphoryl bromide forms solvates with both Lewis acids and bases. Solid solvates such as $TiBr_4$, $2POBr_3$, $FeBr_2 \cdot 2POBr_3$[202]; $BCl_3 \cdot POBr_3$, $BBr_3 \cdot POBr_3$, $AlCl_3 \cdot POBr_3$, $AlBr_3 \cdot POBr_3$,[198] $TiCl_4 \cdot 2POBr_3$, $AlCl_3 \cdot 2POBr_3$, $AlBr_3 \cdot 2POBr_3$, and $SbCl_5 \cdot POBr_3$[236] have been reported. Infrared spectra of all these compounds have been interpreted in terms of coordination through the oxygen atom of phosphoryl bromide.[198,202,236] Phase diagram studies[235] show that phosphorylbromide reacts with gallium(III) bromide to form a 1 : 1 addition compound melting at 154°C. This complex is less stable than $GaCl_3 \cdot POCl_3$, but otherwise has similar properties and is considered to contain the cation $POBr_2^+$.

Sulphur trioxide, tin(IV) bromide, tin(IV) chloride, and antimony(III) bromide are miscible with molten phosphoryl bromide in all proportions. Plots of specific conductances vs. the composition of solutions of these Lewis acids in phosphoryl bromide indicate the formation of adducts with the acid to phosphoryl bromide molar ratio of 1 : 1 and 2 : 1 for SO_3, 1 : 1 and 1 : 2 for $SnBr_4$ and $SnCl_4$, and 1 : 1 for $SbBr_3$.[236] Three of these adducts, $2SO_3 \cdot POBr_3$, $SnBr_4 \cdot 2POBr_3$, and $SnCl_4 \cdot 2POBr_3$ have actually been isolated.

The specific conductivities of most of the phosphoryl bromide adducts with Lewis acids is considerably higher than those of either the solvent or the solutes (Table XII). Measurements of specific conductivity, density, and viscosity of $2SO_3 \cdot POBr_3$ and $SnBr_4 \cdot 2POBr_3$ at various temperatures give the energies of activation of ionic migration as 6.708 and 1.72 kcal/mole and energies of activation of viscous flow as 4.9 and 3.67 kcal/mole, respectively.[236]

Lithium bromide forms a yellow crystalline solvate $LiBr \cdot POBr_3$.[236] Quinoline, isoquinoline, α-picoline, γ-picoline, diethylaniline, and dimethylaniline form solid solvates in the base : $POBr_3$ molar ratio of 3 : 1, pyridine in the ratio of 4 : 1, and dimethylformamide in ratio of 5 : 1, 4 : 1, 3 : 1, and 2 : 1.[236] Phosphoryl bromide solutions of these bases are highly conducting[236] but the solid solvates are reported to be nonelectrolytes in nitrobenzene.[230] Infrared spectra of the adducts show that the bases are coordinated to the phosphorus atom of phosphoryl bromide.[230]

TABLE XII

SPECIFIC CONDUCTANCE[a] OF PHOSPHORYL BROMIDE SOLVATES OF LEWIS ACIDS[b]

Solvate	Specific conductance ($ohm^{-1}\ cm^{-1}$)	Temperature (°C)
$SO_3 \cdot POBr_3$	0.7×10^{-4}	60
$2SO_3 \cdot POBr_3$	8.3×10^{-4}	60
$SnBr_4 \cdot POBr_3$	3.95×10^{-5}	60
$SnBr_4 \cdot 2POBr_3$	4.9×10^{-5}	60
$SnCl_4 \cdot 2POBr_3$	6.8×10^{-5}	100
$SbCl_5 \cdot POBr_3$	4.48×10^{-4}	150
$GaBr_3 \cdot POBr_3$[c]	1.0×10^{-3}	161.2

[a] Specific cond. of pure $POBr_3$ is $4 \times 10^{-8}\ ohm^{-1}\ cm^{-1}$ at 56°C.

[b] R. C. Paul and S. K. Vasisht, *J. Indian Chem. Soc.* **43**, 141 (1966).

[c] N. N. Greenwood and I. J. Worrall, *J. Inorg. Nucl. Chem.* **6**, 34 (1958).

Neutralization reactions between tin(IV) bromide and α-picoline, dimethylaniline, and lithium bromide have been studied conductometrically.[236] The titration curves show breaks at an $SnBr_4$: base molar ratio of 1 : 1 and 1 : 2 for α-picoline, dimethylaniline, and lithium bromide, and 1 : 2 for quinoline. The reactions have been interpreted in terms of the equilibrium

$$2POBr_3 \rightleftharpoons POBr_2^+ + POBr_4^- \quad (73)$$

C. Thiophosphoryl Halides

In contrast to phosphoryl halides, thiophosphoryl halides have received little attention as far as investigations of their potential as solvents or in the formation of adducts with metal halides are concerned.

Thiophosphoryl chloride is a transparent, colorless liquid which can be prepared by the action of phosphorus(III) chloride on sulfur in the presence of anhydrous aluminum chloride.[237] Thiophosphoryl bromide is a yellow, crystalline solid with a pungent odor and can be synthesized by digesting phosphorus(V) sulfide with phosphorus(V) bromide until the reaction is complete or by the action of phosphorus(III) bromide on sulfur in the presence of aluminum(III) bromide.[238,239] Important physical constants of both the thiohalides were listed in Table VIII.

Thiophosphoryl chloride dissolves aluminum(III), antimony(III), iron-(III), and bismuth(III) chlorides to the extent of 30.7, 125.3, 7.3, and 0.5 g/100g of the solvent, respectively, at 35°C. Other metal chlorides and nitrogen bases have solubilities similar to the corresponding bromides in thiophosphoryl bromide.[240]

Qualitative determination of solubilities[241,242] in molten $PSBr_3$ at 50°C indicates that ionic halides such as alkali, alkaline earth, and transition metal bromides are insoluble; Al(III), Bi(III), Hg(II), Cu(II), and Cd(II) bromides are slightly soluble; and Sb(III), As(III), Sn(IV), and P(III) bromides are miscible in all proportions. Tertiary nitrogen bases such as pyridine, quinoline, and piperidine react with $PSBr_3$ to form solid solvates which are almost completely insoluble in $PSBr_3$. However, diethylaniline and dimethylaniline are soluble in it. Of the quaternary ammonium bromides, those with bulky cations such as dimethylphenylbenzylammonium bromide and diethylphenylbenzylammonium bromide are appreciably soluble in molten $PSBr_3$.

Thiophosphoryl halides are much weaker donors toward metal chlorides than phosphoryl halides. Only antimony(V) chloride, aluminum chloride and bromide, sulfur trioxide, and iron(III) chloride have been reported to form solid solvates with $PSCl_3$ and $PSBr_3$. Infrared spectra[242–244] of all these solvates indicate that coordination in the adducts is through the sulfur atom of the thiophosphoryl halides. Existence of some other solvates have been indicated through conductance measurements.[242] Several metal chlorides, which form addition compounds with phosphoryl chloride, were found not to do so with thiophosphoryl chloride[240,243,245] and thiophosphoryl bromide.[241]

Antimony(V) chloride forms $SbCl_5 \cdot PSCl_3$[240,243] and $SbCl_5 \cdot PSBr_3$[242,244,246] with thiophosphoryl chloride and bromide. The latter is light yellow when freshly prepared. However, it is unstable and changes to reddish yellow on standing. Infrared spectra of the compound taken over a period of time indicate halogen exchange between phosphorus and antimony.[241]

Two solvates of aluminum chloride with thiophosphoryl chloride, i.e., $PSCl_3 \cdot AlCl_3$[240] and $PSCl_3 \cdot 2AlCl_3$[247] have been reported. Later work by Van der Veer and Jellinek,[243] however, shows that $PSCl_3 \cdot AlCl_3$, obtained by cooling a warm solution of $AlCl_3$ in $PSCl_3$, is a mixture and not a pure compound. Maier[247] found mixed solutions of $AlCl_3$ and $PSCl_3$ in dichloromethane to be good conductors of electricity. The NMR signal of the ^{31}P nucleus was observed at much lower fields than in solution of $PSCl_3$ only. Maier interpreted these observations in terms of the structures $PSCl_2^+ \cdot AlCl_4^-$ and $PSCl_2^+ \cdot Al_2Cl_7^-$ for the solvates $PSCl_3 \cdot AlCl_3$ and $PSCl_3 \cdot 2AlCl_3$. Maier's experimental results have been confirmed by Vriezen and Jellinek.[248] From the mixed solutions of $AlCl_3$ and $PSCl_3$ in CH_2Cl_2, needles of a colorless compound were obtained which contained $AlCl_3$, $PSCl_3$, and CH_2Cl_2. If CH_2Cl_2 was removed from the complex by prolonged evaporation *in vacuo*, $Cl_3PS \cdot (AlCl_3)_q$ was formed with partial decomposition into $PSCl_3$ and $AlCl_3$. This indicates that the solvent takes part in the reaction of $AlCl_3$ and $PSCl_3$ and, therefore, Maier's interpretation of his results was considered as an oversimplification. Aluminum chloride and bromide give $AlCl_3 \cdot PSBr_3$[242,243] and $AlBr_3 \cdot PSBr_3$[242,244,246] with thiophosphoryl bromide. The former is unstable at room temperature and is also formed as an intermediate in the reaction of aluminum bromide with thiophosphoryl chloride which takes place with exchange of the halogen atoms.[243,244]

Sulfur trioxide forms monosolvates $SO_3 \cdot PSCl_3$[243] and $SO_3 \cdot PSBr_3$[242,244] which are unstable and very hygroscopic. Iron(III) chloride has also been reported to form monosolvates with both $PSBr_3$ and $PSCl_3$.[241] Electrolysis[249] of $PSCl_3 \cdot {}^*FeCl_3$ in nitromethane suggests ionization in the manner of $PSCl_2^+ + FeCl_4^-$. Such ionization is in agreement with the observation that aluminum chloride catalyzes the chemical reaction between benzene and thiophosphoryl chloride in which two chlorine atoms are replaced by phenyl groups. This reaction supports the presence of the $PSCl_2^+$ ion which is then involved in the electrophilic substitution. Surprisingly, ferric chloride does not catalyze the thiophosphorylation reaction.[249]

Tertiary nitrogen bases[241,242] such as quinoline, isoquinoline, α- and γ-picolines, morpholine, and aniline form solid solvates with both $PSCl_3$ and $PSBr_3$ in the base to thiophosphoryl halide molar ratio of 3 : 1. Pyridine forms solvates in the molar ratio 4 : 1 with both thiophosphoryl halides, piperidine forms a 4 : 1 compound with $PSBr_3$ and a 3 : 1 compound with $PSCl_3$, while dimethylformamide forms a 4 : 1 compound with $PSBr_3$.

Pure thiophosphoryl bromide has a specific conductivity of 7.8×10^{-8} ohm^{-1} cm^{-1} at 50°C, which increases considerably on addition of Lewis acids and bases.[241,242] Dimethylphenylbenzylammonium bromide, diethylphenylbenzylammonium bromide, and diethylaniline can be titrated conductometrically against arsenic(III), antimony(III), phosphorus(III), tin(IV), and titanium(IV) bromides. The titration curves show sharp breaks at the acid to base molar ratio of 1 : 1 and 1 : 2 in the case of $SnBr_4$ and $TiBr_4$ and of 1 : 1 in the case of other acids. Conductometric and potentiometric titrations have also been performed in thiophosphoryl chloride as solvent with similar results.[241,242]

Several acid–base neutralization complexes of the type $(B \cdot PSCl_2)SbCl_6$, $(B \cdot PSCl_2)AlCl_4$, $(B \cdot PSCl_2)_2SnCl_6$, $B_2SnCl_4 \cdot PSCl_3$, $(B \cdot PSCl_2)TiCl_5$, and $B_2 \cdot TiCl_4 \cdot PSCl_3$ (where B = quinoline, pyridine, and α- or γ-picoline) have been isolated in thiophosphorylchloride medium.[240]

VI. SELENINYL CHLORIDE

Seleninyl chloride or selenium(IV) oxychloride was first prepared by Weber[250] by heating together the vapors of selenium dioxide and selenium tetrachloride. More conveniently, it can be prepared by the dehydration of dichloroseleneous acid ($SeO_2 \cdot 2HCl$) with concentrated sulfuric acid or by the direct interaction of $SeCl_4$ and SeO_2 in carbon tetrachloride as solvent.[251] Fractional crystallization and/or distillation under reduced pressure yields a pure product which is almost colorless. Its important physical properties are summarized in Table XIII.

The wide liquid range of $SeOCl_2$ indicates that the solvent has a good potential as a medium of chemical reactions over a range of temperatures, However, at higher temperatures it is very reactive and quite frequently behaves as a powerful oxidizing agent.[252-254] Sulfur, selenium, and tellurium are soluble in the cold, but at elevated temperatures they react with the solvent. Bromine and iodine are also readily soluble in the cold. Boron, carbon, and silicon are insoluble and unreactive. Copper, lead, cobalt,

TABLE XIII

PHYSICAL CONSTANTS OF SELENINYL CHLORIDE

Melting point (°C)	10[a]
Boiling point (°C)	179.6[b]
Specific conductance	$2.3 \pm 0.3 \times 10^{-5}$ ohm^{-1} cm^{-1} at 25°C[c]
Dielectric constant	46.2 ± 1 at 20°C[d]
Density	2.427 g/cc at 25°C[b]
Viscosity	9.203 sp[b]
Surface tension	31.6 dyn/cm at bp[b]
Heat of evaporation	11.2 kcal/mole[b]

[a] W. J. R. Henley and S. Sugden, *J. Chem. Soc.* p. 1058 (1929).
[b] L. A. Niselson, K. V. Tretyakova, E. P. Paremuzov, and E. N. Torhina, *Izv. Akad. Nauk SSSR, Neorg. Mater.* **7**, 792 (1971).
[c] A. P. Julien, *J. Am. Chem. Soc.* **47**, 1799 (1925).
[d] J. E. Wildish, *J. Am. Chem. Soc.* **42**, 2607 (1920).

nickel, iron, tin, and silver react to give the corresponding metal chlorides, selenium monochloride, and selenium dioxide.

$$3Cu + 4SeOCl_2 \rightarrow 3CuCl_2 + Se_2Cl_2 + 2SeO_2 \quad (74)$$

Seleninyl chloride reacts with chromium(III) chloride and potassium dichromate on heating to give chromyl chloride and with potassium chlorate to give chlorine. Carbonates of the alkali metals are attacked by $SeOCl_2$ to liberate carbon dioxide, whereas those of calcium, strontium, copper, nickel, cobalt, and iron(II) are not attacked by the dry solvent. Many oxides and sulfides also undergo solvolysis with the formation of the metal chlorides.

Seleninyl chloride has been usefully employed as an ionizing solvent for a variety of reactions. Most of the earlier work along these lines has been reviewed by Smith.[255] A quantitative estimate of the solubilities of metal chlorides in $SeOCl_2$ has been given by Wise[256] (Table XIV). With the exception of barium chloride, the solubility of most of the chlorides increases with increasing temperatures. It should be noted that the solubilities of predominantly covalent chlorides are usually higher than those of the ionic chlorides and that solvate formation generally accompanies high solubility.

Conductance of pure $SeOCl_2$ has been attributed by Smith[255] to its autoionization.

$$2SeOCl_2 \rightleftharpoons (SeOCl \cdot SeOCl_2)^+ + Cl^- \quad (75)$$

Electrolysis of the pure liquid yields chlorine at the anode and selenium monochloride is probably formed at the cathode. A solution of potassium chloride or tin(IV) chloride in $SeOCl_2$ when electrolyzed gave chlorine at the anode and selenium dioxide and selenium monochloride at the cathode.

TABLE XIV

SOLUBILITIES OF METAL CHLORIDES IN SELENINYL CHLORIDE AT 25°C[a]

Metal	Solubility[b]	Metal	Solubility
Li	3.21	Cu(II)	Insoluble
Na	0.57	Ag	Insoluble
K	2.89	Ti(IV)	0.75
Rb	3.56	Sn(IV)	13.73
Cs	3.83	Pb(II)	Insoluble
Mg	4.96	As(III)	Miscible in all proportions
Ca	6.11	Sb(V)	38.64
Sr	5.17	Cr(III)	Insoluble
Ba	3.95	Mn(II)	0.16
Zn	1.10	Fe(III)	23.40
Cd	0.15	Ni(II)	0.15
Hg(II)	0.89	Co(II)	0.17

[a] C. R. Wise, *J. Am. Chem. Soc.* **45**, 1233 (1923).

[b] Solubilities are expressed as grams of solute per 100 grams of the solvent.

The following equations have been formulated for the electrolytic decomposition of the solvent:

$$\text{anodic } 2Cl^- \rightarrow Cl_2 + 2e$$

$$\text{cathodic } 6(SeOCl)^+ + 6e \rightarrow 6(SeOCl) \tag{76}$$

$$6(SeOCl) \rightarrow 2SeO_2 + Se_2Cl_2 + 2SeOCl_2$$

The suggested autoionization of $SeOCl_2$, which has been supported by Masters *et al.*[257] by radio chlorine exchange reactions in the solvent, has proved to be extremely helpful in understanding the behavior of a large number of substances in terms of acid–base phenomena in the solvent. The Lewis acids $SnCl_4$, $TiCl_4$, $FeCl_3$, $AsCl_3$, and $ZrCl_4$ and the bases pyridine, quinoline, and isoquinoline give highly conducting solutions in $SeOCl_2$.[255] This has been explained on the basis of the ionization of the corresponding solvoacids and solvobases:

$$SnCl_4 + 2SeOCl_2 \rightleftharpoons 2(SeOCl)^+ + SnCl_6^{2-} \tag{77}$$

$$C_5H_5N + SeOCl_2 \rightleftharpoons C_5H_5N \cdot SeOCl^+ + Cl^- \tag{78}$$

Several conductometric titrations involving acid–base neutralization reactions (Table XV) have been performed in $SeOCl_2$.[182,255] The equation for a typical conductometric titration between tin(IV) chloride and pyridine in $SeOCl_2$ has been formulated as

$$2(C_5H_5N \cdot SeOCl)Cl + (SeOCl)_2SnCl_6 \rightarrow (C_5H_5N \cdot SeOCl)_2SnCl_6 + 2SeOCl_2 \tag{79}$$

TABLE XV

CONDUCTOMETRIC ACID–BASE TITRATIONS IN $SeOCl_2$[a,b]

System	Stoichiometric ratios indicated
$C_5H_5N-SnCl_4$	3 : 1, 2 : 1, 1 : 1, 1 : 2
$KCl-SnCl_4$	3 : 2, 1 : 1
$CaCl_2-SnCl_4$	3 : 2, 1 : 1, 1 : 2
$C_5H_5N-SO_3$	2 : 1, 1 : 1
$C_5H_5N-AsCl_3$	2 : 1, 1 : 1, 1 : 2
$C_5H_5N-FeCl_3$	3 : 1, 2 : 1, 1 : 1
$KCl-ZrCl_4$	3 : 1, 2 : 1, 1 : 1

[a] G. B. L. Smith, *Chem. Rev.* **23**, 165 (1938).
[b] V. Gutmann and R. Himml, *Z. Anorg. Allg. Chem.* **287**, 199 (1956).

Potentiometry has been employed to study neutralization reactions in $SeOCl_2$.[258] In Fig. 11 the titration curves for the titrations of potassium chloride, pyridine, quinoline, and isoquinoline with sulfur trioxide are shown. The order of relative base and acid strengths as determined by such potentiometric titrations are as follows: (1) relative acid strengths: $SO_3 > FeCl_3 > SnCl_4$, (2) relative base strengths: isoquinoline > quinoline > pyridine > potassium chloride.

Solvates of several substances with $SeOCl_2$ are known. Wise[256] has reported the formation of $KCl \cdot SeOCl_2$, $RbCl \cdot SeOCl_2$, $MgCl_2 \cdot 3SeOCl_2$, $TiCl_4 \cdot 2SeOCl_2$, $SnCl_4 \cdot 2SeOCl_2$, $FeCl_3 \cdot 2SeOCl_2$, and $SbCl_5 \cdot 2SeOCl_2$. The formation of the last solvate, however, has not been confirmed by Agermann *et al.*[70] Their melting point diagram for the $SbCl_5-SeOCl_2$ system indicated the formation of only $SbCl_5 \cdot SeOCl_2$. The crystal structure of the solvate consists of discrete $SbCl_5 \cdot SeOCl_2$ molecules.[259] The Sb atom is octahedrally coordinated by five Cl atoms and one oxygen atom. The $SeOCl_2$ part of the molecule has a pyramidal shape. Hermodsson has carried out and reviewed x-ray structural studies on complexes of $SeOCl_2$ with Lewis acids.[260,261] The data indicate that $SeOCl_2$ always coordinates via the oxygen atom in the solid state.

Sheldon and Tyree[262] have observed that pure $TiCl_4 \cdot 2SeOCl_2$ is obtained when $TiCl_4$ and $SeOCl_2$ solutions are mixed in CCl_4. However, mixing of pure $TiCl_4$ and $SeOCl_2$ gives a mixture of $SeCl_4$ and TiO_2 instead of the pure adduct. A similar reaction has been shown to occur between $SiCl_4$ and $SeOCl_2$. This has been supported by the work of Frazer[263] who has shown that both BCl_3 and $AlCl_3$ react with $SeOCl_2$ to give a solid compound of composition $3SeOCl_2 \cdot 2MCl_3$ or $M_2O_3 \cdot 3SeCl_4$. On sublimation, the solid product gives $SeCl_4$ and the infrared spectrum indicates the presence of M_2O_3 and $SeCl_4$ mixture rather than the adduct.

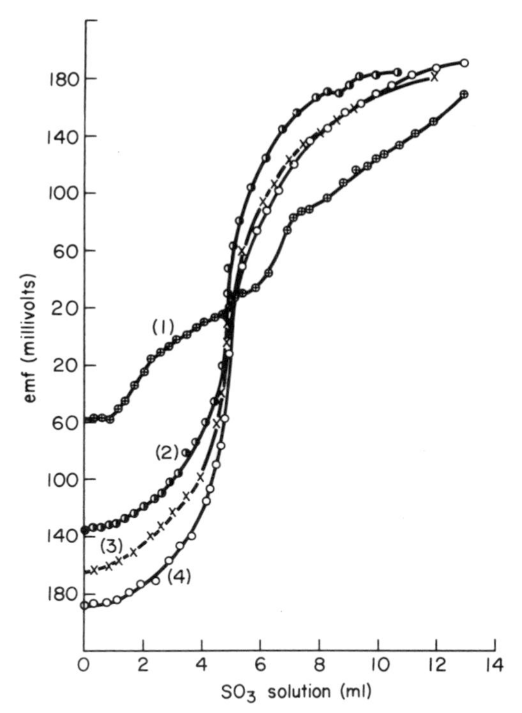

FIG. 11. Potentiometric titration curves of (1) potassium chloride, (2) pyridine, (3) quinoline, and (4) isoquinoline with SO_3 in $SeOCl_2$.

Two other solvates, $ZrCl_4 \cdot 2SeOCl_2$[182] and $SO_3 \cdot SeOCl_2$[264] have also been reported. Study of the infrared spectra of the latter favors coordination through oxygen and gives no evidence for the existence of SO_3Cl^-. Seleninyl chloride gets protonated in disulphuric acid[265]

$$SeOCl_2 + 2H_2S_2O_7 \rightarrow SeOH^+Cl_2 + HS_3O_{10}^- + H_2SO_4 \tag{80}$$

whereas a compound capable of furnishing Cl^- is expected to react differently

$$MCl + 3H_2S_2O_7 \rightarrow M^+ + HSO_3Cl + HS_3O_{10}^- + 2H_2SO_4 \tag{81}$$

The two modes of reaction represented by equations (79) and (80) can easily be distinguished with conductance and cryoscopic investigations in $H_2S_2O_7$.

Devynck and Termillion[266] have carried out a voltometric study with rotating disk microelectrodes and potentiometric measurements with a chloride electrode. Pyridine acts as a strong base having an electrochemical behavior similar to tetramethylammonium chloride. The cationic entity,

$SeOCl^+$, is produced by a constant current coulometric oxidation of the solvent. The value of ionic product, $K_i = [SeOCl]^+[Cl]^-$, has been determined to be $10^{-9.7 \pm 0.2}$ (mole/liter)^2 at an ionic strength of 0.5 and temperature of 25°C.

Birchall *et al.*[267] have studied the ^{77}Se-NMR spectra of pure seleninyl halides as well as their mixtures. It has been found that a mixture of $SeOF_2$ and $SeOCl_2$ contains SeOClF and that rapid selenium exchange occurs between $SeOCl_2$ and $SeOBr_2$, presumably via the intermediate, SeOClBr. It has been further observed by these workers that addition of $SbCl_5$ to $SeOCl_2$ gives a high-field shift that increases linearly with the $SbCl_5$ concentration, thus substantiating the formation of $SeOCl^+$. A small shift, such as that for $AsCl_3$, indicates it to be a weak acid. The large shift produced by $SnCl_4$, on the other hand, shows that it is ionized to a large extent according to the Eq. 77. The two bases quinoline and potassium chloride give high field shifts of the same order of magnitude as those produced by acids. The larger shift for KCl indicates a strong interaction of the chloride ions with the solvent. This can be attributed to the formation of the complex ion $SeOCl_3^-$ so that the self-ionization of the solvent could be represented as

$$2SeOCl_2 \rightleftharpoons SeOCl^+ + SeOCl_3^- \tag{82}$$

Data for chemical shifts of the acids and bases are listed in Table XVI.

Agermann *et al.*[70] have reported that three intermediate phases, $Me_4NCl \cdot 5SeOCl_2$, $Me_4NCl \cdot 3SeOCl_2$, and $Me_4NCl \cdot 2SeOCl_2$ occur in the $(CH_3)_4NCl$–$SeOCl_2$ system. The crystal structure[268] of $Me_4NCl \cdot 5SeOCl_2$ consists of discrete tetramethylammonium ions and solvated chloride ions. Each anion is octahedrally surrounded by six $SeOCl_2$ molecules, two of which are shared by an adjacent octahedron.

Seleninyl chloride forms well-defined stable adducts of the type

TABLE XVI

CHEMICAL SHIFTS FOR SOME ACIDS AND BASES IN $SeOCl_2$[a]

Acid or base	Moles of acid or base/moles $SeOCl_2$	$SeOCl_2$ (ppm)	Calc. for mole ratio = $\frac{1}{3}$
$AsCl_3$	1 : 2	2.5	3.2
$AsCl_3$	1 : 1	7.0	2.3
$SnCl_4$	1 : 6	26.2	52.4
$SbCl_5$	1 : 2.9	35.0	33.6
Quinoline	1 : 7	10.3	24.0
KCl	1 : 15	5.1	25.3

[a] T. Birchall, R. J. Gillespie, and S. L. Vekris, *Can. J. Chem.* **43**, 1672 (1965).

$SeOCl_2 \cdot nB$ ($n = 1$ or 2 and B = pyridine, quinoline, or isoquinoline) with Lewis bases.[269,270] The disolvates are not decomposed at 61°C in the case of quinoline and isoquinoline, whereas the pyridine adduct is slowly converted to the monosolvate at this temperature. Edgington and Firth[270] have reported that the compound $SeOCl_2 \cdot 2Py$ decomposes at 145°C giving Py_2SeCl_4 and Py_2SeCl_2. An x-ray analysis[271] of the compound $SeOCl_2 \cdot 2Py$ shows that one selenium–chlorine bond distance is abnormally long. In solution this chlorine atom could easily undergo ionization and establish an equilibrium according to Eq. 78.

VII. Sulfur Halides and Oxyhalides

Sulfur halides and oxyhalides which have been studied as solvents to some extent are disulfur dichloride, thionyl chloride, thionyl bromide, and sulfuryl chloride. Most thoroughly investigated among these, however, is thionyl chloride.

A. Disulfur Dichloride

Disulfur dichloride or sulfur monochloride is prepared by the action of dry chlorine with solid or molten sulfur at temperatures ranging from 50°–130°C. The product may contain sulfur dichloride and other sulfur halides. A chlorine carrier, e.g., iron, iodine, or a trace of iron(III) chloride facilitates the reaction.[272] It can be purified by distillation over activated bone charcoal (1%) and sulfur (2%). The final product is a light-yellow liquid (bp 137°C/760 mm with decomposition). It can be distilled under reduced pressure at temperatures less than 60°C with no decomposition.

Electron diffraction studies[273–275] favor the right-angled structure of the H_2O_2 type for S_2Cl_2. Calculations of bond dissociation energies show that there is a possibility of the existence of an ionic structure of the type $[S{=}S{-}Cl]^+Cl^-$ in addition to the homopolar structures.[276] The specific electrical conductivity (1.3×10^{-10} ohm^{-1} cm^{-1} at 20°C) of the pure solvent,[277] and the reactions carried out in it as solvent,[278] indicate that S_2Cl_2 dissociates (although very slightly) as

$$S_2Cl_2 \rightleftharpoons S_2Cl^+ + Cl^- \qquad (83)$$

if the possibility of further solvation of ions is overlooked. Disulfur dichloride rapidly exchanges radiochlorine with tetraethylammonium chloride, antimony(III) chloride, and antimony(V) chloride, but exchanges very slowly with thionyl chloride.[279] The results suggest ready dissociation of S_2Cl_2 in the presence of a strong chloride acceptor. Some important physical properties of S_2Cl_2 are recorded in Table XVII.

TABLE XVII

PHYSICAL CONSTANTS OF DISULFUR DICHLORIDE[a]

Melting point	−76.5°C
Boiling point at 760 mm	137.1°C
Density at 20°C	1.6828 g/ml
Molecular volume	90.8 ml
Dielectric constant at 22°C	4.9
Dipole moment at 25°C	1.60 Debye
Specific electrical conductivity at 20°C	1.3×10^{-10} ohm^{-1} cm^{-1}
Cryoscopic constant	5.36°C/mole
Ebullioscopic constant	5.02°C/mole

[a] V. H. Spandau and H. Hattwig, *Z. Anorg. Allg. Chem.* **311**, 32 (1961).

Spandau and Hattwig[278] have made a qualitative study of the solubilities of various substances in S_2Cl_2 at room temperature and at 120°C. Alkali, alkaline earth metals, and ammonium chlorides are insoluble in S_2Cl_2. Several predominantly homopolar chlorides such as CCl_4, $TiCl_4$, and PCl_3 are completely miscible with the solvent without increasing the conductance, while $POCl_3$, $AsCl_3$, $SnCl_4$, $SOCl_2$, and SO_2Cl_2 dissolve to give conducting solutions. Chlorides such as Hg_2Cl_2, $HgCl_2$, $(C_2H_5)_4NCl$, and $(C_2H_5)_3NHCl$ are only slightly soluble at room temperature but possess remarkably large positive temperature coefficients of solubility. For example, Hg_2Cl_2 is practically insoluble at room temperature but has a solubility of 14.08 g/liter at 102°C, and $HgCl_2$, with a solubility of 1.14 g/liter at 20°C, has a solubility of 13.94 g/liter at 102°C. Of the organic compounds, pyridine, quinoline, and cyclohexanone are miscible in S_2Cl_2 to give solutions of relatively high specific conductances.

Several investigators have examined the action of S_2Cl_2 on metals and metal oxides and the best conditions for the preparation of chlorides have also been worked out.[280–285] Several other salts,[286] and sulfide and oxide minerals[287–289] react with S_2Cl_2 to yield the corresponding chlorides.

Disulfur dichloride forms adducts with both inorganic and organic compounds and several solvates of metal chlorides, oxides, acetates, sulfates, and organic bases such as pyridine, quinoline, α-picoline, morpholine, piperidine, diethylamine, triethylamine, and ethylenediamine have been reported.[290–292] Substances like $(C_2H_5)_4NCl$, $(C_2H_5)_3NHCl$, $(C_6H_5)_3CCl$, $HgCl_2$, $POCl_3$, PCl_5, pyridine, quinoline, and cyclohexanone act as bases in S_2Cl_2, while chlorides such as $SbCl_3$, $SnCl_4$, and $AsCl_3$ act as acids in S_2Cl_2. Spandau and Hattwig have determined the specific and molar conductances of these acids and bases in S_2Cl_2.[278] The low conductance of the solutions has been ascribed principally to the smaller dissociation constant of S_2Cl_2. Acid–base neutralization reactions in S_2Cl_2 have been studied by conductometric titrations and the products of the reactions have been identified analytically.[278] All these neutralization reactions have been explained on the basis of the ionization of the solvent and the existence of solvoacids and solvobases specific to it. Figure 12 shows the conductance curves obtained in the titrations of $(C_6H_5)_3CCl$ with $AsCl_3$ and $SnCl_4$. The neutralization reactions proceeds according to equations of the type

$$AsCl_3 + S_2Cl_2 \rightleftharpoons S_2Cl^+ + AsCl_4^- \quad (84)$$

$$(C_6H_5)_3CCl \rightleftharpoons (C_6H_5)_3C^+ + Cl^- \quad (85)$$

$$(C_6H_5)_3C^+ + Cl^- + S_2Cl^+ + AsCl_4^- \rightarrow [(C_6H_5)_3C]AsCl_4 + S_2Cl_2 \quad (86)$$

Infrared spectra of the solvates[292] $BCl_3 \cdot S_2Cl_2$ and $SbCl_5 \cdot S_2Cl_2$ have been interpreted in terms of the structures $S_2Cl^+ \cdot BCl_4^-$ and $S_2Cl^+ \cdot SbCl_6^-$. Similar studies on the solvates of organic tertiary bases with S_2Cl_2 indicate that the nitrogen of the bases coordinates to the sulfur atom of the sulfur chloride.[292]

B. Thionyl Chloride

Thionyl chloride is a fairly good ionizing medium for chemical reactions, but its usefulness as a nonaqueous solvent is slightly limited because of its high chemical reactivity and its inability to dissolve salts to an appreciable extent.

Crude thionyl chloride is usually contaminated with chlorides of sulfur and can be easily purified to give a product of over 99% purity by treatment with flowers of sulfur (3–5%) in the presence of a trace of an iron-containing

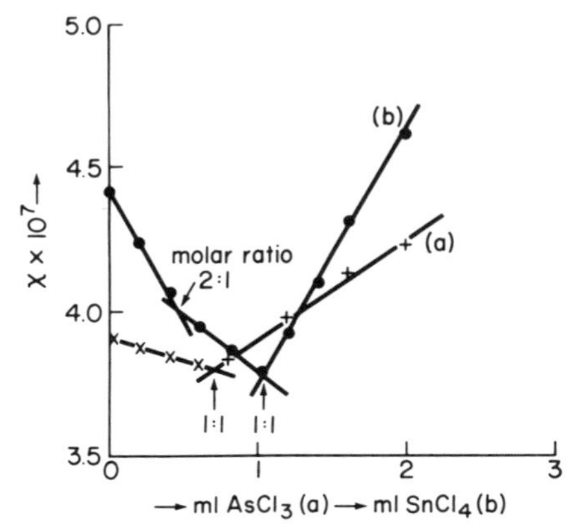

FIG. 12. Conductometric titrations of $(C_6H_5)_3CCl$ with $AsCl_3$ (curve a) and $SnCl_4$ (curve b) in S_2Cl_2 at 20°C.

catalyst such as iron stearate, iron(III) oxide, or iron(III) chloride and fractionating the mixture.[293] Sulfuryl chloride, if present, can be removed by refluxing with anhydrous aluminum chloride and naphthalene.[294] Thionyl chloride is conveniently obtained in a high state of purity by atmospheric distillation over 10 weight % of triphenylphosphite. Two distillations give pure material spectroscopically and gas chromatographically.[295] It can also be purified by treating with 2% quinoline and fractionating in a glass apparatus.[296] Important physical constants of thionyl chloride are listed in Table XVIII.

Spandau and Brunneck[297] have investigated the solubilities of numerous compounds in thionyl chloride. These results show that, in general, predominantly ionic compounds are insoluble while covalent compounds are fairly soluble. Action of thionyl chloride on metals,[298–300] metalloids,[298,301,302] and oxides[303–305] have been extensively studied. All the substances solvolyze to yield the corresponding metal chlorides. If the metal exists in two oxidation states, the products formed depend on whether the metal or the thionyl chloride is in excess. In the former case, lower metal chloride, sulfide, and sulfur dioxide are formed, and in the latter case higher metal chloride, sulfur dioxide, and sulfur monochloride are formed. Precipitated sulfides[306] and sulfide minerals[307] are also solvolyzed by thionyl chloride to yield the corresponding metal chloride.

The readiness with which thionyl chloride reacts with water is responsible for its application as a dehydrating agent in the preparation of anhydrous metal halides. Hydrated chlorides of thorium, lanthanides, and several other elements can be readily dehydrated by refluxing with thionyl chloride.[308] Anhydrous metal halides are also formed when nitrates, carbonates, formates, acetates, and propionates of alkali and alkaline metals, Cu, Ni, Co, Zn, Cd, and Mn are treated with thionyl chloride at room temperature or at the boiling point.[309]

TABLE XVIII

PHYSICAL CONSTANTS OF SULFUR OXYHALIDES[a,b]

	$SOCl_2$	$SOBr_2$	SO_2Cl_2
Melting point (°C)	−104.5	−50	−54.1
Boiling point (°C)	75.6	138/773 mm dec.	69.2
Dielectric constant	9.05 at 22°C	9.06 at 20°C	10.0 at 21.5°C
Dipole moment	1.58 D	1.47 Debye	1.79 Debye
Density	1.656 at 14.5°C	2.688 at 20°C	1.658 at 25°C
Specific conductance	3.5×10^{-9} ohm^{-1} cm^{-1}	1×10^{-7} ohm^{-1} cm^{-1}	2×10^{-8} ohm^{-1} cm^{-1}

[a] V. H. Spandau and E. Brunneck, *Z. Anorg. Allg. Chem.* **270**, 201 (1952).
[b] R. C. Paul, M. Singh, and S. K. Vasisht, *J. Indian Chem. Soc.* **47**, 635 (1970).

Uranium(V) chloride,[305] antimony(V) chloride,[310] titanium(IV) chloride,[262] zirconium(IV) chloride,[182] and carbonyl chloride[311] form monosolvates with thionyl chloride. Aluminum chloride forms $AlCl_3 \cdot SOCl_2$ and $2AlCl_3 \cdot SOCl_2$.[312] A comparison of the Raman spectra of the solutions of aluminum chloride in thionyl chloride and the two solvates show that the solutions consist of the essentially unassociated 1 : 1 complex in excess of thionyl chloride and that the 1 : 1 complex is of the donor–acceptor type with coordination through the oxygen atom.[312] The 2 : 1 complex is probably a 1 : 1 complex with a loosely attached second $AlCl_3$ molecule. Organic tertiary bases such as pyridine, β-picoline, γ-picoline, quinoline, isoquinoline, and piperidine form solvates of the type $2B–SOCl_2$ whereas trimethylamine forms a monosolvate.[313,314] Infrared spectra[313] of the solvates with tertiary bases has been interpreted to indicate the coordination of the base molecule through its nitrogen atom to the sulfur atom of thionyl halide.

Thionyl chloride has a specific conductance of 3.5×10^{-9} ohm^{-1} cm^{-1} which has been attributed to its slight ionization.

$$SOCl_2 \rightleftharpoons SOCl^+ + Cl^- \tag{87}$$

Compounds such as antimony(III) chloride, antimony(V) chloride, iron(III) chloride, tin(IV) chloride, aluminum chloride, and sulfur dioxide dissolve in thionyl chloride to give solutions which show higher conductance than either the solvent or the solute. These compounds have been classified as solvoacids of the solvosystem thionyl chloride because of their tendency to form chlorometallate anions, thereby increasing the relative concentration of $SOCl^+$. The formation of the complex anions has been explained on the basis of the chloride ion transfer from the solvent to the solute.

$$AlCl_3 + SOCl_2 \rightleftharpoons [SOCl][AlCl_4] \rightleftharpoons SOCl^+ + AlCl_4^- \tag{88}$$

$$SbCl_5 + SOCl_2 \rightleftharpoons [SOCl][SbCl_6] \rightleftharpoons SOCl^+ + SbCl_6^- \tag{89}$$

Organic bases triethylamine, pyridine, quinoline, diethylaniline, and acetone react with thionyl chloride to increase the concentration of chloride ions, thus acting as solvobases of the system[297,315]

$$C_5H_5N + SOCl_2 \rightleftharpoons (C_5H_5N \cdot SOCl)^+ + Cl^- \tag{90}$$

Substituted ammonium chlorides ionize in thionyl chloride to give chloride ions directly. Titanium(IV) chloride and phosphorus(V) chloride show amphoteric behavior in $SOCl_2$.

Neutralization reactions between acids and bases characteristic of the solvent system thionyl chloride have been investigated conductometrically[63,296,297,315–318] and potentiometrically. Potentiometric titrations of organic bases with $SbCl_5$, $AlCl_3$, $SnCl_4$, $TiCl_4$, and $SbCl_3$ as titrants have been performed in $SOCl_2$.[315,319] Titration of $SbCl_5$ with Et_4NCl is more satisfactory with Mo electrodes than with Ag, Cu, Pt, or Ni electrodes. The titration curve shows a break at a 1 : 1 molar ratio. With molybdenum electrodes and Et_4NCl as a titrant, breaks in the potentiometric curves at the molar ratios for MCl_x : Et_4NCl are 1 : 1 with $AlCl_3$ and PCl_5, 1 : 1 and 1 : 2 with $SnCl_4$ and $TiCl_4$; and 1 : 1, 1 : 2, and 1 : 3 with $SbCl_3$.[315]

Tetramethylammonium chloride can be titrated conductometrically against the solvoacids $SnCl_4$, $SbCl_5$, $AlCl_3$, $FeCl_3$, and SO_3.[297] The conductance curves show breaks at acid to base molar ratios of 1 : 1 and 1 : 2 in the case of $SnCl_4$ and SO_3, 1 : 1 in the case of $SbCl_5$ and $AlCl_3$, and 1 : 1, 1 : 2, and 1 : 3 in the case of $FeCl_3$. The two breaks at the acid to base molar ratio of 1 : 2 and 1 : 1 for a titration with $SnCl_4$ have been attributed to the reactions

$$2Et_4NCl + [SOCl]_2SnCl_6 \rightarrow [Et_4N]_2[SnCl_6] + 2SOCl_2 \quad (91)$$

$$[Et_4N]_2[SnCl_6] + [SOCl]_2SnCl_6 \rightarrow 2[Et_4N][SOCl][SnCl_6] \quad (92)$$

Other substituted ammonium chlorides have been titrated conductometrically against $TiCl_4$, $ZrCl_4$, and $SnCl_4$ in thionyl chloride by Sandhu *et al.*[296] The titration curves show sharp breaks at an acid to base molar ratio of 1 : 2 and solid complexes of the formula $(R_1 \cdot R_2 \cdot R_3 \cdot R_4N)_2MCl_6$ have been isolated in each case. Titanium(IV) chloride is amphoteric and can be titrated against Et_4NCl or $SbCl_5$. Mercury(II) chloride is a base and can be titrated against $SbCl_5$.[297]

Conductometric titrations of triethylamine show breaks at the acid to base molar ratio of 1 : 1 with $SbCl_5$; 1 : 1 and 1 : 2 with $SnCl_4$; and 1 : 1, 2 : 3, and 1 : 3 with $FeCl_3$.[315] A potentiometric titration with $SbCl_5$ also shows a single break at a 1 : 1 molar ratio. This indicates that triethylamine acts as a monoacidic base in thionyl chloride.

$$[(C_2H_5)_3N \cdot SOCl]Cl + [SOCl][SbCl_6] \rightarrow [(C_2H_5)_3N \cdot SOCl]SbCl_6 + SOCl_2 \quad (93)$$

Pyridine, quinoline, aniline, diethylaniline, and diphenylamine, however, react differently with acids in thionyl chloride and form basic, neutral, and acid salts.[315] For example, potentiometric and conductometric titrations of pyridine against tin(IV) chloride show four breaks at the acid to base molar ratios of 1 : 1, 1 : 2, 1 : 3, and 1 : 4. Spandau and Brunneck[315] explain these breaks on the basis of the formation of the solvoacid salt $[(C_5H_5N)_2SO] \times [SOCl]_2[SnCl_6]_2$ at the molar ratio of 1 : 1, neutral salt $[(C_5H_5N)_2SO]SnCl_6$ at the molar ratio of 1 : 2, and solvobasic salts $[(C_5H_5N)_2SO]_3Cl_2[SnCl_6]_2$ and $[(C_5H_5N)_2SO]_2Cl_2SnCl_6$ at 1 : 3 and 1 : 4. Pease and Luder,[318] who detected two breaks at the $SnCl_4$: C_5H_5N molar ratio of 1 : 1 and 1 : 2 in a conductometric titration curve, have suggested the following reactions to account for the two breaks:

$$C_5H_5N + SnCl_4 \rightleftharpoons (C_5H_5N \cdot SnCl_3)^+ + Cl^- \tag{94}$$

$$C_5H_5N + (C_5H_5N \cdot SnCl_3)^+ + Cl^- \rightleftharpoons (C_5H_5N)_2SnCl_4 \tag{95}$$

According to Sandhu,[320] when a solution of pyridine in thionyl chloride is added to a tin(IV) chloride solution in thionyl chloride, the first break at the acid to base molar ratio of 1 : 1 is due to the formation of the acid salt $[C_5H_5N \cdot SOCl][SOCl][SnCl_6]$. Further addition of pyridine solution results in the formation of the neutral salt $[C_5H_5N \cdot SOCl]_2SnCl_6$ at a molar ratio of 1 : 2. At this stage, when the simple neutralization reaction is complete, an equilibrium exists between the following species:

$$\begin{array}{c} [C_5H_5N \cdot SOCl]_2SnCl_6 \rightleftharpoons [C_5H_5N \cdot SOCl]^+[C_5H_5N \cdot SnCl_5]^- + SOCl_2 \\ \upharpoonleft\downharpoonright \\ [(C_5H_5N)_2]SnCl_4 + 2SOCl_2 \end{array} \tag{96}$$

Isolation of solid complexes with the general formula $B_2MCl_4 \cdot xSOCl_2$ (x varies between zero and one), as a result of neutralization reactions between the solvobases pyridine, quinoline, and isoquinoline and the solvoacids tin(IV) chloride, titanium(IV) chloride, and zirconium(IV) chloride, has been presented as an evidence for the above argument. Molar conductance measurements[321] on solutions of these complexes in nitrobenzene show that they exist as uniunivalent electrolytes and might be represented as $(B \cdot SOCl)^+(B \cdot MCl_5)^-$. The addition of thionyl chloride to the solution of these complexes in nitrobenzene converts them into unibivalent complexes of the type $(B \cdot SOCl)_2{}^+MCl_6{}^{2-}$. Sandhu[320] has represented the two other breaks at the molar ratio 1 : 3 and 1 : 4 by the equations

$$(C_5H_5N)_2SnCl_4 + C_5H_5N \rightleftharpoons (C_5H_5N)_3SnCl_3{}^+ + Cl^- \tag{97}$$

$$(C_5H_5N)_3SnCl_3{}^+ + C_5H_5N \rightleftharpoons (C_5H_5N)_4SnCl_2{}^{2+} + Cl^- \tag{98}$$

In these two reactions, thionyl chloride acts as a simple solvent which does not influence the interacting species except to provide a medium for them.

The solvent system approach has proved to be of great value in correlating the data from conductance measurements and conductometric and potentiometric titrations in $SOCl_2$, but so far there does not seem to be any direct evidence for the existence of $SOCl^+$ cations in solutions of Lewis acids in thionyl chloride. Although radiochlorine and radiosulfur exchange studies in thionyl chloride[257,322,323] and experiments on the electrolysis of thionyl chloride solutions[188] are in agreement with the postulated self-ionization of thionyl chloride, they do not prove unambiguously that such ionization actually occurs. The solute–solvent interaction in thionyl chloride can also be explained on the basis of the coordination model, which does not preclude the ionization of the solvent. For example, the conductance of solutions of $SbCl_5$ and $AlCl_3$ in $SOCl_2$ could be attributed to the ionizations

$$SbCl_5 + SOCl_2 \rightleftharpoons SbCl_5 \cdot SOCl_2 \rightleftharpoons [SbCl_4 \cdot 2SOCl_2]^+ + SbCl_6^- \quad (99)$$

$$AlCl_3 + SOCl_2 \rightleftharpoons AlCl_3 \cdot SOCl_2 \rightleftharpoons [AlCl_2 \cdot 2SOCl_2]^+ + AlCl_4^- \quad (100)$$

and the break occurring at the acid to base molar ratio of 1 : 2 in the conductometric titration of Et_4NCl and $SnCl_4$ in thionyl chloride could be explained on the basis of the equation[324]

$$3SnCl_4 + 6SOCl_2 \rightleftharpoons 3[SnCl_4 \cdot 2SOCl_2] \rightleftharpoons 2[SnCl_3 \cdot 3SOCl_2]^+ + SnCl_6^{2-} \xrightarrow{+6Et_4NCl} 3\,[Et_4N]_2[SnCl_6] + 6SOCl_2. \quad (101)$$

Visual titrations in thionyl chloride between the acids tin(IV) chloride, titanium(IV) chloride, and aluminum chloride and the bases pyridine and quinoline, using internal indicators such as crystal violet, melachite green, and benzanthrone, have been reported.[86,87,325,326] These indicators are reversible and give fairly good results.

C. Thionyl Bromide

Paul, Singh, and Vasisht[327] have recently studied thionyl bromide as a medium for chemical reactions. Most of the work reported below has been done by them.

Thionyl bromide can be prepared by passing dry hydrogen bromide through thionyl chloride at 0°C for about a week. The liquid thus formed is repeatedly fractionally distilled under reduced pressure. Thionyl bromide is a yellowish-orange, fuming liquid which is easily hydrolyzed to hydrogen bromide and sulfur dioxide. It is stable at room temperature but, when heated above 80°C, decomposes into sulfur dioxide, bromine, and sulfur monobromide. Some of its physical constants were given in Table XVIII.

Thionyl bromide reacts with a large number of metals, metal oxides, carbonates, iodides, formates, and acetates to yield anhydrous metal bromides. Most of these substances react completely at room temperature, while others have to be heated to 70°C to complete the reaction.

Monosolvates of aluminum bromide, aluminum chloride, and boron bromide have been isolated while the existence of disolvates of $AlBr_3$, $AlCl_3$, and $SbBr_3$ has been indicated conductometrically. Tin(IV), titanium(IV), and antimony(V) chlorides and bromides and tellurium(IV) and phosphorus(III) bromides are soluble in $SOBr_2$ without significant change in the specific conductance, and no solid solvate of these compounds has been isolated. Organic bases such as pyridine, α-picoline, quinoline, isoquinoline, benzylamine, morpholine, and dimethyaniline form solvates of the type 2 base · $SOBr_2$. Molar conductance of these solvates in nitrobenzene shows that they are uniunivalent electrolytes. Infrared spectra of the solvates indicate coordination of the base nitrogen atom to the sulfur atom of thionyl bromide.[313] Tetramethylammonium bromide and phenylbenzyldimethylammonium bromide form yellow 1 : 1 solvates whereas tetraethylammonium bromide and phenylbenzyldiethylammonium bromide form 2 : 1 orange-red solvates with thionyl bromide.[327] The molar conductance measurements in nitrobenzene indicate that the former complexes are uniunivalent and the latter are unidivalent electrolytes.

Acid–base titrations between pyridine, α-picoline, quinoline, piperidine, dimethylaniline, tetramethyl, and tetraethylammonium bromides against $SnBr_4$, $TiBr_4$, $AlBr_3$, and $SbBr_3$ have been performed in $SOBr_2$. The equations for the neutralization reactions have been formulated on the basis of the solvent system concept and by analogy with reactions in $SOCl_2$. Solid neutralization complexes of the composition $B \cdot MBr_4 \cdot SOBr_2$, B_2MBr_4 (where B = pyridine, isoquinoline, and M = Sn or Ti) and $Q \cdot MBr_3$ and Q_2MBr_4 (where Q = tetramethyl or tetraethylammonium bromide and M = Al, Sb, or Ti) have been isolated.

D. Sulfuryl Chloride

The solvent system sulfuryl chloride has been investigated by Gutmann[63,328,329] and Paul *et al.*[330,331] It is commercially available as a pale yellow liquid. It can be purified by keeping over mercury for a few hours followed by fractional distillation.[330] It can also be purified by stirring it over lithium under argon followed by distillation.[332] Pure SO_2Cl_2 is a colorless liquid with an extremely pungent smell. Some of its physical constants were listed in Table XVIII.

Anhydrous SO_2Cl_2 is a poor solvent for ionic compounds but dissolves numerous covalent compounds.[63,328,329] Sulfuryl chloride does not react with sulfur at room temperature, but at 150°–180°C a reaction takes place with the formation of sulfur dioxide and sulfur monochloride. Similar reaction takes place with red phosphorus.[333] Selenium[302] and tellurium[301] react with SO_2Cl_2 to form their respective tetrachlorides. In a current of SO_2Cl_2 vapors diluted with CO_2, selenium and tellurium are first converted to Se_2Cl_2 and Te_2Cl_2 at room temperature and then to $SeCl_4$ and $TeCl_4$ with the passage of time.[334] Precipitated sulfides, sulfide minerals,[334] and metal oxides[335] solvolyze in SO_2Cl_2 to give the corresponding metal chlorides. Compounds such as $SnSO_4$, $TiSO_4$, $V_2O(SO_4)_4$, $Sb_2O(SO_4)_4$, $CrO(SO_4)_2$, $MoO(SO_4)_2$, $WO(SO_4)_2$, and $U(SO_4)_2$, and the addition compounds of $SeCl_4$, $SeOCl_2$, and $TeCl_4$ with SO_3, can be conveniently prepared by treating solutions of corresponding metal halides or oxyhalides with SO_3 using SO_2Cl_2 as the solvent.[336]

Two binary adducts, $SO_3 \cdot SO_2Cl_2$[337] and $SbCl_5 \cdot SO_2Cl_2$,[310] and a ternary adduct, $SO_2Cl_2 \cdot AlCl_3 \cdot S_2Cl_2$,[338] have been reported. Organic tertiary bases, pyridine, β and γ picolines, quinoline, isoquinoline, and piperidine form solvates of the type $2B \cdot SO_2Cl_2$.[331] Monosolvates of dimethylformamide[339] and triphenyl phosphine[340] are also known. Infrared spectra[331] of the adduct $2B \cdot SO_2Cl_2$ indicate that the coordination of the tertiary amines takes place through the nitrogen atom to the sulfur atom of sulfuryl chloride.

The specific conductance of sulfuryl chloride (2×10^{-8} ohm^{-1} cm^{-1} at 20°C) has been attributed to its autoionization as[328]:

$$SO_2Cl_2 \rightleftharpoons SO_2Cl^+ + Cl^- \quad (102)$$

Radiochlorine exchange between pyridinium chloride and sulfuryl chloride in chloroform takes place to the extent of 97% at 20°C. The exchange may be explained either in terms of ionization of SO_2Cl_2 or by the intermediate formation of an addition complex between the chloride ion and SO_2Cl_2.[323] The chlorides, $SbCl_5$, $SbCl_3$, $AsCl_3$, $TiCl_4$, VCl_4, and $SnCl_4$ act as acids and the substituted ammonium chlorides, PCl_5, $TeCl_4$, pyridine, quinoline, the picolines, and dimethylaniline act as bases of the sulfuryl chloride solvosystem. Acid–base neutralization reactions in the solvent have been studied conductometrically.[63,328,329,341] A number of solid acid–base neutralization complexes of the type $[\text{Base} \cdot SO_2Cl]SbCl_6$, $[\text{Base} \cdot SO_2Cl]_2SnCl_6$, and $(\text{Base})_2SnCl_4 \cdot SO_2Cl_2$ have been isolated.[330] The neutralization reactions have been formulated on the basis of the solvent system concept as

$$SbCl_5 + SO_2Cl_2 \rightleftharpoons SbCl_5 \cdot SO_2Cl_2 \rightleftharpoons SO_2Cl^+ + SbCl_6^- \quad (103)$$

$$\text{Amine} + SO_2Cl_2 \rightleftharpoons \text{amine} \cdot SO_2Cl_2 \rightleftharpoons \text{amine} \cdot SO_2Cl^+ + Cl^- \quad (104)$$

$$SO_2Cl^+ + SbCl_6^- + [\text{amine} \cdot SO_2Cl]^+ + Cl \rightarrow [\text{amine} \cdot SO_2Cl]SbCl_6 + SO_2Cl_2 \quad (105)$$

Acid–base neutralization reactions have also been studied by visual titrations using benzanthrone as a reversible internal indicator. Pyridine, quinoline, stannic chloride, and titanium tetrachloride have been used as titrants.[87]

VIII. Tin(IV) Chloride

Although tin(IV) chloride has long been used as a Lewis acid in various nonaqueous solvents, as an ionizing solvent it has been examined only by Spandau and Hattwig[342] and unless otherwise specified the work reported in this section has been done by these workers.

Tin(IV) chloride is purified by repeated distillation in a carbon dioxide atmosphere followed by distillation over phosphorus(V) oxide. Some of its important physical constants are given in Table XIX.

TABLE XIX

PHYSICAL CONSTANTS OF LIQUID $SnCl_4$[a]

Boiling point (°C)	113
Freezing point (°C)	−33
Density	2.232 g/cm³ at 20°C
Molecular volume	116.7 cm³
Dielectric constant	3.2 at 22°C
Dipole moment	0 Debye
Specific electrical conductance	6.3×10^{-11} ohm^{-1} cm^{-1} at 20°C

[a] V. H. Spandau and H. Hattwig, *Z. Anorg. Allg. Chem.* **295**, 281 (1958).

Tin(IV) chloride dissolves only a limited number of inorganic compounds. The solubilities of various substances in $SnCl_4$ have been measured qualitatively. Ionic chlorides are almost insoluble. Iron(III) chloride, phosphorus(V) chloride, triphenylmethyl chloride, triethylammonium chloride, and diethylaniline have slight solubilities and show very little change in the conductances of these solutions. Hydrogen chloride, acetone, and ethylmethyl ketone are slightly soluble but show large changes in conductance of the solutions. However, I_2, Cl_2, $TiCl_4$, PCl_3, and S_2Cl_2 have large solubilities but show no change in conductance of the solutions and $POCl_3$, $AsCl_3$, $SbCl_3$, $SbCl_5$, $SOCl_2$, SO_2Cl_2, ICl, and cyclohexanone have large solubilities but their solutions show large changes in conductances.

Tin(IV) chloride has only a very slight tendency toward self-dissociation. The small specific conductance of pure tin(IV) chloride has been attributed to the autoionization of the solvent:

$$3SnCl_4 \rightleftharpoons 2SnCl_3^+ + SnCl_6^{2-} \tag{106}$$

Chlorides such as $SbCl_5$, $AsCl_3$, and ICl act as solvo acids in $SnCl_4$ whereas $SOCl_2$, $POCl_3$, HCl, and cyclohexanone act as solvo-bases.

$$SbCl_5 + SnCl_4 \rightleftharpoons SnCl_3^+ + SbCl_6^- \tag{107}$$

$$AsCl_3 + SnCl_4 \rightleftharpoons SnCl_3^+ + AsCl_4^- \tag{108}$$

$$2SOCl_2 + SnCl_4 \rightleftharpoons 2SOCl^+ + SnCl_6^{2-} \tag{109}$$

Neutralization reactions between these solvoacids and solvobases have been studied by conductometric titrations and preparative analytical investigations. A conductometric titration between thionyl chloride and antimony(V) chloride in tin(IV) chloride solutions shows a break in the curve at the $SOCl_2 : SbCl_5$ molar ratio of 1 : 1. The reaction has been postulated as

$$2SOCl^+ + SnCl_6^{2-} + 2SnCl_3^+ + 2SbCl_6^- \rightarrow 3SnCl_4 + 2[SOCl]SbCl_6 \tag{110}$$

The reaction product, $[SOCl]SbCl_6$, is soluble in $SnCl_4$ and shows greater

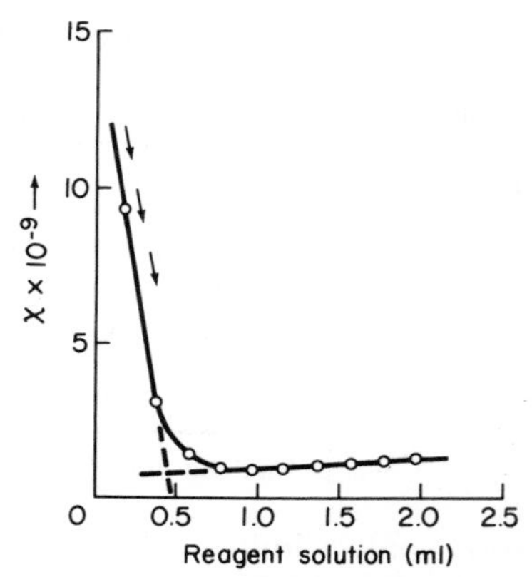

FIG. 13. Conductometric titration of a saturated solution of HCl with a 0.15 *M* $SbCl_5$ solution in $SnCl_4$.

conductance than the two compounds $SOCl_2$ and $SbCl_5$. The behavior of $POCl_3$ and $AsCl_3$ in stannic chloride is similar to that of $SOCl_2$ and $SbCl_5$, respectively.

The behavior of dry HCl toward $SnCl_4$ is particularly interesting. By bubbling dry HCl in $SnCl_4$ only a small quantity of the gas dissolves. At 20°C a saturated solution is about 5×10^{-3} molar, but the solution shows a very good conductivity due to the reaction

$$2HCl + SnCl_4 \rightleftharpoons H_2SnCl_6 \rightleftharpoons 2H^+ + SnCl_6^{2-} \tag{111}$$

By titrating a saturated solution of HCl in $SnCl_4$ with dilute $SbCl_5$ solution (Fig. 13) a white precipitate of $HSbCl_6$ separates out and, simultaneously, the conductance decreases sharply until the equivalence point is reached. The reaction has been formulated as

$$2H^+ + SnCl_6^{2-} + 2SnCl_3^+ + 2SbCl_6^- \rightarrow 3SnCl_4 + 2HSbCl_6 \tag{112}$$

A solution of HCl can also be titrated against a solution of ICl in $SnCl_4$. The conductometric titration curve is very much like that in Fig. 12.

$$2H^+ + SnCl_6^{2-} + 2SnCl_3^+ + 2ICl_2^- \rightarrow 3SnCl_4 + 2HICl_2 \tag{113}$$

A titration of cyclohexanone solution with antimony(V) chloride solution in $SnCl_4$ also shows a break at the 1 : 1 molar ratio in the titration curve.

REFERENCES

1. P. Walden, *Z. Anorg. Chem.* **25**, 209 (1900).
2. S. Tolloczko, *Bull. Int. Acad. Pol. Sci. Lett., Cl. Sci. Math. Nat.* **1**, 1 (1901).
3. E. Beckmann, *Z. Anorg. Chem.* **51**, 111 (1906).
4. K. Frycz and S. Tolloczko, *Festschr. Univ. Lwow* **1**, 1 (1912).

5. K. Frycz and S. Tolloczko, *Chem. Zentralbl.* **1**, 91 (1913).
6. Z. Klemensiewicz, *Bull. Int. Acad. Pol. Sci. Lett., Cl. Sci. Math. Nat.* **6**, 418 (1908).
7. Z. Klemensiewicz, *Z. Phys. Chem.* **113**, 28 (1924).
8. Z. Klemensiewicz and Z. Balowna, *Rocz. Chem.* **10**, 481 (1930).
9. Z. Klemensiewicz and Z. Balowna, *Rocz. Chem.* **11**, 683 (1931).
10. Z. Klemensiewicz and A. Zebrowska, *Rocz. Chem.* **14**, 14 (1934).
11. G. Jander and K. H. Swart, *Z. Anorg. Allg. Chem.* **299**, 252 (1959).
12. G. Jander and K. H. Swart, *Z. Anorg. Allg. Chem.* **301**, 54 (1959).
13. G. Jander and K. H. Swart, *Z. Anorg. Allg. Chem.* **301**, 80 (1959).
14. G. B. Porter and E. C. Baughan, *J. Chem. Soc.* p. 744 (1958).
15. A. G. Davies and E. C. Baughan, *J. Chem. Soc.* p. 1711 (1961).
16. E. C. Baughan, T. P. Jones, and L. G. Stoodley, *Proc. Chem. Soc., London* p. 274 (1963).
17. J. R. Lacher, D. E. Campion, and J. D. Park, *Science* **110**, 300 (1949).
18. J. R. Lacher, V. D. Croy, A. Kianpour, and J. D. Park, *J. Phys. Chem.* **58**, 206 (1954).
19. J. R. Lacher, J. L. Bitner, and J. D. Park, *J. Phys. Chem.* **59**, 610 (1955).
20. J. R. Lacher, J. L. Bitner, D. J. Emery, M. E. Seffl, and J. D. Park, *J. Phys. Chem.* **59**, 615 (1955).
21. H. Szymanski, K. Broda, J. May, W. Collins, and D. Bakalik, *Anal. Chem.* **37**, 617 (1965).
22. H. Szymanski, A. Bluemle, and W. Collins, *Appl. Spectrosc.* **19**, 137 (1965); *C. A.* **64**, 5947 (1966).
23. G. Jander and J. Weis, *Z. Elektrochem.* **61**, 1275 (1957); **62**, 850 (1958).
24. G. Jander and J. Weis, *Z. Elektrochem.* **63**, 1037 (1959).
25. E. C. Puente, *An. R. Soc. Esp. Fis. Quim.* **57**, 259 (1961); *C. A.* **56**, 11200 (1962).
26. P. Ehrlich and G. Dietz, *Z. Anorg. Allg. Chem.* **305**, 158 (1960).
27. C. G. Vonk and E. H. Wiebenga, *Recl. Trav. Chim. Pays-Bas* **78**, 913 (1959).
28. E. Pietsch, Ed., "Gmelin Handbuch der anorganischen Chemie," Part 18B. Verlag Chemie, Weinheim, 1949.
29. V. S. Grechishkin and I. A. Kyuntsel, *Opt. Spectrosk.* **16**, 161 (1964); *C. A.* **61**, 199 (1964).
30. S. S. Raskin, *Fiz. Sb., Lvov. Gos. Univ.* **3**, 203 (1957); *C. A.* **55**, 16143 (1961).
31. M. Zackrisson and K. I. Alden, *Acta Chem. Scand.* **14**, 994 (1960).
32. I. L. Abarbarchuk and G. G. Rusin, *Zh. Obshch. Khim.* **35**, 1902 (1965).
33. V. S. Grechishkin and I. A. Kyuntsel, *Zh. Strukt. Khim.* **4**, 269 (1963).
34. P. D. Simova and R. Angelova, *Izv. Bulg. Akad. Nauk, Otd. Fiz.-Mat. Tekh. Nauki, Ser. Fiz.* **7**, 333 (1959).
35. S. Prashad, G. B. Punna Rao, S. Kumar, V. R. Reddy, and K. P. Kaeker, *J. Indian Chem. Soc.* **36**, 129 (1929).
36. H. Funk and H. Koehler, *J. Prakt. Chem.* [4] **14**, 226 (1961).
37. K. C. Malhotra, G. K. Chawla, and S. C. Chaudhry, *J. Indian Chem. Soc.* **51**, 1035 (1974).
38. B. A. Voitovich, *Zh. Neorg. Khim.* **6**, 1914 (1961).
39. A. Chrétien and J. C. Couturier, *Rev. Chim. Miner.* **2**, 487 (1965); *C. A.* **64**, 1411 (1966).
40. L. Kolditz, *Z. Anorg. Allg. Chem.* **289**, 118 (1957).
41. J. R. Atkinson, E. C. Baughan, and B. Dacre, *J. Chem. Soc. A* p. 1377 (1970).
42. J. R. Atkinson, T. P. Jones, and E. C. Baughan, *J. Chem. Soc.* p. 5808 (1964).
43. P. V. Johnson and E. C. Baughan, *J. Chem. Soc. A* p. 2686 (1969).
44. G. B. Porter, J. Simpson, and E. C. Baughan, *J. Chem. Soc. A* p. 2806 (1970).
45. P. Texier, *Bull. Soc. Chim. Fr.* p. 4716 (1968).
46. P. Texier, *Bull. Soc. Chim. Fr.* p. 4315 (1968).
47. J. Badoz-Lambling, D. Bauer, and P. Texier, *Anal. Lett.* **2**, 411 (1969).
48. P. Texier, *J. Electroanal. Chem. Interfacial Electrochem.* **29**, 343 (1971).
49. E. V. Gorenbein, *J. Gen. Chem. USSR (Engl. Transl.)* **15**, 729 (1945).

50. T. C. Waddington, ed., "Non-Aqueous Solvent Systems," p. 303, Academic Press, New York, 1965.
51. J. D. Corbett, S. V. Winbush, and F. C. Albers, *J. Am. Chem. Soc.* **79**, 3020 (1957).
52. J. D. Corbett and F. C. Albers, *J. Am. Chem. Soc.* **82**, 533 (1960).
53. B. L. Bruner and J. D. Corbett, *J. Inorg. Nucl. Chem.* **20**, 62 (1961).
54. R. D. Whealy and J. B. Blackstock, *J. Inorg. Nucl. Chem.* **26**, 243 (1964).
55. T. Bjorvatten, O. Hassel, and A. Lindheim, *Acta Chem. Scand.* **17**, 689 (1963).
56. P. Walden, *Z. Anorg. Chem.* **25**, 214 (1900).
57. P. Walden, *Z. Anorg. Chem.* **29**, 371 (1902).
58. V. Gutmann, *Z. Anorg. Allg. Chem.* **266**, 331 (1951).
59. V. Gutmann, *Z. Anorg. Allg. Chem.* **264**, 151 (1951).
60. V. Gutmann, *Monatsh. Chem.* **83**, 159 (1952).
61. V. Gutmann, *Monatsh. Chem.* **83**, 583 (1952).
62. V. Gutmann, *Monatsh. Chem.* **84**, 1191 (1953).
63. V. Gutmann, *Monatsh. Chem.* **85**, 286 (1954).
64. V. Gutmann, *Monatsh. Chem.* **85**, 491 (1954).
65. V. Gutmann, *Sven. Kem. Tidskr.* **68**, 1 (1956).
66. V. Gutmann and H. Spandau, *Angew Chem.* **64**, 93 (1952).
67. I. Lindqvist and L. H. Andersson, *Acta Chem. Scand.* **8**, 128 (1954).
68. I. Lindqvist, *Acta Chem. Scand.* **9**, 73 (1955).
69. I. Lindqvist and L. H. Andersson, *Acta Chem. Scand.* **9**, 79 (1955).
70. M. Agerman, L. H. Andersson, I. Lindqvist, and M. Zackrisson, *Acta Chem. Scand.* **12**, 477 (1958).
71. G. Jander and K. Gunther, *Z. Anorg. Allg. Chem.* **297**, 81 (1958).
72. G. Jander and K. Gunther, *Z. Anorg. Allg. Chem.* **298**, 241 (1959).
73. G. Jander and K. Gunther, *Z. Anorg. Allg. Chem.* **302**, 155 (1959).
74. J. Lewis and D. B. Sowerby, *J. Chem. Soc.* p. 336 (1957).
75. H. L. Wheeler, *Am. J. Sci.* **46**, 90 (1893).
76. H. L. Wheeler, *Z. Anorg. Chem.* **4**, 452 (1893).
77. W. B. Shirey, *J. Am. Chem. Soc.* **52**, 1720 (1930).
78. E. Montignie, *Bull. Soc. Chim. Fr.* **2**, 1365 (1935).
79. O. Dafert and Z. A. Melinski, *Ber. Dtsch. Chem. Ges. B* **59**, 788 (1926).
80. C. S. Gibson, J. D. A. Johnson, and D. C. Vining, *J. Chem. Soc.* p. 1710 (1930).
81. L. Kolditz, *Z. Anorg. Allg. Chem.* **284**, 144 (1956).
82. L. Kolditz, *Z. Anorg. Allg. Chem.* **286**, 307 (1956).
83. L. Kolditz and A. Felts, *Z. Anorg. Allg. Chem.* **293**, 286 (1957).
84. L. Kolditz and W. Schmidt, *Z. Anorg. Allg. Chem.* **296**, 188 (1958).
85. F. J. J. Birkmann and H. Gerding, *Rev. Chim. Miner.* **7**, 729 (1970).
86. R. C. Paul, J. Singh, and S. S. Sandhu, *J. Indian Chem. Soc.* **36**, 305 (1959).
87. R. C. Paul, J. Singh, and S. S. Sandhu, *Anal. Chem.* **31**, 1495 (1959).
88. E. G. Brame, R. C. Ferguson, and G. J. Thomas, *Anal. Chem.* **39**, 517 (1967).
89. J. W. Rettgers, *Z. Phys. Chem.* **11**, 328 (1893).
90. N. A. Pusin and J. Makuc, *Z. Anorg. Allg. Chem.* **237**, 177 (1938).
91. W. Isbekow and W. Plotnikow, *Z. Anorg. Chem.* **71**, 332 (1911).
92. J. Kendall, E. D. Crittenden, and H. K. Miller, *J. Am. Chem. Soc.* **45**, 963 (1923).
93. N. A. Pusin and S. Lowy, *Z. Anorg. Allg. Chem.* **150**, 167 (1926).
94. A. A. Woolf and N. N. Greenwood, *J. Chem. Soc.* p. 2200 (1950).
95. E. L. Muetterties and W. D. Phillips, *J. Am. Chem. Soc.* **79**, 3686 (1957).
96. L. Kolditz, *Z. Anorg. Allg. Chem.* **289**, 128 (1957).
97. L. Kolditz, *Z. Anorg. Allg. Chem.* **280**, 313 (1955).

98. H. M. Dess, R. W. Parry, and G. L. Vidale, *J. Am. Chem. Soc.* **78**, 5730 (1956).
99. H. M. Dess and R. W. Parry, *J. Am. Chem. Soc.* **78**, 5735 (1956).
100. L. Kolditz and W. Schaefer, *Z. Anorg. Allg. Chem.* **315**, 35 (1962).
101. L. J. Beckham, W. A. Fessler, and M. A. Kise, *Chem. Rev.* **48**, 319 (1951).
102. E. Briner and Z. Pylkoff, *J. Chim. Phys.* **10**, 640 (1912).
103. A. F. Scott and C. R. Johnson, *J. Phys. Chem.* **33**, 1975 (1929).
104. C. W. Whittaker, F. O. Lundstrom, and A. R. Merz, *Ind. Eng. Chem.* **23**, 1410 (1931).
105. G. H. Coleman, G. A. Litlis, and G. E. Goheen, "Inorganic Synthesis" (H. S. Booth, ed.) Vol. I, p. 55. McGraw-Hill, New York, 1939.
106. A. B. Burg and D. E. McKenzie, *J. Am. Chem. Soc.* **74**, 3143 (1952).
107. J. A. Ketelaar and K. J. Palmer, *J. Am. Chem. Soc.* **59**, 2629 (1937).
108. J. A. Ketelaar, *Recl. Trav. Chim. Pays-Bas* **62**, 289 (1943).
109. A. B. Burg and G. W. Campbell, *J. Am. Chem. Soc.* **70**, 1964 (1948).
110. H. Houtgraaf and A. M. de Roos, *Recl. Trav. Chim. Pays-Bas* **72**, 963 (1953).
111. V. Gutmann, *J. Phys. Chem.* **63**, 378 (1959).
112. P. A. Guye and G. Gluss, *J. Chim. Phys.* **6**, 732 (1908).
113. J. McMorris and D. M. Yost, *J. Am. Chem. Soc.* **54**, 2247 (1932).
114. J. R. Partington and A. L. Whynes, *J. Chem. Soc.* p. 3135 (1949).
115. R. Perrot, *C. R. Hebd. Seances Acad. Sci.* **201**, 275 (1935).
116. F. Seel, R. Schmutzler, and K. Wasem, *Angew. Chem.* **71**, 340 (1959).
117. T. C. Waddington and F. Klanberg, *Z. Anorg. Allg. Chem.* **304**, 185 (1960).
118. J. MacCordick, *Experientia* **26**, 1289 (1970).
119. J. MacCordick, C. Devin, R. Perrot, and R. Rohmer, *C. R. Hebd. Seances Acad. Sci.* **274**, 278 (1972).
120. H. Gerding and H. Houtgraaf, *Recl. Trav. Chim. Pays-Bas* **72**, 21 (1953).
121. H. Gerding, *Recl. Trav. Chim. Pays-Bas* **75**, 589 (1956).
122. F. Seel, *Z. Anorg. Allg. Chem.* **252**, 24 (1943).
123. F. Seel and H. Bauer, *Z. Naturforsch.* **126**, 397 (1947).
124. H. Klinkenberg, *Recl. Trav. Chim. Pays-Bas* **56**, 759 (1937); *Chem. Weekbl.* **35**, 197 (1938).
125. R. W. Asmussen, *Z. Anorg. Allg. Chem.* **243**, 127 (1939).
126. B. Vandorpe and J. Heubel, *C. R. Hebd. Seances Acad. Sci.* **260**, 6619 (1965).
127. B. Vandorpe, *Rev. Chim. Miner.* **4**, 589 (1967).
128. H. Gerding, T. Hohle, and K. van Schaik, *Rev. Chim. Miner.* **3**, 617 (1966).
129. J. Lewis and R. G. Wilkins, *J. Chem. Soc.* p. 56 (1955).
130. J. Lewis and D. B. Sowerby, *J. Chem. Soc.* p. 1617 (1957).
131. J. Lewis and D. B. Sowerby, *Chem. Soc., Spec. Publ.* **10**, 123 (1957).
132. J. Lewis and D. B. Sowerby, *Recl. Trav. Chim. Pays-Bas* **75**, 615 (1956).
133. J. Lewis and D. B. Sowerby, *J. Chem. Soc.* p. 150 (1956).
134. A. E. Comyus, *J. Chem. Soc.* p. 1557 (1955).
135. R. C. Paul and S. L. Chadha, *Indian J. Chem.* **6**, 272 (1968).
136. J. R. Geichman, E. A. Smith, S. S. Trond, and P. R. Ogle, *Inorg. Chem.* **1**, 661 (1962).
137. A. A. Woolf, *J. Chem. Soc.* p. 1053 (1950).
138. A. G. Sharp and A. A. Woolf, *J. Chem. Soc.* p. 798 (1951).
139. F. Seel, *Z. Anorg. Chem.* **250**, 331 (1943).
140. F. Seel and T. Gossl, *Z. Anorg. Allg. Chem.* **263**, 253 (1950).
141. J. H. Holloway and H. Selig, *J. Inorg. Nucl. Chem.* **30**, 473 (1968).
142. N. Bartlett, S. P. Beaton, and N. K. Jha, *Chem. Commun.* p. 168 (1966).
143. F. P. Gortsema and R. H. Toeniskoetter, *Inorg. Chem.* **5**, 1217 (1966).
144. H. C. Clark and H. J. Emeléus, *J. Chem. Soc.* p. 190 (1958).
145. F. Seel, W. Birukraut, and D. Werner, *Chem. Ber.* **95**, 1264 (1962).

146. J. R. Geichman, E. A. Smith, and P. R. Ogle, *Inorg. Chem.* **2**, 1012 (1963).
147. G. J. Moody and H. Selig, *Inorg. Nucl. Chem. Lett.* **2**, 319 (1966).
148. M. S. Toy and W. A. Cannon, *J. Electrochem. Soc.* **114**, 940 (1967).
149. E. E. Aynsley, G. Hetherington, and P. L. Robinson, *J. Chem. Soc.* p. 1119 (1954).
150. A. C. Legon and D. J. Millen, *J. Chem. Soc.* p. 1736 (1968).
151. S. J. Kuhn, *Can. J. Chem.* **45**, 3207 (1967).
152. G. Brauer, ed., "Handbook of Preparative Inorganic Chemistry," 2nd ed., Vol. 1, p. 513. Academic Press, New York, 1963.
153. R. C. Paul, D. Singh, and K. C. Malhotra, *J. Chem. Soc. A* p. 1396 (1969).
154. V. Gutmann, *Monatsh. Chem.* **83**, 164 (1952).
155. V. Gutmann, *Monatsh. Chem.* **83**, 279 (1952).
156. V. Gutmann, *Z. Anorg. Allg. Chem.* **269**, 279 (1952).
157. V. Gutmann, *Z. Anorg. Allg. Chem.* **270**, 179 (1952).
158. V. Gutmann, *Monatsh. Chem.* **85**, 1077 (1954).
159. A. Maschka, V. Gutmann, and R. Sponer, *Monatsh. Chem.* **86**, 52 (1955).
160. V. Gutmann and R. Himml, *Z. Phys. Chem.* (*Frankfurt am Main*) **4**, 157 (1955).
161. V. Gutmann and F. Mairinger, *Z. Anorg. Allg. Chem.* **289**, 279 (1957).
162. V. Gutmann and F. Mairinger, *Monatsh. Chem.* **89**, 724 (1958).
163. V. Gutmann and M. Baaz, *Z. Anorg. Allg. Chem.* **298**, 121 (1958).
164. V. Gutmann and M. Baaz, *Monatsh. Chem.* **90**, 239 (1959).
165. M. Baaz and V. Gutmann, *Monatsh. Chem.* **90**, 256 (1959).
166. V. Gutmann and M. Baaz, *Monatsh. Chem.* **90**, 271 (1959).
167. M. Baaz and V. Gutmann, *Monatsh. Chem.* **90**, 276 (1959).
168. M. Baaz and V. Gutmann, *Monatsh. Chem.* **90**, 426 (1959).
169. V. Gutmann and M. Baaz, *Monatsh. Chem.* **90**, 729 (1959).
170. M. Baaz and V. Gutmann, *Monatsh. Chem.* **90**, 744 (1959).
171. V. Gutmann and F. Mairinger, *Monatsh. Chem.* **91**, 529 (1960).
172. M. Baaz, V. Gutmann, and L. Hubner, *Monatsh. Chem.* **91**, 537 (1960).
173. M. Baaz, V. Gutmann, and L. Hubner, *J. Inorg. Nucl. Chem.* **18**, 276 (1961).
174. M. Baaz, V. Gutmann, and M. Y. A. Talaat, *Monatsh. Chem.* **91**, 548 (1960).
175. M. Baaz, V. Gutmann, and L. Hubner, *Monatsh. Chem.* **91**, 694 (1960).
176. M. Baaz, V. Gutmann, and L. Hubner, *Monatsh. Chem.* **92**, 272 (1961).
177. M. Baaz, V. Gutmann, and L. Hubner, *Monatsh. Chem.* **92**, 707 (1961).
178. M. Baaz, V. Gutmann, and M. Y. A. Talaat, *Monatsh. Chem.* **92**, 714 (1961).
179. V. Gutmann and F. Mairinger, *Monatsh. Chem.* **92**, 720 (1961).
180. V. Gutmann, *Oesterr. Chem.-Ztg.* **56**, 126 (1955).
181. V. Gutmann, *Recl. Trav. Chim. Pays-Bas* **75**, 603 (1956).
182. V. Gutmann and R. Himml, *Z. Anorg. Allg. Chem.* **287**, 199 (1956).
183. V. Gutmann and M. Baaz, *Electrochim. Acta* **3**, 115 (1960).
184. V. Gutmann, *Oesterr. Chem.-Ztg.* **62**, 326 (1961).
185. M. Baaz, V. Gutmann, and J. R. Masagner, *Monatsh. Chem.* **92**, 590 (1961).
186. V. Gutmann, H. Hubacek, and A. Steininger, *Monatsh. Chem.* **95**, 678 (1964).
187. V. Gutmann, F. Mairinger, and H. Winkler, *Monatsh. Chem.* **96**, 574 (1965).
188. H. Spandau, A. Beyer, and F. Preugschat, *Z. Anorg. Allg. Chem.* **306**, 13 (1960).
189. D. S. Payne, *in* "Non-Aqueous Solvent Systems" (T. C. Waddington, ed.), p. 334. Academic Press, New York, 1965.
190. V. Gutmann, *in* "Halogen Chemistry" (V. Gutmann, ed.), Vol. 2, p. 403. Academic Press, New York, 1967.
191. G. Oddo and M. Tealdi, *Gazz. Chim. Ital.* **33**, 427 (1903).
192. H. P. Cady and R. Taft, *J. Phys. Chem.* **29**, 1057 (1925).

193. I. Lindqvist and C. I. Branden, *Acta Crystallogr.* **12**, 642 (1959).
194. C. I. Branden and I. Lindqvist, *Acta Chem. Scand.* **17**, 353 (1963).
195. C. I. Branden, *Acta Chem. Scand.* **16**, 1806 (1962).
196. C. I. Branden and I. Lindqvist, *Acta Chem. Scand.* **14**, 726 (1960).
197. C. I. Branden, *Acta Chem. Scand.* **17**, 759 (1963).
198. E. W. Wartenberg and J. Goubeau, *Z. Anorg. Allg. Chem.* **329**, 269 (1964).
199. H. Gerding, J. A. Koningstein, and E. R. Van der Worm, *Spectrochim. Acta* **16**, 881 (1960).
200. T. C. Waddington and F. Klanberg, *J. Chem. Soc.* p. 2339 (1960).
201. M. E. Peach and T. C. Waddington, *J. Chem. Soc.* p. 3450 (1962).
202. J. C. Scheldon and S. Y. Tyree, *J. Am. Chem. Soc.* **80**, 4775 (1958).
203. D. W. Meek and R. S. Drago, *J. Am. Chem. Soc.* **83**, 4322 (1961).
204. R. S. Drago and K. F. Purcell, *Prog. Inorg. Chem.* **6**, 271 (1964).
205. G. Leman and G. Tridot, *C. R. Hebd. Seances Acad. Sci.* **248**, 3439 (1959).
206. B. A. Voitovich, A. S. Barabanova, and G. N. Novitskaya, *Ukr. Khim. Zh.* **33**, 271 (1967).
207. I. A. Sheka, B. A. Voitovich, and L. A. Niselson, *Zh. Neorg. Khim.* **4**, 1803 (1959).
208. A. S. Barabanova, *Ukr. Khim. Zh.* **37**, 123 (1971).
209. W. L. Groeneveld, J. W. Van Sponsen, and H. W. Kouwenhoven, *Recl. Trav. Chim. Pays-Bas* **72**, 950 (1953).
210. G. Adlafsson, R. Bryntse, and I. Lindqvist, *Acta Chem. Scand.* **14**, 949 (1960).
211. B. A. Voitovich and N. F. Lozovskaya, *Ukr. Khim. Zh.* **31**, 1136 (1965).
212. B. F. Markov, A. S. Barabanova, and B. A. Voitovich, *Ukr. Khim. Zh.* **29**, 1035 (1963).
213. V. V. Dadape and M. R. A. Rao, *J. Am. Chem. Soc.* **77**, 6192 (1955).
214. O. Ruff and H. Einbeck, *Ber. Dtsch. Chem. Ges.* **37**, 4518 (1904).
215. M. Baaz, V. Gutmann, L. Hubner, F. Mairinger, and T. S. West, *Z. Anorg. Allg. Chem.* **311**, 302 (1961).
216. M. J. Frazer, W. Gerrard, and J. K. Patel, *J. Chem. Soc.* p. 726 (1960).
217. W. Gerrard, E. F. Mooney, and H. A. Willis, *J. Chem. Soc.* p. 4255 (1961).
218. R. H. Herber, *J. Am. Chem. Soc.* **82**, 792 (1960).
219. A. S. Barabanova and B. A. Voitovich, *Ukr. Khim. Zh.* **30**, 1298 (1964).
220. W. L. Groeneveld and A. P. Zuur, *J. Inorg. Nucl. Chem.* **8**, 241 (1958).
221. K. N. V. Raman and A. R. V. Murthy, *Proc. Indian Acad. Sci., Sect. A* **51**, 270 (1960).
222. B. A. Voitovich, A. S. Barabanova, and G. N. Novitskaya, *Ukr. Khim. Zh.* **33**, 367 (1967).
223. N. N. Greenwood and K. Wade, *J. Chem. Soc.* p. 1516 (1957).
224. N. N. Greenwood and P. G. Perkins, *J. Inorg. Nucl. Chem.* **4**, 291 (1957).
225. N. N. Greenwood, *J. Inorg. Nucl. Chem.* **8**, 234 (1958).
226. L. A. Woodward, G. Garton, and H. L. Roberts, *J. Chem. Soc.* p. 3723 (1956).
227. B. A. Voitovich and E. V. Zvagolskaya, *Ukr. Khim. Zh.* **34**, 778 (1968).
228. M. Revel, J. Navech, and J. P. Vives, *Bull. Soc. Chim. Fr.* p. 2327 (1963).
229. J. Navech and J. P. Vives, *C. R. Hebd. Seances Acad. Sci.* **248**, 1354 (1959).
230. R. C. Paul, H. Khurana, S. K. Vasisht, and S. L. Chadha, *J. Indian Chem. Soc.* **46**, 915 (1969).
231. B. Prajsnar, *Zesz. Nauk Politech. Slask, Chem.* **20**, 148 (1963).
232. V. Gutmann and H. Hubacek, *Monatsh. Chem.* **94**, 1019 (1963).
233. E. Berger, *C. R. Hebd. Seances Acad. Sci.* **146**, 400 (1908); *Bull. Soc. Chim. Fr.* **3**, 721 (1908).
234. W. C. Fernelius, ed., "Inorganic Synthesis," Vol. II, p. 151. McGraw-Hill, New York, 1946.
235. N. N. Greenwood and I. J. Worrall, *J. Inorg. Nucl. Chem.* **6**, 34 (1958).
236. R. C. Paul and S. K. Vasisht, *J. Indian Chem. Soc.* **43**, 141 (1966).
237. J. C. Bailar, Jr., ed., "Inorganic Synthesis," Vol. IV, p. 71. McGraw-Hill, New York, 1953.
238. H. S. Booth and C. A. Seabright, *J. Am. Chem. Soc.* **65**, 1834 (1943).
239. W. C. Fernelius, ed., "Inorganic Synthesis," Vol. II, p. 153. McGraw-Hill, New York, 1946.

240. R. C. Paul, K. C. Malhotra, and G. Singh, *J. Indian Chem. Soc.* **37**, 105 (1960).
241. O. P. Abrol, Ph.D. Thesis, Panjab University, Chandigarh, India (1970).
242. G. Singh and O. P. Abrol, *Chem. Ind.* (*London*) p. 1812 (1968).
243. W. Van der Veer and F. Jellinek, *Recl. Trav. Chim. Pays-Bas* **87**, 365 (1968).
244. W. Van der Veer and F. Jellinek, *Recl. Trav. Chim. Pays-Bas* **89**, 833 (1970).
245. L. A. Niselson, *Zh. Neorg. Khim.* **5**, 1634 (1960).
246. H. Teichmann and G. Hilgetag, *Angew. Chem., Int. Ed. Engl.* **6**, 1013 (1967).
247. L. Maier, *Z. Anorg. Allg. Chem.* **345**, 29 (1966).
248. W. H. N. Vriezen and F. Jellinek, quoted in Jander and Swart.[11]
249. J. A. Cade, M. Kasrai, and I. R. Ashton, *J. Inorg. Nucl. Chem.* **27**, 2375 (1965).
250. I. R. Weber, *Ann. Phys.* (*Leipzig*) [2] **108**, 615 (1859).
251. V. Lenher, *J. Am. Chem. Soc.* **42**, 2498 (1920).
252. V. Lenher, *J. Am. Chem. Soc.* **43**, 29 (1921).
253. V. Lenher, *J. Am. Chem. Soc.* **44**, 1664 (1922).
254. W. L. Ray, *J. Am. Chem. Soc.* **45**, 2090 (1923).
255. G. B. L. Smith, *Chem. Rev.* **23**, 165 (1938).
256. C. R. Wise, *J. Am. Chem. Soc.* **45**, 1233 (1923).
257. B. J. Masters, N. D. Dotter, D. R. Asher, and T. H. Norris, *J. Am. Chem. Soc.* **78**, 4252 (1956).
258. W. S. Paterson, C. J. Heimerheim, and G. B. L. Smith, *J. Am. Chem. Soc.* **65**, 2403 (1943).
259. Y. Hermodsson, *Acta Chem. Scand.* **21**, 1313 (1967).
260. Y. Hermodsson, *Acta Crystallogr.* **13**, 656 (1960).
261. Y. Hermodsson, *Ark. Kemi* **31**, 199 (1969).
262. J. C. Sheldon and S. Y. Tyree, *J. Am. Chem. Soc.* **81**, 2290 (1959).
263. M. J. Frazer, *J. Chem. Soc.* p. 3165 (1961).
264. R. C. Paul, C. L. Arora, and K. C. Malhotra, *Indian J. Chem.* **10**, 92 (1972).
265. R. C. Paul, V. P. Kapila, and K. C. Malhotra, *J. Chem. Soc.* p. 2267 (1970).
266. J. Devynck and B. Termillion, *J. Electroanal. Chem. Interfacial Electrochem.* **23**, 241 (1969).
267. T. Birchall, R. J. Gellespie, and S. L. Vekris, *Can. J. Chem.* **43**, 1672 (1965).
268. Y. Hermodsson, *Acta Chem. Scand.* **21**, 1328 (1967).
269. J. Jackson and G. B. L. Smith, *J. Am. Chem. Soc.* **62**, 544 (1940).
270. B. Edgington and J. B. Firth, *J. Soc. Chem. Ind.* **55**, 192T (1936).
271. I. Lindqvist and G. Nahringbauer, *Acta Crystallogr.* **12**, 638 (1959).
272. L. A. Wiles and Z. S. Ariyan, *Chem. Ind.* (*London*) p. 2102 (1962).
273. O. G. Ackermann and J. E. Mayer, *J. Chem. Phys.* **4**, 377 (1936).
274. K. J. Palmer, *J. Am. Chem. Soc.* **60**, 2360 (1938).
275. E. Hirota, *Bull. Chem. Soc. Jpn.* **31**, 130 (1958); *C. A.* **52**, 13341 (1958).
276. Y. K. Syrkin and M. E. Dyatkina, "Structure of Molecules," p. 260. Butterworth, London, 1950.
277. V. Gutmann and G. Schober, *Monatsh. Chem.* **87**, 792 (1956).
278. V. H. Spandau and H. Hattwig, *Z. Anorg. Allg. Chem.* **311**, 32 (1961).
279. R. R. Wiggle and T. H. Norris, *Inorg. Chem.* **3**, 539 (1964).
280. P. Nicolardot, *C. R. Hebd. Seances Acad. Sci.* **147**, 1304 (1908).
281. F. Bourion, *C. R. Hebd. Seances Acad. Sci.* **148**, 170 (1909).
282. N. Domanicki, *J. Russ. Phys.-Chem. Soc.* **48**, 1724 (1916); *C. A.* **11**, 3184 (1917).
283. A. K. Batalin, *Nauchn. Zap., Ukr. Akad. Sotsialist Zemledeliya Artema* **2**, 125 (1940); *Khim. Ref. Zh.* **4**, 34 (1941).
284. A. Ibanez and R. Uson, *Rev. Acad. Cienc. Exactas Fis.-Quim. Nat. Zaragoza* **12**, 79 (1957); *C. A.* **54**, 11789 (1960).
285. H. Funk, K. H. Berndt, and G. Henze, *Wiss. Z. Martin-Luther-Univ., Halle-Wittenberg, Math.-Naturwiss. Reihe* **6**, 815 (1957); *C. A.* **54**, 12860 (1960).

286. R. E. Kirk and D. F. Othmer, eds., "Encylopedia of Chemical Technology," Vol. 13, p. 401. Wiley (Interscience), New York, 1954.
287. E. F. Smith, *J. Am. Chem. Soc.* **22**, 289 (1898).
288. R. D. Hall, *J. Am. Chem. Soc.* **26**, 1243 (1904).
289. W. B. Hicks, *J. Am. Chem. Soc.* **33**, 1492 (1911).
290. R. Chand, G. S. Hamdard, and K. Lal, *J. Indian Chem. Soc.* **35**, 28 (1958).
291. K. L. Jaura and I. P. Bhatia, *J. Sci. Ind. Res., Sect. B* **20**, 315 (1961).
292. R. C. Paul, J. Kishore, D. Singh, and K. C. Malhotra, *Indian J. Chem.* **8**, 729 (1970).
293. F. C. Trager, U.S. Patent 2,539,679 (1951).
294. W. E. Bissinger, U.S. Patent 2,529,671 (1950).
295. L. Friedman and W. P. Wetter, *J. Chem. Soc. A* p. 36 (1967).
296. S. S. Sandhu, B. S. Chakkal, and G. S. Sandhu, *J. Indian Chem. Soc.* **37**, 329 (1960).
297. V. H. Spandau and E. Brunneck, *Z. Anorg. Allg. Chem.* **270**, 201 (1952).
298. H. B. North and A. M. Hageman, *J. Am. Chem. Soc.* **34**, 890 (1912).
299. R. A. Hubbard and W. F. Luder, *J. Am. Chem. Soc.* **73**, 1327 (1951).
300. P. W. Schenk and H. Platz, *Z. Anorg. Allg. Chem.* **215**, 113 (1933).
301. B. V. Horvath, *Z. Anorg. Chem.* **70**, 408 (1911).
302. V. Lenher and H. B. North, *J. Am. Chem. Soc.* **29**, 33 (1907).
303. H. B. North and A. M. Hageman, *J. Am. Chem. Soc.* **35**, 352 (1913).
304. H. B. North, *J. Am. Chem. Soc.* **32**, 184 (1910).
305. H. Hecht, G. Jander, and H. Schlapmann, *Z. Anorg. Chem.* **254**, 255 (1947).
306. H. B. North and C. B. Conover, *J. Am. Chem. Soc.* **37**, 2486 (1915).
307. H. B. North and C. B. Conover, *Am. J. Sci.* **40**, 640 (1915).
308. J. H. Freeman and M. L. Smith, *J. Inorg. Nucl. Chem.* **7**, 224 (1958).
309. D. Khristov, St. Karaivanov, and V. Kolushki, *God. Sofii. Univ., Fiz.-Mat. Fak. Khim.* **55**, 49 (1960).
310. I. Lindqvist and P. Einarsson, *Acta Chem. Scand.* **13**, 420 (1959).
311. T. W. Saults and J. J. Wimberly, U.S. Patent 3,042,490 (1962).
312. D. A. Long and R. T. Bailey, *Trans. Faraday Soc.* **59**, 594 (1963).
313. R. C. Paul, H. Khurana, S. K. Vasisht, and S. L. Chadha, *Z. Anorg. Allg. Chem.* **370**, 185 (1969).
314. P. W. Schenk and R. Steudel, *Angew. Chem.* **75**, 793 (1963).
315. V. H. Spandau and E. Brunneck, *Z. Anorg. Allg. Chem.* **278**, 197 (1955).
316. C. Bertoglio, T. Soldi, and F. Martinotti, *Ann. Chim. (Rome)* **52**, 1043 (1962).
317. M. C. Henry, J. F. Hazel, and W. M. McNabb, *Anal. Chim. Acta* **15**, 187 (1956).
318. L. E. D. Pease and W. F. Luder, *J. Am. Chem. Soc.* **75**, 5195 (1953).
319. C. Bertoglio and T. Soldi, *Ann. Chim. (Rome)* **50**, 1540 (1960).
320. S. S. Sandhu, *J. Indian Chem. Soc.* **39**, 589 (1962).
321. S. S. Sandhu and A. Singh, *J. Indian Chem. Soc.* **42**, 744 (1965).
322. L. F. Johnson and T. H. Norris, *J. Am. Chem. Soc.* **79**, 1584 (1957).
323. M. J. Frazer, *J. Chem. Soc.* p. 3319 (1957).
324. D. W. Meek, *in* "The Chemistry of Nonaqueous Solvents" (J. J. Lagowski, ed.), Vol. I, p. 51. Academic Press, New York, 1966.
325. E. B. Garber, L. E. D. Pease, and W. F. Luder, *Anal. Chem.* **25**, 581 (1953).
326. V. Gutmann and H. Hubacek, *Monatsh. Chem.* **94**, 1098 (1963).
327. R. C. Paul, M. Singh, and S. K. Vasisht, *J. Indian Chem. Soc.* **47**, 635 (1970).
328. V. Gutmann, *Monatsh. Chem.* **85**, 393 (1954).
329. V. Gutmann, *Monatsh. Chem.* **85**, 404 (1954).
330. R. C. Paul, S. S. Sandhu, and S. B. Vij, *J. Indian Chem. Soc.* **39**, 2 (1962).
331. R. C. Paul, H. Kiran, and S. L. Chadha, *J. Indian Chem. Soc.* **46**, 863 (1969).
332. K. French, P. Cukor, C. Persiani, and J. Auborn, *J. Electrochem. Soc.* **121**, 1045 (1974).

333. H. B. North and J. C. Thomson, *J. Am. Chem. Soc.* **40**, 774 (1918).
334. H. Danneel and F. Schlottman, *Z. Anorg. Allg. Chem.* **212**, 225 (1933).
335. H. Danneel and W. Hesse, *Z. Anorg. Allg. Chem.* **212**, 214 (1933).
336. E. Hayek and A. Engelbrecht, *Monatsh. Chem.* **80**, 640 (1949).
337. G. P. Luchinskii and A. I. Likhacheva, *Zh. Fiz. Khim.* **9**, 65 (1937).
338. O. Hutzinger, S. Safe, and V. Zitko, *Int. J. Environ. Anal. Chem.* **2**, 95 (1972).
339. G. Walter and T. Kojscheff, German (East) Patent, 49,320 (1966).
340. A. J. Banister, B. Bell, and L. F. Moore, *J. Inorg. Nucl. Chem.* **34**, 1161 (1972).
341. A. R. Pray and C. R. McCrosky, *J. Am. Chem. Soc.* **74**, 4719 (1952).
342. V. H. Spandau and H. Hattwig, *Z. Anorg. Allg. Chem.* **295**, 281 (1958).

5

Molten Salts as Nonaqueous Solvents

D. H. Kerridge

Department of Chemistry, The University
Southampton SO9 5NH, U.K.

I. Introduction

Molten salts are evidently a numerous and potentially most significant and useful class of nonaqueous solvents. While "salt" is a widely used though imprecise term, this class can be broadly defined as the liquid state of those compounds which melt, usually above room temperature, to give liquids displaying ionic properties, at least to some degree. While they are probably more numerous than other nonaqueous solvents, molten salts are currently the least-known and least appreciated of the nonaqueous solvents, certainly much less so than their practical importance demands.

This relative obscurity is in part due to the lack of suitable introductory material of sufficient detail. While short chapters or sections can be found in some general books on nonaqueous solvents[1,2] and in a few inorganic chemistry textbooks,[3] almost the only books dealing solely with molten salts are the two original compilations of reviews dealing largely with their physical aspects,[4,5] a slightly later report of conference proceedings,[6] a book by Bloom[7] mainly devoted to methods of physical measurements, and an invaluable compilation of data extracted from the existing literature.[8] A series of "Advances in Molten Salt Chemistry"[9] has also been started, currently reaching Volume 3 and comprising an average of five reviews per volume.

The total number of original research papers now available in the literature on the chemistry of molten salts is obviously much too large for coverage in this chapter. We will therefore have to be highly selective and, in order to maintain adequate depth, the chemistry discussed will be confined mainly to molten alkali metal chlorides and nitrates. The first class of melts is chosen both because of their intrinsic simplicity and because these melts have been the most frequently chosen for study. More than 15% of the papers on all types of melts are concerned with the chemistry of molten alkali metal chlorides. The oxyanions, besides providing numerous different molten salt systems, display chemistry of great and still intriguing complexity, and already include the melts of greatest industrial importance, i.e., sulfates, carbonates, borates, and silicates. Despite significant present-day applications and numerous potential uses, the academic study of these melts is only at a very early stage. The oxyanion class chosen for description is the alkali metal nitrates since most of the research carried out thus far has been in this class of oxyanion melts (roughly half the number of papers devoted to the alkali metal chlorides) and it is therefore the most effectively understood. Where necessary the reasonably detailed account of these two melt classes will be supplemented by information on other halide or oxyanion classes.

II. Halide Melts

A. Use of Alkali Metal Chloride Melts

The selection of alkali metal chloride melts by so many investigators arises chiefly from the desire to begin with the simplest system, in which acid–base and oxidation–reduction reactions are of minimal importance or entirely removed. In addition, much of the early impetus was derived from the measurement of physical properties and it was felt that by using melts consisting solely of simple, spherical, singly charged cations and anions, a theoretical understanding of the experimental data would be most readily obtained. Moreover, it seemed that such simple ions, interacting largely by ionic bonding, would enable a more rapid development of generalized theories of the liquid state which might assist with understanding the behavior of more complex liquids such as water.

In fact development of a quantitative understanding of molten salt systems has been relatively slow. Thermodynamic and transport data have been interpreted on the basis of "lattice models" with and without allowance for the presence of vacant sites, a "significant structures model" where some ions were presumed to be "gas-like" and others to have properties as in a crystal lattice, and a "hole theory" in which spherical cavities are created against a (macroscopic) surface tension. These theories have been reviewed elsewhere.[10,11] Despite obvious defects in the initial hypotheses, some progress has been made in explaining variations in physical properties, once the values of one or more emperical parameters have been decided.

A more fundamental attempt to understand molten salt systems has been made by numerical calculation of the interaction of an assembly of ions by molecular dynamics or by Monte Carlo methods.[12] Encouraging progress has been reported for the simplest system, molten potassium chloride, where the spherical, singly charged cation and anion are of similar size. As the relative ion sizes vary, and particularly when their charges and polarizabilities are different, these methods are at present less effective. For example, with $2KCl/MgCl_2$, trigonal planar assemblages of three chloride anions around each magnesium cation are predicted,[13] whereas spectroscopy (infrared and Raman) suggests the clusters are in fact tetrahedral $MgCl_4^{2-}$.[14] Part of the difficulty lies in choosing the correct form of the pair potential and part in using only a small number of ions (usually ~ 500) and for too small a time interval ($\sim 10^{-10}$ sec). Eventually limitations of computer size and computation cost may be eased and so remove, to some extent, the latter restrictions and thus improve the usefulness of results for other than the simplest chloride melts. However, the impossibility of using these methods at their present level of sophistication, to obtain data on the more practically useful melts which contain polyatomic anions, emphasizes the importance of direct experimental investigation. Such investigation is of course likely to remain forever irreplaceable when information on the behavior of one or more solute species is needed.

The comparatively high melting points of the pure alkali metal chlorides (LiCl, 613°C; NaCl, 801°C; KCl, 776°C; RbCl, 715°C; CsCl, 646°C) imposes sufficient experimental difficulty that many investigators have opted to use the much lower melting eutectic mixtures [LiCl/KCl (60/40 mole%) mp 450°C; NaCl/KCl (50/50 mole%) 658°C; LiCl/NaCl/KCl (43/33/24 mole%) 383°C].[7] The two eutectics with the conveniently low melting points have the drawback of containing lithium cations, which have a sufficiently small radius that considerable interaction occurs with anions. This tendency and also the higher cost of lithium salts have induced other workers to prefer the sodium chloride–potassium chloride eutectic, particularly for reactions with potential industrial applications.

B. The Solubilities and Reactivities of the Elements in Chloride Melts

Despite the considerable number of studies with molten alkali metal chlorides already reported, data on solubilities is by no means complete and comparatively little information is available on the solubilities of the elements despite interest in such solutions. Of the nonmetallic elements only those of argon, sulfur, and the halogens have been investigated.

1. Nonmetallic Elements

The solubilities of the latter group of elements have generally been found to be very small, most workers giving values near 10^{-3} *M*. Only the earliest references to the solubility of chlorine have been given in Janz's "Molten Salts Handbook"[8]: that of Greenberg and Sundheim,[15] who suggested a value of 10^{-4}–10^{-3} molal, (hereafter represented as *m*) in LiCl/KCl at 400°C, while Ryabukin[16] has made a number of more recent measurements refining his values and showing that the solubility of chlorine increases steadily with temperature, with increasing radius of the alkali metal cation (LiCl < NaCl < KCl < CsCl),[17] and as the melt becomes more potassium rich in NaCl/KCl mixtures.[18] Earlier still, Wartenberg[19] had reported values of 10^{-3} *m* for various alkali metal chlorides, while more recently Leonova and Ukshe[20] have given values of 10^{-3} *m* in rubidium chloride (2.4×10^{-3} *m* at 781°C and 3.7 at 1002°C) and Nakajima *et al.*[21] 10^{-4} *m* at 450°C in LiCl/KCl. Outside this general agreement Kostin *et al.*[22] have reported much higher values (e.g., 1.115 g/cm^3 in LiCl/KCl at 400°C), while conversely Olander and Camahart[23] found an upper limit of 10^{-6} *m* for the same eutectic and temperature. At present no detailed explanation can be given for these differences although an increase in the solubility with temperature is generally agreed upon, however. For example, van Norman and Tivers[24] (1.26×10^{-4} *m* at 400°C, 2.05×10^{-4} *m* at 550°C) give a positive heat of solution (ΔH^{-0} + 3.7 kcal/mole) which would suggest a largely physical solution as is usually found for inert gaseous solutes in molten salts where it is postulated that the size of the "holes" in the structure increases with temperature.

In contrast, evidence for chemical solution has been advanced from spectroscopic studies where a charge transfer band at 31,000 cm^{-1} in LiCl/KCl at 400°C has been assigned as resulting from the trichloride anion,[15] though the validity of the observation has been criticized and the apparent maximum attributed to stray light. Additionally, however, the diffusion coefficient has been measured as some ten times larger than expected and thus chain conduction of the Grotthus type has been suggested[24]

$$Cl{-}Cl{-}Cl^- + Cl^- \rightarrow Cl^- + Cl{-}Cl{-}Cl^- \quad (1)$$

and isotope exchange studies in $KCl/PbCl_2$ have also been interpreted on the basis of the formation of the trichloride anion.[25]

By contrast, bromine has been reported to dissolve more freely in LiBr, values of an order of magnitude higher being quoted.[24] Similar values were given by Wartenberg for bromine in alkali metal bromides and iodine in alkali metal iodides, a negative heat of solution (ΔH^{-0} – 9.5 kcal/mole)[24] being taken to indicate a chemical solution. The tribromide anion was also postulated from spectroscopic studies.[15] These studies also resulted in the suggested formation of the diiodochloride anion when iodine was dissolved in LiCl/KCl melts at 400°C,[15] though Leroy[26] found iodide anions and iodinium(I) cations as well as iodine molecules in chloride melts (LiCl/KCl eutectic at 450°C).

Solutions of sulfur in alkali metal chloride melts which show a very characteristic bright blue color have been found to have a more complicated chemistry and have been the subject of many investigations. These solutions (in LiCl/KCl at 420°C) are reported to be reduced to sulfide at the cathode and oxidized to disulfur dichloride at the anode, though the latter compound was not trapped out as a volatile product.[27,28] The first assumption that the blue solution contained sulfur dimer units was advanced by Greenberg, Sundheim, and Gruen[29] on the basis of spectroscopic measurements. However, after more detailed measurements, Giggenbach[30] postulated that the 17,000 cm^{-1} charge transfer band (molar absorptivity ε 2400 liter $mole^{-1}$ cm^{-1}) was due to the singly charged disulfide anion (i.e., S_2^-, analogous to the superoxide anion) and found that sulfur was not soluble in highly purified lithium chloride–potassium chloride eutectic, solution only occurring when traces of hydroxide or water were present. Addition of Lux–Flood acids ($NaPO_3$, $K_2S_2O_7$ or NH_4Cl) caused the formation of a yellow suspension of sulfur which reverted to a yellow or orange solution on addition of potassium carbonate or hydroxide, with absorption bands at 25,000 and 32,000 cm^{-1} and postulated first a disproportionation

$$nS + nO^{2-} \rightleftharpoons SO_n^{2-} + (n-1)S^{2-} \tag{2}$$

where $n = 3$ or 4

although no evidence was advanced for the formation of sulfite or sulfate, which was followed by polysulfide formation

$$3S + S^{2-} \rightleftharpoons \underset{\text{blue}}{2S_2^-} \overset{<400^\circ C}{\rightleftharpoons} \underset{\text{yellow}}{S_4^{2-}} \tag{3}$$

Most recently the elegant investigation by Gruen *et al.*[31] resulted in the assignment of the 17,000 cm^{-1} band ($\varepsilon \sim 10{,}000$ l $mole^{-1}$ cm^{-1}) to the trisulfide radical anion S_3^-, the 25,600 cm^{-1} band to the disulfide anion, and the 31,440 cm^{-1} band to sulfide, via the equilibria

$$2S^{2-} + S_2 \rightleftharpoons 2S_2^{2-} \tag{4}$$

$$2S^{2-} + 5S_2 \rightleftharpoons 4S_3^- \tag{5}$$

Confirmation of the radical anion, which is also the chromophore in lapis lazuli, was obtained by EPR, endor, Raman, and infrared measurements. A slightly later spectroscopic and electrochemical investigation resulted in less precise but broadly similar assignments. Bernard *et al.*,[32] working with solutions of calcium sulfide in lithium chloride–potassium chloride at 400°C, showed the 17,000 cm^{-1} band to initially increase in intensity with time and then to diminish after 90 minutes. They assigned this band to a radical anion (S_2^- or S_3^-), the intermediate band at 25,600 cm^{-1} to a polysulfide (S_x^{2-}, where $x < 5$), and the high-energy band (31,300 cm^{-1}) also to the sulfide anion. In the presence of oxide, oxidation of sulfur to sulfur dioxide or sulfate can take place, though without oxide disulfur dichloride is the product[33]:

$$S + 2O^{2-} + 4Cu^{2+} \rightarrow 4Cu + SO_2\uparrow \tag{6}$$

$$S + 4O^{2-} + 6Cu^{2} \rightarrow 6Cu^+ + SO_4^{2-} \tag{7}$$

$$2S + 2Cl^- + 2Au^+ \rightarrow 2Au + S_2Cl_2 \tag{8}$$

In contrast to nitrate melts, little work has so far been reported on the solubilities of rare gases in the higher melting alkali metal halides, the only value quoted being that for argon.[34]

2. Metallic Elements

In complete contrast to the rare gases the solubilities of the alkali metals have been the subject of many investigations. The well-known phase diagrams were established largely by the group led by Bredig[35] at Oak Ridge on the basis of thermal analysis and of chemical analysis of two liquid layers separated by differences of density in "two-legged" stainless steel vessels. The immiscible liquids formed above the melting point of the halide showed increasing metal solubility with temperature.[36] The miscibility increases from lithium to cesium (Fig. 1a) and also from iodide to fluoride (Fig. 1b), though the latter is largely brought about by the higher melting temperatures resulting from the reduction in anion size. Solutions dilute in alkali metal are blue in color and show many properties arising from the presence of free electrons, similar to the better-known solutions of alkali metals in liquid ammonia (see Chapter 6 of Volume II of this treatise). For example, electrical conductivity showed a smooth variation with composition, even through the immiscibility gap, from ionic to metallic values over several orders of magnitude. Earlier electron spin resonance and magnetic measurements were inconclusive but later magnetic susceptibility indicated ionization and were interpreted on a free-electron model.[37,38] Electronic

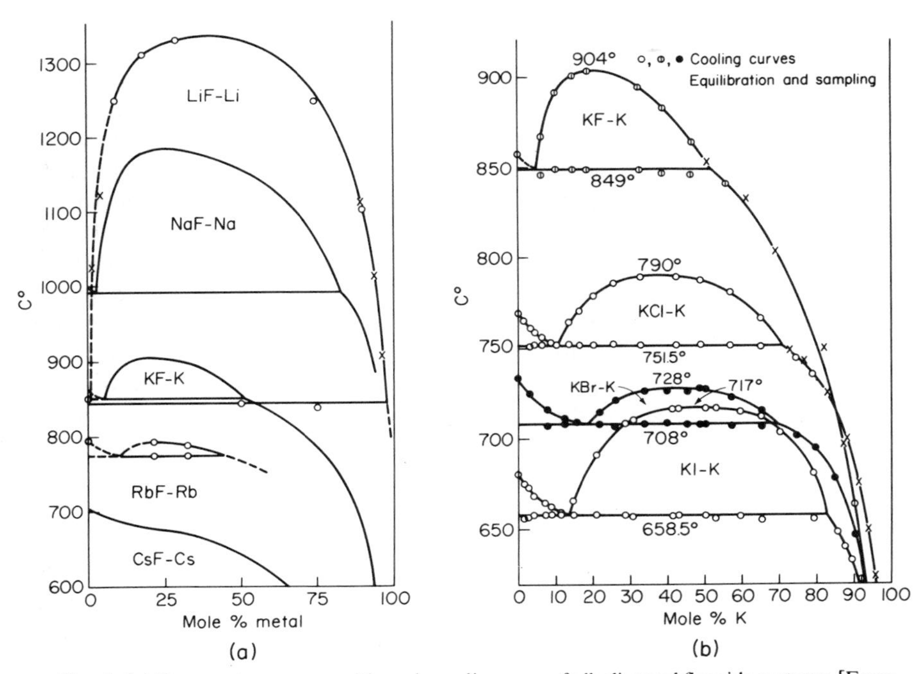

FIG. 1. (a) Temperature-composition phase diagrams of alkali metal fluoride systems. [From A. S. Dworkin, H. R. Bronstein, and M. A. Bredig, *J. Phys. Chem.* **66**, 572 (1962).] (b) Temperature-composition phase diagrams of potassium-potassium halide systems. [From J. W. Johnson and M. A. Bredig, *J. Phys. Chem.* **62**, 604 (1958).]

spectroscopy has shown the absorption maximum of the broad band of alkali metals in their own halides (e.g., maximum at 13,000 cm^{-1} for Na in NaCl at 809°C) to vary smoothly with the molar volume of the halide and this has been taken to support the existence of electrons trapped at anion vacancies, analogous to the F (German farbe) centers in solid alkali metal halides doped with excess metal,[39,40] a theory which can also account for the susceptibility measurements.[37] However, as has been pointed out, the differences between the various postulated environments of the electron are relative rather than absolute, since an association between an F center, an anion vacancy, and two intervening cations is equivalent to the lower oxidation state cation M_2^+ suggested by certain Russian workers on the basis of potentiometric and conductometric measurements after reacting halide melts with alkali metal containing alloys (NaCl with Na/Pb at 850°C and LiCl/KCl with Li/Hg at 400°–800°C).[41] An entity consisting of an F center and one neighboring cation is equivalent to the atomic species postulated by Rice.[42]

Earlier workers had reported very different spectroscopic results and had even found difficulty in obtaining blue solutions when potassium dissolved in lithium chloride–potassium chloride eutectic unless potassium hydroxide, potassium oxide, or lithium oxide additions were made, when an absorption band at 13,900 cm^{-1} was obtained. However, no additions were found necessary to obtain a 10,600 cm^{-1} band from potassium in potassium chloride at 500°C (or a 8600 cm^{-1} band for Cs in CsCl and 18,200 cm^{-1} for Li in LiCl.[43]) In contrast, when lithium, sodium, or potassium metals were added to the lithium chloride–potassium chloride eutectic, an identical band at 18,900 cm^{-1} was reported. However, a fairly rapid fall of optical density with time indicated reaction, possibly of the metal with water absorbed on the quartz surfaces, which could have provided the solubilizing anions if such are indeed necessary with the eutectic.

In other cases postulation of the formation of lower oxidation state cations is generally acceptable in those cases where a metal has been found to react and dissolve in alkali metal halide melts containing the halide of that metal in solution. A number of examples are given later in Section II,F,2.

The more reactive metals will of course reduce the alkali metal cations themselves. Addition of calcium to lithium chloride–potassium chloride eutectic resulting in the distillation of an unspecified alkali metal,[44] while several lanthanide metals (Sm, Eu, Yb) displace lithium from this melt at 450°C[45] and lanthanum metal reacts with sodium chloride–potassium chloride at 690°–800°C.[46]

Certain other metals, in the form of intermetallic compounds, are very soluble in halide melts and appear to be present in anionic form. This is illustrated by the lithium chloride–lithium fluoride eutectic in which pure bismuth had only a low solubility (0.9×10^{-3} wt % at 600°C). However, in the form of the lithium alloy (Li_3Bi) it dissolved to 0.3 wt % at 600°C, the Li/Bi ratio of the solute in the red solution (absorption edge 16,300 cm^{-1}) was 3.09, suggesting the presence of bismuthide anions (Bi^{3-}).[47] Very similar results were reported for pure lithium chloride but at somewhat higher temperatures.[48] In molten sodium or potassium iodides even higher solubilities were found. Various sodium–bismuth alloys dissolved (e.g., a 26% Bi alloy to 2.25 mole % Bi at 605°C in NaI/NaCl eutectic with a dissolved Na/Bi ratio of 3) suggesting that the same bismuthide anion was present, while sodium–antimony (Na_3Sb) was soluble in the same eutectic, to the extent of 20 mole % at 700°C. Cryometric measurements on this solute in sodium iodide gave a cryoscopic number (number of foreign ions per solute molecule) of 1.04, suggesting Sb^{3-}, and of 0.34 when NaSb was dissolved, suggesting the more polymerized anion $Sb_3{}^{3-}$. Alloys of sodium with thallium, tin, lead, and gold gave much smaller solubilities with Na/metal ratios varying with temperature and it was assumed solution occurred as atoms.[49]

C. Solubility and Reactivity of Oxides and Oxyacids

1. Oxides

Of the gaseous oxides investigated carbon monoxide had a low solubility (of 10^{-4} *m* in NaCl/KCl at 680°–800°C)[50] which increased with temperature and pressure, indicating physical solution. Similar results were obtained for carbon dioxide (in NaCl, KCl, and CsCl at 900° and 950°C)[51] but the values were sensitive, as expected, to the presence of oxide anions. Even an impurity of nickel(II) oxide increased the value by an order of magnitude due to the formation of carbonate.

The solubility of water is very small, but this impurity is troublesome since hydrolysis can occur[52,53]

$$H_2O + 2Cl^- \rightarrow 2HCl\uparrow + O^{2-} \tag{9}$$

The purification of alkali metal chlorides is therefore of importance, particularly the removal of water, and strict regimes of purging the melts with streams of hydrogen chloride and chlorine and finally with an inert gas are recommended, after thorough pumping off of removable water first from the solid salts and then from the melt mixture.

The phase diagram of NaCl–Na_2O has been published showing a high solubility of the oxide in the chloride melt.[54] However, many solid oxides appear to have much lower solubilities, for example, beryllium oxide (2.1×10^{-6} *m*; see Table 1), though this increased with the presence of fluoride

TABLE I

SOLUBILITIES OF METAL OXIDES IN ALKALI METAL CHLORIDE MELTS

Solute oxide	Solvent	Temperature (°C)	Solubility (m)	Reference
Na_2O	NaCl	700	5.73	54
BeO	NaCl/KCl	1100	2.1×10^{-6}	55
MgO	LiCl	688	$< 1.2 \times 10^{-2}$	56
MgO	NaCl	900	$< 1.2 \times 10^{-2}$	56
MgO	KCl	900	$< 1.2 \times 10^{-2}$	56
MgO	LiCl/KCl	480	3.0×10^{-3}	57
FeO	LiCl/KCl	480	3.0×10^{-3}	57
CoO	LiCl/KCl	480	6.0×10^{-3}	57
NiO	LiCl/KCl	480	2.1×10^{-3}	57
Sb_2O_3	LiCl/KCl	600	$\geq 5.4 \times 10^{-3}$	58
UO_3	LiCl/KCl	480	3.0×10^{-3}	56

anions probably because of compound formation. The full variation of solubility at 1100°C (LiCl < NaCl < KCl < CsCl < NaF < KF < $Na_4P_2O_7$ < $NaPO_3$ < $Na_2B_4O_7$) indicated the effects of cation size, compound formation, and acid–base behavior.[55] The solubilities of a number of other metal oxides have been reported with some precision (Table I).[54–58] In addition, the relative approximate solubilities of these and many further oxides have been given (Table II) together with five higher oxides which decomposed, liberating oxygen.[59]

Certain insoluble oxides may be dissolved by addition of oxidizing agents:

$$NiO + 2Cu^{2+} \rightarrow Ni^{2+} + 2Cu^{+} + \tfrac{1}{2}O_2\uparrow \tag{10}$$

$$FeO + \tfrac{3}{2}Cl_2 \rightarrow Fe^{3+} + 3Cl^{-} + \tfrac{1}{2}O_2\uparrow \tag{11}$$

$$Fe_3O_4 + \tfrac{9}{2}Cl_2 \rightarrow 3Fe^{3+} + 9Cl^{-} + 2O_2\uparrow \tag{12}$$

While other less soluble oxides may be brought into solution by acid–base reactions, examples utilizing acids of increasing strength are:

$$ZnO + CO_2 \rightarrow Zn^{2+} + CO_3^{2} \tag{13}$$

$$CoO + CO_2 \rightarrow Co^{2+} + CO_3^{2-} \tag{14}$$

$$Fe_3O_4 + 8HCl \rightarrow Fe^{2+} + 2Fe^{3+} + 8Cl^{-} + 4H_2O\uparrow \tag{15}$$

$$SnO_2 + 4HCl \rightarrow SnCl_4\uparrow + 2H_2O\uparrow \tag{16}$$

$$NiO + K_2S_2O_7 \rightarrow Ni^{2+} + 2K^{+} + 2SO_4^{2-} \tag{17}$$

Analogous solubilization reactions have also been reported with acidic oxides rather than basic transition metal oxides. For example, boric oxide is

TABLE II

RELATIVE SOLUBILITIES OF METAL OXIDES IN LiCl/KCl AT 450°C[59]

Solubility or reaction	Oxides
Soluble and dissociated	CaO, SrO, BaO, CdO, HgO, Ag_2O, Sb_2O_3, Bi_2O_3, MnO, Cu_2O, Tl_2O, PbO
Slightly soluble	CoO, ZnO
Insoluble	MgO, Al_2O_3, Ce_2O_3, SiO_2, TiO_2, ZrO_2, NiO, SnO, SnO_2, Sb_2O_5, Mn_2O_3, FeO, Fe_3O_4, Cr_2O_3, PdO, PtO
Decomposed (forming stated lower oxidation state product)	$CuO(Cu^+)$, $Tl_2O_3(Tl^+)$, $Fe_2O_3(Fe_3O_4)$, $PbO_2(Pb^{2+})$, $Pb_3O_4(Pb^{2+})$

neutralized by alkali or alkaline earth carbonates to form borates.[60] Surprisingly a whole series of different products were found by potentiometric titration (Table III). Variations in the stoichiometry occurred even when changing from adding the carbonate to a melt initially containing boric acid (dehydration is assumed), termed "backward titration," to finding the glass composition causing rapid change in oxygen electrode potential of chloride melts containing identical quantities of borate glasses (formed by firing various amounts of boric acid and carbonate at 1000°–1100°C) the so called "forward titration."

Boric oxide in a sodium chloride melt facilitated the oxidation of chloride anions by oxygen[61] supposedly by the reaction

$$2NaCl + \tfrac{1}{2}O_2 \xrightarrow{B_2O_3} Na_2O + Cl_2\uparrow \tag{18}$$

The oxide initially produced presumably formed borate with excess boric oxide. However, the more acidic phosphorus pentoxide has been reported to form some chlorine apparently without additional oxygen (in LiCl–NaCl–KCl at 350°–500°C)[62]

$$P_4O_{10} + Cl^- \xrightarrow{\text{vacuum}} POCl_3 + Cl_2 + Li_2O\ (?) \tag{19}$$

though the reduced product was not identified. Perhaps similarly, tungsten(VI) oxide has been stated to oxidize chloride anions.[63] Uranium(VI) oxide and uranyl(VI) cations can also oxidize chloride anions:

$$UO_2^{2+} + Cl^- \rightleftharpoons UO_2^{+} + \tfrac{1}{2}Cl_2 \tag{20}$$

though the equilibrium lies well to the left except at high temperatures (650°C) or low chlorine pressures (0.7 mm).[64-66] Eq. 20 is in fact only one of many equilibria which have been found to interconnect the four oxidation states of uranium in chloride melts (a Littlewood diagram has been given).[65] Further equilibria have been determined when the uranium species are also in the presence of hydrogen, oxygen, hydrogen chloride,[66] etc.

TABLE III

PRODUCTS OF POTENTIOMETRIC TITRATION OF BORIC ACID WITH CARBONATES IN NaCl/KCl MELT AT 750°C

Carbonate	Forward titration		Backward titration	
	CO_3^{2-}/B_2O_3 ratio	Product	CO_3^{2-}/B_2O_3 ratio	Product
Li_2CO_3	2 : 1	$B_2O_5^{4-}$ or $BO_2^- \cdot BO_3^{3-}$	1 : 1	BO_2^-
Na_2CO_3	3 : 2	$B_4O_9^{6-}$ or $3BO_2^- \cdot BO_3^{3-}$	1 : 2	$B_4O_7^{2-}$
			and 1 : 1	$B_4O_8^{4-}$ or $4BO_2^-$
K_2CO_3	3 : 2	$B_4O_9^{6-}$ or $3BO_2^- \cdot BO_3^{3-}$	1 : 2	$B_4O_7^{2-}$
			and 1 : 1	$B_4O_8^{4-}$ or $4BO_2^-$
$CaCO_3$	4 : 1	$B_2O_7^{8-}$ or $2BO_3^{3-} \cdot O^{2-}$	3 : 2	$B_4O_9^{6-}$ or $3BO_2^- \cdot BO_3^-$
$SrCO_3$	2 : 1	$B_2O_5^{4-}$ or $BO_2^- \cdot BO_3^{3-}$	2 : 3	$B_6O_{11}^{4-}$ or $B_4O_7^{2-} \cdot 2BO_2^-$
			and 1 : 1	$B_6O_{12}^{6-}$ or $6BO_2^-$
$BaCO_3$	3 : 2	$B_4O_9^{6-}$ or $3BO_2^- \cdot BO_3^{3-}$	1 : 2	$B_4O_7^{2-}$
			and 1 : 1	$4BO_2^-$

Acidic properties can also be shown by oxides of transition metals when in the highest oxidation states:

$$V_2O_5 + CO_3^{2-} \rightarrow 2VO_3^- + CO_2\uparrow \text{ (Ref. 57)} \tag{21}$$

and can also act as oxidizing agents in the presence of acids[57]:

$$V_2O_5 + 2I^- + 2HSO_4^- \rightarrow 2VO^{2+} + 2SO_4^{2-} + I_2 + H_2O\uparrow \tag{22}$$

$$V_2O_5 + 2I^- + S_2O_7^{2-} \rightarrow 2VO^{2+} + 2SO_4^{2-} + I_2 \tag{23}$$

Vanadium(IV) oxide, itself insoluble, can be oxidized to the soluble vanadium(V) cation:

$$VO_2 + Cu^{2+} \rightarrow VO_2^+ + Cu^+ \tag{24}$$

Other oxides of transition metals in lower oxidation states show only a basic character. For example, the solubilities of the manganese oxides are said to vary in the order $MnO_2 < Mn_2O_3 < MnO$ and to increase with decreasing size of alkali cation ($KCl < NaCl < LiCl$) and with temperature.[67]

Peroxide anions have been found to be stable in LiCl/KCl at 400°–500°C[68] but not in NaCl/KCl at 700°C.[69,70] More surprisingly, peroxide has been found in melts to which only oxide anions were originally added.[71,72] The equilibria (in LiCl/KCl, NaCl/KCl, and LiCl/NaCl/KCl at 585°C)

$$O_2 + Cl^- \rightleftharpoons O_2^- + \tfrac{1}{2}Cl_2 \tag{25}$$

$$\Updownarrow O_2$$

$$O_2 + 2Cl^- \rightleftharpoons O_2^{2-} + Cl_2 \tag{26}$$

were postulated by Martinot,[73] the oxygen anions being characterized by polarography, chronopotentiometry, and infrared spectroscopy, and may provide an alternative hypothesis for chlorine production (see Eq. 18).

2. Oxyanions

A report of the electroreduction of nitrate (in LiCl/KCl at 400°–500°C) implies that both this anion and the nitrite anion formed as a product are stable in these melts.[74] Phosphate anions are certainly stable, having been the subject of acid–base titrations,[75]

$$PO_3^- + CO_3^{2-} \rightarrow PO_4^{3-} + CO_2\uparrow \tag{27}$$

electroreduction reactions (e.g., at 780°C in NaCl/KCl),[76]

$$PO_4^{3-} + 5e \rightarrow P + 4O^{2-} \tag{28}$$

and reactions with oxidizing gases, e.g., the chlorination of apatite in sodium chloride at 850°C[77]

$$Ca_3(PO_4)_2 + Cl_2 \rightarrow CaCl_2 + Ca_2(P_2O_7) + Ca_5(P_3O_{10}) + P_2O_5 \text{ etc.} \tag{29}$$

The sulfites of several cations (Mg^{2+}, Ca^{2+}, Sn^{2+}, Ba^{2+}, Mn^{2+}, Tl^{+}) are soluble in lithium chloride–potassium chloride eutectic giving colorless solutions which decompose, either by liberating sulfur dioxide or by disproportionation to sulfide and sulfate anions. The latter reaction is favored if a cation forming an insoluble sulfide is present, and the former is favored if a particularly insoluble oxide may be formed (e.g., with Pd^{2+} and Pt^{2+}).[33] With oxidizing cations (Cu^{2+} or Au^{+}) sulfur dioxide and oxygen were formed, while with mercury(II) cations oxidation was to sulfur dioxide and sulfate

$$2SO_3^{2-} + Hg^{2+} \rightarrow SO_4^{2-} + Hg\uparrow + SO_2\uparrow \quad (30)$$

In the presence of oxide, oxidation was entirely to sulfate

$$SO_3^{2-} + O^{2-} + 2Cu^{2+} \rightarrow SO_4^{2-} + 2Cu^{+} \quad (31)$$

$$SO_3^{2-} + CrO_3 \rightarrow SO_4^{2-} + Cr_2O_3 \quad (32)$$

while a kinetic study has recently shown oxidation to sulfate was first-order with respect to the partial pressure of oxygen.[78] Reduction occurred with metals that formed insoluble sulfides

$$SO_3^{2-} + 3Zn \rightarrow 2ZnO + ZnS + O^{2-} \quad (33)$$

$$SO_3^{2-} + 6Cu \rightarrow 4Cu^{+} + 3O^{2-} + Cu_2S \quad (34)$$

Hydrogen sulfite, thiosulfate, and dithionate anions were all unstable, giving the products listed, though stoichiometries were not given

$$HSO_3^{-} \rightarrow SO_4^{2-} + SO_3^{2-} + S + SO_2 \quad (35)$$

$$S_2O_3^{2-} \rightarrow SO_4^{2-} + SO_3^{2-} + S \quad (36)$$

$$S_4O_6^{2-} \rightarrow SO_4^{2-} + SO_3^{2-} + S \quad (37)$$

By contrast, and as would be expected, most sulfates were found to be stable, in conformity with the reports of several phase diagram[79] and infrared studies.[80] The more acidic cations were found to release sulfur dioxide

$$Al_2(SO_4)_3 \rightarrow Al_2O_3 + SO_3\uparrow \quad (38)$$

$$Ti(SO_4)_2 \rightarrow TiO_2 + SO_3\uparrow \quad (39)$$

The self-ionization constant of sulfate in chloride melts has been found to be small[81] ($K < 10^{-2}$ in LiCl/KCl at 750°C). Hydrogen sulfate, disulfate (pyrosulfate), and peroxydisulfate anions were all reactive, as acids, thermally unstable compounds, and oxidizing agents.[33]

$$NaHSO_4 + Cl^{-} \rightarrow SO_4^{2-} + HCl\uparrow \quad (40)$$

$$Na_2S_2O_7 \rightarrow SO_4^{2-} + SO_3 \quad (41)$$

$$K_2S_2O_8 + 2Cl^{-} \rightarrow 2SO_4^{2-} + Cl_2\uparrow \quad (42)$$

Until recently no work had been done on the reactions of halates or perhalates in chloride melts, but now a rich reaction chemistry has begun to be uncovered by Slama and co-workers, in which iodide can be oxidized by iodate or by oxygen, the more acidic lithium cations apparently assisting the reaction and precipitating the sparingly soluble hexoxoperiodate.[82,83]

$$3I^- + 2IO_3^- + 5Li^+ \rightarrow Li_5IO_6 + 2I_2 \tag{43}$$

$$9I^- + 4O_2 + 5Li^+ \rightarrow Li_5IO_6 + 2O^{2-} + 4I_2 \tag{44}$$

However, in a stream of carbon dioxide or with phosphorus pentoxide, the stronger acid takes the oxidation reaction further[83]

$$Li_5IO_6 + 7I^- + 6CO_2 \rightarrow 6CO_3^{2-} + 5Li^+ + 4I_2 \tag{45}$$

The formation of metavanadate anions has been mentioned above (Section II,C) but further acid–base reduction has been reported[57,75]

$$VO_3^- + O^{2-} \rightleftharpoons (Li, K)_3VO_4 \tag{46}$$

Lithium orthovanadate is the least soluble product. Metavanadate may also be electrolytically reduced[84] (in LiCl/KCl at 450°C):

$$\begin{array}{l} LiVO_3 \xrightarrow{e} LiV_2O_5 \xrightarrow{e} LiVO_2 \xrightarrow[\text{or } -e]{LiVO_3} LiV_2O_4 \\ \qquad\qquad + Li_3VO_4\downarrow \quad + Li_3VO_4\downarrow \end{array} \tag{47}$$

Potassium chromate is both soluble and stable, showing the same charge-transfer band (in LiCl/KCl at 370°C) as found in aqueous solution.[85] Dichromate was also stable when it was the only solute but in the presence of acids or bases reaction occurred,[75] the suggested products being:

$$K_2Cr_2O_7 + HCl \longrightarrow Cr^{3+} + H_2O + Cl^- \tag{48}$$

$$+ PO_3^- \longrightarrow Cr_2O_3 + 2PO_4^{3-} \tag{49}$$

$$+ CO_3^{2-} \xrightarrow{400^\circ C} 2CrO_4^{2-} + CO_2 \tag{50}$$

Chromate(V) anions can be readily reduced electrolytically to insoluble chromate(III) in pure lithium chloride–potassium chloride at 400°C. The product had the stoichiometry Li_5CrO_4, but with some magnesium chloride in solution this became $Li_xMg_yCrO_4$ where $x = 0.3$–0.5, and $x + 2y = 5$. With nickel(II) chloride in solution, $Li_xNi_yCrO_4$ ($x = 0.6$–0.9)[86] was formed, as was a similar stoichiometry with $x = 1$ when dissolved zinc chloride was present, the structure suggested being

```
   O     O
  / \   / \
Zn   Cr    Zn
  \ /   \ /
   O     O
```

With cobalt(II) chloride in high concentration and at low potentials, Co_2CrO_4 was formed containing equal numbers of cobalt(II) and cobalt(III) cations, but at high potentials and lower cobalt(II) concentration the more usual product, $LiCo_2CrO_4$, was formed containing only cobalt(II).[87] The presence of lithium for this type of product seems essential because in sodium chloride–potassium chloride at 700°C only chromium(III) oxide was reported as a reduction product.[88] Molybdate(VI) and tungstate(VI) have been the subject of crystal growth and phase diagram studies and appear to be stable.[89–92] The solubilities of neodynium molybdates have been stated to be smaller than for tungstates, and to decrease with the size of the alkali metal cation in the solvent,[93] while lanthanide oxides have been extracted from their titanoniobate compounds at 900°–1000°.[94]

D. Solubility and Reactivity of Sulfides

Solubilities of six sulfides have been reported quantitatively (Table IV)[33,95,96] and a number of others have been reported qualitatively (the sulfides of Mg^{2+}, Ca^{2+}, Sr^{2+}, Ba^{2+}, Mn^{2+}, and Tl^{+} are stated to be soluble in LiCl/KCl at 500°C, and those of Sn^{2+}, Pb^{2+}, Sb^{3+}, Bi^{3+}, Fe^{2+}, Co^{2+}, Ni^{2+}, Pd^{2+}, Pt^{2+}, Cu^{+}, Ag^{+}, Zn^{2+}, Cd^{2+}, and Ce^{3+} to be insoluble).[33] Sodium sulfide has also been stated to be insoluble, but sodium polysulfide ($Na_2S_{2.2-5.9}$) is soluble though decomposing liberating sulfur (in LiCl/KCl at 420°C).[97] Additions of sodium sulfide will increase the solubility of certain other sulfides, (PbS, Sb_2S_3, Bi_2S_3, and Cu_2S) apparently by the formation of thioanions.[96]

The sulfide anion may be oxidized to sulfur by copper(II), iron(III), gold(I), mercury(II), and tin(IV) cations and by iodine

$$SnS_2 \rightleftharpoons Sn^{4+} + 2S^{2-} \rightarrow SnS + S \quad (51)$$

$$S^{2-} + I_2 \rightarrow 2I^- + S \quad (52)$$

and can thus bring cations into solution from insoluble sulfides,

$$CdS + HgCl_2 \rightarrow Cd^{2+} + S + Hg\uparrow + 2Cl^- \quad (53)$$

a process also achieved by addition of cyanide, which is accompanied by oxidation of the cation though the reduction product was not specified[33]

$$Cu_2S + KCN \rightarrow Cu(CN)_4{}^{2-} \quad (54)$$

$$FeS + KCN \rightarrow Fe(CN)_6{}^{3-} \quad (55)$$

Hydrogen sulfide has a limited solubility, but if bubbled through lithium chloride/potassium chloride at 400°–500°C displaces hydrogen chloride

$$H_2S + Cl^- \rightleftharpoons HS^- + HCl \quad (56)$$

The resulting anion has a higher solubility.[98] Hydrogen chloride itself has only a limited solubility (e.g., 1.4×10^{-4} *m* at 900°C in NaCl) that decreases

TABLE IV

SOLUBILITY OF SULFIDES IN ALKALI METAL CHLORIDE MELTS

Sulfide	Melt	Temperature (°C)	Solubility (*m*)	Reference
Li_2S	LiCl/KCl	375	0.008	95
Li_2S	LiCl/KCl	475	0.028	95
CaS	LiCl	700	0.5	33
PbS	NaCl/KCl	700	0.024	96
PbS	NaCl/KCl	850	0.11	96
Sb_2S_3	NaCl/KCl	700	0.029	96
Sb_2S_3	NaCl/KCl	950	0.28	96
Bi_2S_3	NaCl/KCl	700	0.0015	96
Bi_2S_3	NaCl/KCl	950	0.02	96
NiS	LiCl/KCl	335	1.1×10^{-8}	95
NiS	LiCl/KCl	475	2.1×10^{-7}	95
Cu_2S	NaCl/KCl	800	0.0075	96

with temperature, and decreasing size of alkali metal cation. Formation of complex anions (HCl_2^-), is suggested from infrared spectroscopy.[99,100]

E. Solubility and Reactivity of Other Compounds

1. BINARY COMPOUNDS

Molten chlorides are well known as being good solvents for a number of compounds which are otherwise almost unknown in solution. Among these are the hydrides (LiH is very soluble in LiCl, the phase diagram indicating a eutectic at 495°C with 36% LiH, i.e., 13.2 *m*)[101] and carbides [LiC and CaC_2 dissolve in LiCl at 820°C giving 33.6 and 12.0 mole% (or 12.0 and 3.2 *m*) solutions, respectively].[102] Such solutions are unstable in sodium chloride and form carbon and sodium.[103] However, in molten calcium chloride, such solutions can react with steel surfaces giving a pearlitic layer with enhancement of mechanical properties.[104]

Ionic nitrides are also soluble [Li_3N 15 mole% (4.1 *m*) in LiCl at 650°C decreasing as KCl was added, Ca_3N_2 7 mole% (1.8 *m*)],[105] but two other insoluble nitrides have been prepared by the reaction of ammonia

$$B_2O_3 + NH_3 \rightarrow BN \tag{57}$$

in NaCl/KCl at 850°–950°C[106]

$$UCl_4 + NH_3 \rightarrow UNCl \tag{58}$$

in NaCl/KCl at 800°C.[107]

2. Other Anions

Potassium cyanide is very soluble in molten chloride (e.g., $>2.3\ m$ in LiCl/KCl at 450°C[108]) and is oxidized by chlorine and quantitatively by oxygen at 400°C

$$2KCN + Cl_2 \rightarrow (CN)_2 + 2KCl \tag{59}$$

$$KCN + Cl_2 \rightarrow CNCl + KCl \tag{60}$$

$$2KCN + O_2 \rightarrow 2KCNO \tag{61}$$

Potassium cyanate and thiocyanate are also very soluble ($>1.9\ m$ and $>0.7\ m$ in LiCl/KCl at 400°C.[108,109]) Sodium hydroxide is unstable, losing water in sodium chloride–potassium chloride at 800°C and forming red solutions attributed to the disproportionation reaction[52]

$$4NaOH \rightarrow O_2^{2-} + 2Na_2^{+} + 2H_2O \tag{62}$$

F. Solutions of Metal Chlorides

1. Coordination Chemistry

As would be expected, most metal chlorides are very soluble and many of the more strongly acidic ones form distinct complexes in chloride melts. Perhaps the best-known example is aluminum trichloride where the tetrachloroaluminate anion forms (Fig. 2). The sodium salt (which can also be called the 1 : 1 mixture) has the conveniently low melting point of 175°C and is often used as a melt for organic reactions (c.f. Section II,H) and for preparing lower oxidation state compounds (c.f. Section II,F,2). This melt shows a further ionization

$$2AlCl_4^- \rightleftharpoons Al_2Cl_7^- + Cl^- \tag{63}$$

which can be expressed in acid–base terms otherwise not possible in pure alkali metal chloride melts. This equilibrium is displaced to the right by addition of lithium cations and by increasing temperature.[110] Chloride ions, defined as the basic species, are produced by the addition of the strongly basic oxide ion, water, and the weakly basic fluoride ion

$$2AlCl_4^- + O^{2-} \rightleftharpoons Al_2OCl_5^- + 3Cl^- \tag{64}$$

$$2AlCl_4^- + H_2O \rightleftharpoons Al_2OCl_5^- + Cl^- + 2HCl \tag{65}$$

$$2AlCl_4^- + F^- \rightleftharpoons Al_2Cl_6F^- + Cl^- \tag{66}$$

Protons increase the concentration of the acidic polymeric anion

$$2AlCl_4^- + H^+ \rightleftharpoons Al_2Cl_7^- + HCl \tag{67}$$

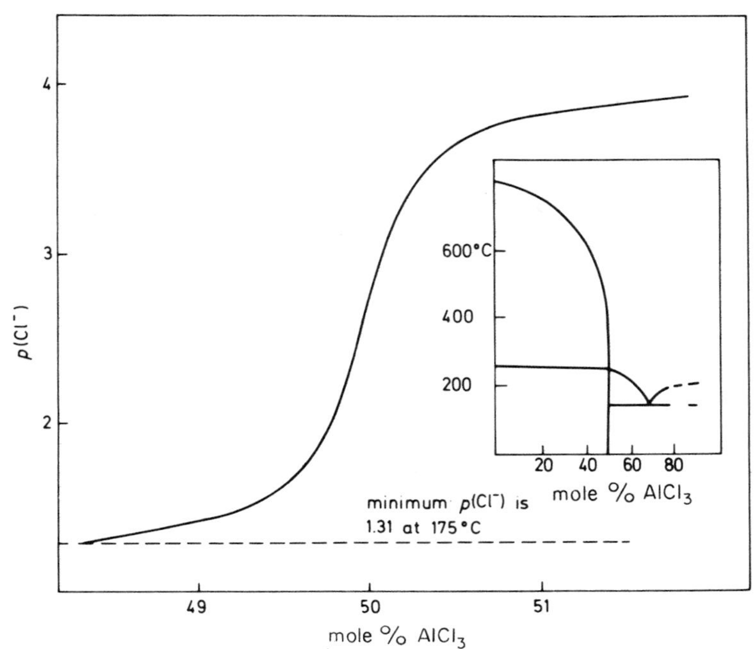

FIG. 2. Changes in the chloride activity with composition for the solvent system NaCl–$AlCl_3$ at 175°C [adapted from B. Tremillon and G. Letisse, *J. Electroanal. Chem.* **17**, 371 (1968)]. The insert shows the phase diagram for this system [W. Fischer and A. L. Simon, *Z. Anorg. Allg. Chem.* **306**, 1 (1960)].

as does addition of aluminum trichloride. Figure 2 indicates how the activity of the chloride anion is heavily dependent on the precise ratio of sodium chloride–aluminum chloride.

With the larger group IIIB metals the coordination number is higher, indium(III) chloride being six coordinated octahedral (from Raman spectroscopy).[111] With Group IVB halides complex formation is inferred from electrochemical[112] and Raman results,[113] though Braunstein[114] has shown vapor pressure data that give no positive evidence. The coordination involved is more doubtful, tetrahedral $PbCl_4^{2-}$ being suggested, but there is agreement that the chain structures of tin(II)[115] and lead(II)[113] chlorides are broken up on addition of potassium chloride to form individual anions. Raman measurements have also suggested that antimony(III) and bismuth(III) chlorides dissolve in potassium chloride forming complexes ($SbCl_4^-$, $BiCl_4^-$, and $BiCl_6^{3-}$ are suggested),[116] while phase diagrams of tellurium(IV) chloride have shown compound formation in the solid phase (K_2TeCl_6, Cs_2TeCl_6, $CsTeCl_5$) which may also occur in solution.[117]

Pentachlorotellurate(IV) anions are also postulated from conductometric measurements in potassium chloride.[118]

The coordination chemistry of transition metal cations in chloride melts has been the subject of much study. The early work was reviewed by Gruen[119] who showed the deviations of oxidation state (II) and (III) cations of the first row transition elements from octahedral coordination, by distortion and formation of tetrahedral geometries, to be those expected on the basis of ligand field stabilization energies. These results and those for second and third row elements are listed in Table V.[120-134]

Lest the study of coordination appear simple or even trivial, it must be pointed out that for the most-investigated cations an impressive number and variety of equilibria have been discovered. In the case of nickel(II) the three bands were initially assigned to a distorted tetrahedral coordination[130] though in the light of more detailed measurements over a temperature range there are both octahedral and tetrahedral coordinations and the ratio of tetrahedral/octahedral centers increases with the potassium chloride concentration (in LiCl/KCl mixtures) as well as with increasing temperature.[132] The dominance of the tetrahedral geometry appears complete in cesium chloride melts at higher temperatures, while at a concentration of nickel(II) chloride above 20 mole% a further, unidentified geometry appeared.[131] When magnesium chloride was added to molten potassium chloride, the coordination became octahedral,[135] probably because the chloride ions that are nearest neighbors to magnesium contribute less charge to the nickel cations.

With potassium chloride–zinc(II) chloride melts a series of equilibria were established. The change from tetrahedral to octahedral and then back to tetrahedral coordination as the zinc chloride concentration was increased is clearly shown by the spectral curves of Fig. 3. Increasing temperature also caused a change to tetrahedral coordination, though both a two-species equilibrium and a continuous distortion mechanism have been suggested to operate at different concentrations[136] (Table VI). When cesium chloride is substituted for potassium chloride even more complex behavior has been found[137] and is summarized in Table VII. Four different environments for nickel cations have been deduced in this particularly careful study: octahedral$_N$ and tetrahedral$_N$ indicate sites in a network of zinc chloride sharing six and four chlorides, respectively, while octahedral$_P$ and tetrahedral$_P$ indicate the nickel associated with the smaller, but probably still polymeric, entities formed as the zinc chloride network was broken up on addition of the cesium chloride. Computer calculations indicated that some equilibria were internally linear (i.e., two-species), indicated by straight lines in Table VII, and others shown by wavy lines to be of the continuous distortion type.

TABLE V

SPLITTING AND STEREOCHEMISTRY OF TRANSITION METAL CATIONS IN MOLTEN ALKALI METAL CHLORIDES

Solute	Melt	Temperature (°C)	Absorption maxima (cm^{-1})	Coordination	Reference
Ti(III)	LiCl/KCl	400	13,000; 10,000(sh)	Dist. oct.	120
V(III)	LiCl/KCl	400	18,030; 11,000	Oct.	121
		1,000		Dist. tet.	
V(II)	LiCl/KCl	400	19,000; 12,000; 7,200	Oct.	120
Cr(III)	LiCl/KCl	400	18,500; 12,500	Oct.	122
Cr(III)	CsCl	655–765	17,200; 12,050	Oct.	122
Cr(II)	LiCl/KCl	400–1,000	9,800	Dist. tet.	123
Mn(II)	LiCl/KCl	400–1,000	28,000; 22,000	Tet.	120, 124
Re(IV)	LiCl/KCl	400–600	16,150; 15,100; 7,950; 7,000	Dist. oct.	125
Fe(III)	LiCl/KCl	370–450	25,000(sh)	Prob. tet.	126
Fe(II)	LiCl/KCl	400	5,100	Tet.	123
	LiCl/KCl	1,000	6,000	Dist. tet.	123
Ru(III)	LiCl/KCl	400	28,700; 24,700; 18,300	Oct.	127
Os(IV)	LiCl/KCl	450	18,000(sh); 11,000; 5,300	Oct.	128
Co(II)	LiCl/KCl	447–480	16,700; 15,150; 14,700	Tet.	124
Rh(III)	LiCl/KCl	400	38,800; 23,500; 18,400	Oct.	127
	LiCl/KCl	450	23,500; 18,200; 14,400	Oct.	129
Ir(III)	LiCl/KCl	450	28,000; 23,400; 18,800; 15,400	Oct.	129
Ni(II)	LiCl/KCl	700–1,000	15,300; 14,200; 8,000	Dist. tet.	130
	CsCl	864	15,700; 14,000; 7,500	Tet.	131
	LiCl/KCl	363	19,700; 16,000; 14,300; 8,200	Tet., oct.	132
	LiCl/KCl	898	17,600(sh); 16,000; 14,100; 8,000	Tet.	132
Pd(II)	LiCl/KCl	450		Dist. tet.	129
	LiCl/KCl	400–600	20,100	Sq. pl.	133
Pt(II)	LiCl/KCl	400–450	28,400; 25,600; 19,000	Sq. pl.	133
	LiCl/CsCl	<400	28,900; 25,200; 19,000	Sq. pl.	134
Cu(II)	LiCl/KCl	400	9,500	Tet.	123

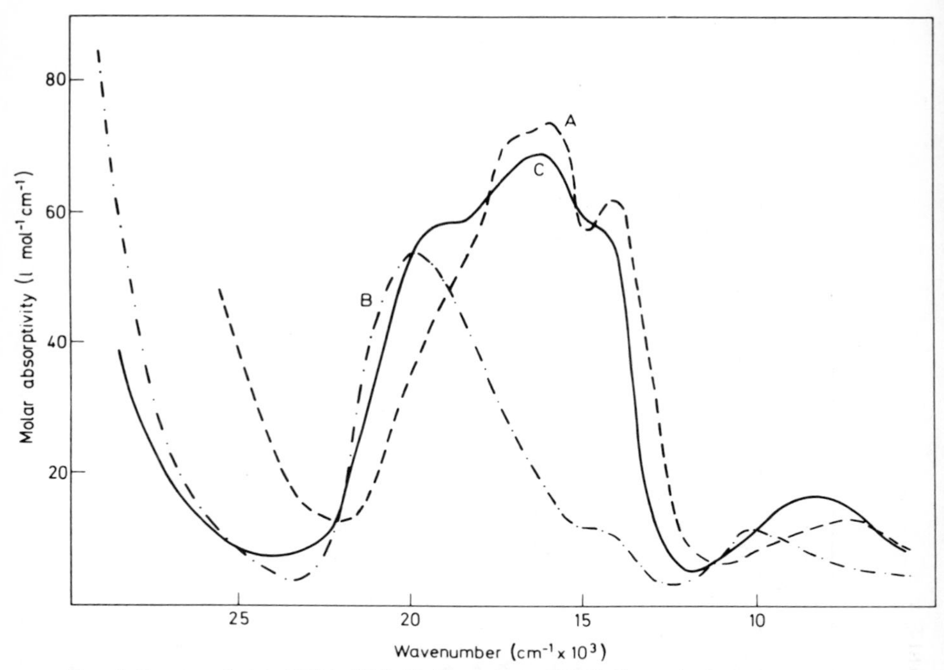

FIG. 3. Spectra of nickel(II) in KCl–$ZnCl_2$ melts at 320°C. Curve A: 58% KCl, 42% $ZnCl_2$; curve B: 45% KCl, 55% $ZnCl_2$; curve C: 0% KCl, 100% $ZnCl_2$ [adapted from C. A. Angell and D. M. Gruen, *J. Phys. Chem.* **70**, 1601 (1966)].

Further coordination geometries have been found in other halide melts. For example, in pure aluminum trichloride nickel(II) has been found in octahedral sites, possibly achieved by sharing the triangular faces of two $Al_2Cl_7^-$ anions,[138] and another new presently undefined coordination, that is neither tetrahedral nor octahedral, has been postulated in molten cesium tetrachloroaluminate.[139] The influence of anion radius is clearly shown in the variation from octahedral coordination in fluoride melts (in LiF/NaF/KF and LiF/BeF_2 at 550°C[140]) to largely tetrahedral in bromide melts (in LiBr/KBr at 335°C, though the shoulder at 19,000 cm^{-1} is considered to indicate some octahedral coordination) to entirely tetrahedral coordination in molten iodides (LiI/KI) even at the low temperature of 295°C.[141] Other ligands have also been found to complex nickel(II) in chloride melts, bromide anions coordinating slightly more strongly and cyanide anions much more strongly (in LiCl/KCl eutectic at 450°C), the latter forming four coordinate complexes of uncertain geometry.[142]

The electronic spectra of nine lanthanide(III) cations (Ce, Pr, Nd, Sm, Eu, Tb, Dy, Tm, and Yb) in molten alkali metal chloride have been measured

TABLE VI

COORDINATION GEOMETRIES AND EQUILIBRIA OF NICKEL(II) IN MOLTEN $KCl/ZnCl_2$

Melt composition (mole% $ZnCl_2$)	Temperature (°C)		
	320°C		700°C
0–42	Mainly tetrahedral		Mainly tetrahedral
47.5	Octahedral	← 2-species equilibria →	Tetrahedral
54.7	Octahedral	← continuous distortion →	Tetrahedral
93.5	Mainly octahedral		
100	Tetrahedral		Octahedral + tetrahedral

and, as expected, found not to be unlike those in aqueous solution. Many bands have been assigned, the sharp bands to $f \to f$ transitions and the broader to $f \to d$ transitions.[143,144] Three lanthanide(II) cations have also been examined in melt solutions and the spectra again were similar to those of aqueous solutions.[145,146] The cations of oxidation state (III) are however coordinated by chloride, octahedral stereochemistry being indicated by electrical conductivity measurements, along with some evidence for other coordinations[147–149] (e.g., $LaCl_4^-$, $NdCl_5^{2-}$, $Sm_2Cl_7^-$, and $Dy_3Cl_{10}^-$) all of which have also been found in phase diagram studies.[150]

Thorium(IV) cations likewise show high coordination in chloride melts. Octahedral complexes ($ThCl_6^{2-}$) are suggested in molten sodium chloride on the basis of density, surface tension measurements,[151] and thermodynamic measurements.[152] Three geometries ($ThCl_6^{2-}$, $ThCl_7^{3-}$, and $ThCl_5^-$) are suggested in potassium chloride on the basis of EMF,[153,154] molar volume, and equivalent conductivity measurements,[154] the coordination varying (to $ThCl_8^{4-}$ and $Th_2Cl_{11}^{3-}$) with further variation of alkali metal cations.[155]

With uranium(V) the electronic spectrum (similar to that of the dioxocations) showed a slight red shift with increase in size of the alkali metal cation and was probably due to the pseudo-octahedral anion ($UO_2Cl_4^{3-}$).[64] Uranium(IV) (in LiCl/KCl eutectic) has a similar spectrum to that in aqueous solution and octahedral coordination has been suggested,[156] while five coordination has been put forward on the basis of density determinations of uranium(III) solutions at 800°–1000°C (in NaCl and KCl).[157] Somewhat similar coordinations are put forward for neptunium on the basis of spectroscopic measurements [$NpO_2Cl_4^{2-}$ for Np(VI) an unspecified chloro complex for Np(V) and $NpCl_6^{2-}$ for Np(IV)[158,159]], for plutonium ($PuO_2Cl_4^{2-}$ and $PuCl_6^{2-}$),[160] and for americium ($AmCl_6^{3-}$).[161]

TABLE VII

COORDINATION GEOMETRIES AND EQUILIBRIA OF NICKEL(II) IN MOLTEN $CsCl/ZnCl_2$

Melt composition (mole % $ZnCl_2$)	Temperature		
	350°C		600°C
0	Tetrahedral $NiCl_4^{2-}$		
50	Tetrahedral$_P$		
	↕		
72	Octahedral$_P$	⟵〰〰⟶	More tetrahedral
92	Octahedral$_N$ + little tetrahedral$_N$		
	↕		
100	Tetrahedral$_N$ + little octahedral$_N$	⟵〰〰⟶	New tetrahedral and octahedral geometry

2. FORMATION OF LOWER OXIDATION STATES

As was mentioned in Section II,B,2, cations of lower oxidation state may be formed when a metal reacts with a higher oxidation state cation of the same metal, present as the halide, in solution in an alkali metal halide melt. As an example the formation of the singly charged beryllium cation (Be^+) is indicated by potentiometric and polarographic measurements on solutions of beryllium(II) chloride (in LiCl/KCl[162] or LiCl/NaCl/KCl[163]) which have been in contact with beryllium metal. Similarly, indium(III) chloride as a solute in lithium chloride–potassium chloride eutectic at 450°C is not stable in the presence of indium metal and forms deep red solutions[164] from which $InCl_{1.2} \cdot 4(Li/K)Cl$ has been separated at 370°C.[111] Raman spectroscopy indicated some disproportionation into indium(I) and indium(III) (as $InCl_4^-$) while a 170 cm^{-1} band suggested some indium–indium bonding.[111]

The higher oxidation state cations of several transition metals have also been found to be reduced. Hafnium(IV) chloride is reduced largely to hafnium(II) in lithium chloride at 820°C, the extent of reduction decreasing with temperature and with size of the alkali metal cation (i.e., LiCl > NaCl > KCl > CsCl) owing to the relatively greater stabilization of the hafnium(IV) chloro complex.[165,166] The earlier reports were less clear-cut because molecular hafnium(IV) chloride is volatile and the lower oxidation state thermally unstable to disproportionation. The same group of workers[167] had carried out analogous potentiometric measurements with iron and showed a reaction with iron(II), postulating the formation of the little-known iron(I) cation (Fe^+). Again the relative stability of the lower oxidation state decreased with increasing size of alkali metal cation but, conversely to hafnium, increased with temperature.

In contrast the much better known cadmium(I) cation (as Cd_2^{2+} rather than Cd^+) has also been found to form when cadmium(II) cations reacted with cadmium in lithium chloride–potassium chloride eutectic at 600°C on the basis of polarographic measurements.[168]

Finally some doubt remains as to whether thorium(IV) is reduced to thorium(II) by thorium metal. The original claim[169] that thorium(II) was formed (in NaCl/KCl at 680°–825°C) was counterbalanced by the failure to find any reduction below 500°C in LiCl/KCl,[170] but later reinforced in two further papers[171,172] reporting reduction in a variety of alkali metal halide melts.

The formation of lower oxidation state cations has been shown to be greatly facilitated by the presence of strong Lewis acids such as aluminum trichloride. As above mentioned (Section II,F,1) tetrachloroaluminate anions can be formed and such large anions assist in the stabilization of larger, sometimes polymeric, lower oxidation state cations, and also ensure reduced concentrations of the much smaller, more strongly polarizing chloride anions.

The element most studied in this connection is bismuth, where Bi^+, Bi_5^{3+} (of trigonal bipyramidal structure) and Bi_8^{2+} (square antiprismatic) cations have been crystallized from such melts.[173] Earlier electrochemical studies indicated metal cations could be reduced (in NaCl/$AlCl_3$ at 277°C)

$$mM^{2+} + mne^- = M_n^{m(2-n)+} \tag{68}$$

where $mn = 3$ for tin(II), 1 for lead(II), and 2 for cadmium(II), probably corresponding to the formation of Sn_2^+, Pb^+ or Pb_2^{3+}, and Cd_2^{2+} cations.[174] Studies on the reduction of selenium(IV) chloride (in NaCl/$AlCl_3$, 37/67 mole%) at 150°C by selenium have given evidence for reduced species (Se_4^{2+} and Se_8^{2+} and possibly Se_{12}^{2+} and Se_{16}^{2+})[175,176] while with tellurium(IV) chloride reduced by tellurium spectrophotometric and potentiometric evidence for six cations has been obtained (Te^{2+}, Te_2^{2+}, Te_3^+, Te_4^{2+}, Te_6^{2+}, and Te_8^{2+}),[177,178] and silver and mercury formed polymeric cations (Ag_q^{p+} [179] and Hg_3^{2+}).[180]

Other lower oxidation state cations have been produced by reduction with dissimilar metals. For example, zirconium(IV) can be reduced with aluminum/copper or nickel alloys (in NaCl/KCl at 700°–900°C[181]); niobium(III) with iron[182] or indium,[183] and molybdenum(III) with iron,[182] lead,[184] copper/nickel,[185] and aluminum/nickel[186] alloys; while uranium(IV) has been reduced with aluminum[156] and lead,[187] uranium(III) with magnesium[170] and nepturium(IV) with aluminum.[156] A variety of lanthanide and actinide cations have been found to be partially or completely reduced by liquid lithium/bismuth alloy or pure liquid bismuth in the course of developing solvent extraction methods for nuclear processing.[188] Uranium(IV) is also reduced by zirconium(II)[189] and, in addition, lower oxidation states can be formed by disproportionation reaction, e.g., molybdenum(III) (present as $Mo_2Cl_9^{3-}$ in LiCl/KCl) forms molybdenum(V) and molybdenum metal,[190] and rhenium(III) (as Re_3Cl_9 in LiCl/KCl) forms rhenium(IV) and rhenium metal.

The extreme case of a lower oxidation state is of course the metal itself and a great deal of research has been devoted to devising suitable processes for the electrowinning of metal from molten salt solutions. One need only mention the extremely large tonnage of aluminum (13×10^6 tons in 1974) made by the electrolytic reduction of aluminum dissolved in molten cryolite (Na_3AlF_6) by the Hall–Heroult process and the large tonnages of magnesium (2.5×10^5 tons)[191] and sodium also formed by electrolysis of the molten chlorides to appreciate the great industrial importance of these methods. Research to develop methods for the extraction of transition metals from such melts is frequently reported. In the case of titanium this has not yet been successful, but niobium, tantalum,[192] silver,[193] and lanthanum[194] have been won from chloride melts. With other melts more processes are possible, for example, titanium,[195] niobium,[196] tantalum,[197] and thorium[198] have been reduced from molten fluoride solution. Such melts have also been proposed for the purification of iron and steel by the electroslag remelting method,[199] while a process for the electrolytic extraction of zinc from zinc sulfide ores via an aqueous zinc chloride melt has been described.[200]

3. OXIDATION OF METALS

A further method of obtaining metal cations, frequently of a low oxidation state, is by oxidation of the metal with oxygen or chlorine, or anodically. These studies are of course important in leading to an understanding of corrosion, corrosion of container materials often being the chief difficulty in scaling up an otherwise attractive process for industrial use.

For example, the reaction of oxygen with nickel–chromium alloy and with stainless steel in chloride melts (NaCl and KCl)[201] has been studied, as has the chlorination of iron–tungsten and iron–molybdenum alloys[202] and even of ferrophosphorus when phosphorus trichloride was volatilized (from NaCl at 730°–950°C).[203] The corrosion of iron (in LiCl/KCl eutectic at 400°–800°C) in both "wet" and "dry" melts (to Fe_2O_3 and Fe_3O_4, respectively) appears to involve water as the oxidizing agent.[204,205] Cobalt corroded more readily (at 450°C in LiCl/KCl) forming a blue-green solution of cobalt(II). The reduction product was not conclusively identified, though potassium was suggested.[142] However, nickel corroded less rapidly than steel (in LiCl/NaCl/KCl at 650°C) in the presence of added fluoride[206] and was chlorinated at 450°–875°C to a decreasing extent as the size of the alkali metal cation decreased.[207] Lanthanide metals were corroded (La in NaCl/KCl at 690°–800°C,[208] and Nd in LiCl/KCl at 400°–800°C).[209]

Several transition metals are not corroded but can be brought into solution by the application of anodic currents. For example, niobium is reported to form niobium(II)[210] or Nb_3Cl_8,[211] tantalum to form tantalum(V) and other species,[212] molybdenum to form molybdenum(III),[211,213] and tungsten to form an average oxidation state of 4.1 (NaCl/KCl at 730°C). The proportion of tungsten(V) to (IV), as WCl_5^-, increases with temperature[214] or addition of 10% sodium fluoride.[215] Rhenium was anodized to rhenium(IV) and (III), the latter product immediately disproportionating to rhenium(IV) and the metal.[216]

Solution of transition metals can also be brought about by reaction with other oxidizing agents ($HgCl_2$, Hg_2Cl_2, and AgCl in LiCl/KCl) forming with mercury(II) chloride, copper(I), nickel(II), cobalt(II), iron(II), manganese(II), chromium(III), titanium(III), and vanadium(III), or (II) in the presence of excess vanadium metal.[217]

G. Exchange Reactions

Many halide and pseudohalide anions are readily exchanged via molten salt solutions, even when they are bonded to organic molecules or to silanes. Such exchanges have potential preparative value, as examples of many such reactions may be quoted

$$Me_3SiCl + KCN \rightarrow Me_3SiCN + KCl \tag{69}$$

using a 2.3 *m* KCN solution in LiCl/KCl at 450°C,[108]

$$MeSiCl_3 + 3KOCN \rightarrow MeSi(OCN)_3 + 3KCl \tag{70}$$

using a 1.9 *m* KOCN solution in LiCl/KCl at 400°C,[108]

$$COCl_2 + 2KSCN \rightarrow CO(NCS)_2 + 2KCl \tag{71}$$

using a 0.7 *m* KSCN solution in LiCl/KCl at 400°C,[169] and

$$Me_2SiCl_2 + 2NaN_3 \rightarrow Me_2Si(N_3)_2 + 2NaCl \tag{72}$$

using a 1–20% suspension of NaN_3 in $KCl/ZnCl_2$ at 250°C[218]

$$Me_3SiCl + CaCN_2 \rightarrow Me_3Si{-}N{=}C{=}N{-}SiMe_3 + CaCl_2 \tag{73}$$

using a 1.6 *m* $CaCN_2$ solution in LiCl/KCl at 400°C[219]

$$Me_3SiCl + CaC_2 \rightarrow Me_3Si{-}C{\equiv}C{-}SiMe_3 + CaCl_2 \tag{74}$$

using a 1.85 *m* CaC_2 solution in LiCl/KCl at 400°C.[219]

With aliphatic chlorides, acetonitrile was obtained when methyl chloride was passed into potassium cyanide or cyanate containing melts (both concentrations 2 *m* in LiCl/KCl eutectic at 420° and 440°C, respectively) though with the latter reactant trimethylisocyanuric acid (the trimeric form of methyl isocyanate) was also formed. The analogous products were obtained using ethyl chloride.[220] Fluoride may also be exchanged, for example, with carbon tetrachloride

$$CCl_4 + F^- \rightarrow CCl_3F,\ C_2Cl_3F,\ C_6Cl_4F_2,\ \text{etc.} \tag{75}$$

(using a KF/KCl melt at 675°C) where, because of the potential importance of the reaction, a careful study has been made of the influence of temperature, flow rate, and concentration of the reactant, etc.[221] Bromides have also been exchanged, where a low melting eutectic was chosen so as to minimize thermal decomposition of the organic reactants and products, e.g., using $NaBr/AlBr_3$ at 170°C[222]

$$CH_2Cl_2 \rightarrow CH_2ClBr + CH_2Br_2 \tag{76}$$

Cl ⟶ Br (77)

Though not strictly a pseudohalide, the analogous hydride anion may also be exchanged

$$SiCl_4 + 4LiH \rightarrow SiH_4 + 4LiCl \tag{78}$$

and a similar reaction may be used for methyl silanes (in LiCl/KCl at 400°C). The process may made continuous by electrolyzing the melt and reacting the metallic lithium with hydrogen to reform lithium hydride, with the chlorine liberated at the anode used in the preparation of further silicon tetrachloride. Thermal decomposition of the silane produced yields high purity silicon, a process used industrially for some years.[44]

H. Organic Reactions

A great number of halogenation, hydrogenation, esterification, ether cleavage, isomerization, elimination, and addition reactions have been reported in a variety of halide melts, frequently of low melting point to avoid too great a loss in yield as a result of various secondary reactions initiated thermally. A number of early reactions in halide melts were reviewed by Sundermeyer[44] and a comprehensive account of organic reactions in halide and in oxyanion melts has been given by Gordon.[223] A specialized account of organic reactions in tetrachloroaluminate melts, including organic electrode reactions, has recently appeared.[224]

Among the recent studies on reactions in molten alkali metal chlorides are the chlorinations of paraffins and olefins by Kunugi *et al.*[225,226] As with most organic reactions, several alternative or consecutive reactions take place resulting in a variety of products, the proportions of which are very dependent on precise experimental conditions. Thus in this review such products will merely be listed, as far as possible, in descending order of importance, and therefore no attempt will be made to represent the stoichiometry of the reaction.

$$CH_4 + Cl_2 \rightarrow MeCl,\ CH_2Cl_2,\ CHCl_3 \tag{79}$$

$$C_2H_4 + Cl_2 \rightarrow \text{1,2-dichloromethane} \tag{80}$$

$$\text{Propane} + Cl_2 \rightarrow \text{2-chloropropane, 1,2-dichloropropane} \tag{81}$$

$$\text{Propene} + Cl_2 \rightarrow \text{1,2-dichloropropane, 3-chloropropene, 1-chloropropene} \tag{82}$$

1,3-butadiene + Cl_2 → 3,4-dichloro-1-butene, *trans*-1,4-dichloro-2-butene,
cis-1,4-dichloro-2-butene, 1-chloro-1,3-butadiene, 4-vinyl-1-cyclohexane, cyclooctadiene (83)

[in KCl/CuCl/$CuCl_2$ (30/40/20 mole %) at 300°–400°C]. From Kikkawa *et al.*[227,228]:

$$\text{Chlorobenzene} + Cl_2 \rightarrow \text{dichlorobenzene} \tag{84}$$

Phenylethane + Cl_2 → styrene, chloroethylbenzenes, 1-chloro-1-phenylethane,
chlorovinylbenzenes, 2-chloro-1-phenyl-ethane 1,2-dichlorophenylethane (85)

[in NaCl/KCl/CuCl or NaCl/KCl/$ZnCl_2$ (both 20/20/60 mole %) at 250°–450°C] the proportions of isomers and the total conversions varying somewhat with temperature.[227]

$$\text{Bromobenzene} + Cl_2 \rightarrow \text{chlorobenzene} + \text{dichlorobenzenes, chlorobromobenzenes} \quad (86)$$

$$\text{Iodiobenzene} + Br_2 \rightarrow \text{bromobenzene} \quad (87)$$

in melts containing copper(I) chloride, and

$$\text{Iodobenzene} + Br_2 \rightarrow \text{bromobenzene, chlorobenzene} \quad (88)$$

in the zinc chloride containing melts.

When the radical inhibitor nitrobenzene was present, reaction proceeded in the zinc chloride-containing melts at 350°C, suggesting that the reaction was ionic and of the Friedel–Crafts type, but the reaction did not proceed in melts containing copper(I) chloride, suggesting a radical mechanism.[228]

Melts containing tin(II) chloride appear to facilitate organic rearrangement reactions, and studies on 16 halohalides have been reported (in $KCl/SnCl_2$ at 280°C). For example, the reaction of 1-bromo-2-methylpropane may be set out schematically as follows[229]:

$$(CH_3)_2CHCH_2Br \longrightarrow (CH_3)_2CHCH_2^+ \rightleftharpoons (CH_3)_3C^+ \quad (89)$$

$$(CH_3)_2CHCH_2Br \underset{+Cl^-}{\overset{-Cl^-}{\rightleftharpoons}} (CH_3)_2CHCH_2Cl$$

$$(CH_3)_2CHCH_2^+ \xrightarrow{-H^+} CH_3CH\begin{matrix}CH_2\\|\\CH_2\end{matrix}$$

$$(CH_3)_2CHCH_2^+ \xrightarrow{-H^+} (CH_3)_2C{=}CH_2 \xleftarrow{-H^+} (CH_3)_3C^+$$

$$(CH_3)_3C^+ \underset{-Cl^-}{\overset{+Cl^-}{\rightleftharpoons}} (CH_3)_3CCl$$

III. Nitrate Melts

A. The Use of Alkali Metal Nitrate Melts

The use of a melt containing polyatomic anions obviously immediately introduces a greater variety of possible chemical reactions for the solvent, including those of coordination, oxidation–reduction, and sometimes acid–base reactions (the last necessarily absent in alkali metal halide melts). In the case of the oxyanion melts systemization of the acid–base type reactions resulted from the work of Lux[230] and of Flood and Forland who defined a "base" as an oxide ion donor and an "acid" as an oxide ion acceptor, a definition which is narrower than that of G. N. Lewis, based on the transfer of a pair of electrons, and analogous to the Brønsted–Lowry definition of the hydrated proton as the characteristic acidic cation in the aqueous system.

The application of this Lux–Flood acid–base definition to molten nitrates has been accompanied by a continuing controversy as to the nature of the acidic and basic species present in this melt which will be described in some detail below (Section III,B). Similar controversy has not yet occurred over the chemistry of other oxyanion melts, possibly because of an intrinsically simpler chemistry but probably more importantly because the number of fundamental investigations of such melts (e.g., molten nitrites, sulfates, carbonates, perchlorates, borates, or even silicates) are much fewer.

Alkali metal nitrates have been selected for so many investigations largely because of their ready availability in a pure form, low cost, and above all conveniently low melting points (e.g., $LiNO_3$ 255°C, $NaNO_3$ 307°C, and KNO_3 334°C). The eutectics provide even lower melting points, e.g., $NaNO_3/KNO_3$ (50 : 50 mole%) 220°C; $LiNO_3/KNO_3$ (43 : 57 mole%) 132°C, and $LiNO_3/NaNO_3/KNO_3$ (27 : 18 : 55 mole%) 125°C.

In contrast to the alkali metal chloride melts, where no review of their chemistry has so far appeared, a systematic survey of reactions in molten alkali metal nitrates was published in 1972,[231] thus enabling this account to be more selective.

B. The Nature of the Acidic Species in Molten Nitrates

Application of the Lux–Flood definition suggests the self-ionization scheme

$$NO_3^- \rightleftharpoons NO_2^+ + O^{2-} \tag{90}$$

This equilibrium was put forward by Duke *et al.* as a result of a series of kinetic[232,233] and emf[234,235] studies. The nitronium cation is thermally unstable, even though it is probably stabilized by ion-pair formation with nitrate anions, thus

$$NO_2^+ + NO_3^- \rightleftharpoons [NO_2^+ \cdot NO_3^-] \rightarrow 2NO_2 + \tfrac{1}{2}O_2 \tag{91}$$

The low values of the equilibrium concentration of the nitronium ion (the equilibrium constants measured for Eq. 90 were 2.7×10^{-26} at 250°C and 5.7×10^{-24} at 300°C in equimolar $NaNO_3/KNO_3$) have prevented its direct identification, even in solutions of Lux–Flood acids, e.g., the weak acid solute, $K_2Cr_2O_7$, or the strong acid solute, $K_2S_2O_7$.

The absence of such identification (by infrared or chronopotentiometry) was partially responsible for the alternative hypothesis, put forward by Topol *et al.* ten years after that of Eq. 90, that nitrogen dioxide was the actual acidic species.[236] Though the electrochemical results certainly showed this compound to be present in all acidic solutions (cf. Eq. 91), a mechanism by which it was produced on the addition of Lux–Flood acids was not elaborated. Indeed it may be noted that in every case mechanistic pathways are simpler if the nitronium cation is involved, and that investigators have made increasing use of the Duke hypothesis in the years after the alternative Topol suggestion was put forward. For example, Bartholomew and Donigan in 1968[237] showed Duke's bromide oxidation kinetics to be compatible with nitrogen dioxide as the acidic species and explicitly supported Topol's hypothesis, but in 1970 Kozlowski, Bartholomew, and Garfinkel[238] proposed a mechanism for another reaction which involved the nitronium ion.

The distinction between the two hypotheses is not clear-cut since besides Eq. 91, nitrogen dioxide dimers are known to ionize in low temperature solvents of high dielectric constant, both unsymmetrically

$$2NO_2 \rightleftharpoons N_2O_4 \rightleftharpoons NO^+ + NO_3^- \tag{92}$$

and, less readily, symmetrically,

$$2NO_2 \rightleftharpoons N_2O_4 \rightleftharpoons NO_2^+ + NO_2^- \tag{93}$$

Such ionizations might be facilitated by a highly ionic medium, though the concentration of the dimers at the melt temperatures would be extremely small.

Nitrogen dioxide has been found to act as an acid in several reactions, e.g., with halide[233,236,239] and halate[239] anions, and with chromate(VI)[236] and manganese(IV) oxide,[240] though usually much more slowly than conventional Lux–Flood acids. However, much of this slowness can be attributed to the difficulty of transport across the gas–melt interface.[237,240]

C. The Nature of the Basic Species in Molten Nitrates

Interestingly both Topol and Duke assumed oxide to be the basic species in nitrate melts, though Duke did suggest it might be present as the orthonitrate anion in the bulk melt in equilibrium with oxide anions at the electrode surface.[235]

The orthonitrate anion (NO_4^{3-}) was originally suggested by Zintl and Morowietz.[241] Kohlmuller[242] postulated its presence in molten nitrates on the basis of phase-diagram studies (that for $NaNO_3/Na_2O$ showed an extremely broad and flat maximum) and some rather unconvincing differences in thermal decomposition, solubilities in liquid ammonia, and reactions with carbon dioxide and oxygen between fused and unfused mixtures of the oxides and nitrates of sodium and potassium. No evidence was found with the lithium compounds. Moreover, the cryoscopic measurements reported were intrinsically incapable of distinguishing any compound formation. However, both Duke[235] and Kust[243] have used the orthonitrate anion to explain the deviations of equilibrium constants from calculated values.

Shams El Din has also used compound formation between nitrate and oxide anions to explain a positive drift of potential with time for an oxygen electrode in molten potassium nitrate at 450°C containing sodium peroxide or hydroxide,[244,245] and an analogous change in the transition time in chronopotentiometric measurements.[246] The stoichiometry of the compound could of course only be determined in an inert solvent, so that the 2 : 1 ratio determined in molten lithium chloride–potassium chloride, i.e., pyronitrate, $N_2O_7^{4-}$, is not necessarily that applicable in a nitrate melt.

A completely different reaction of oxide anions in a nitrate melt was first mentioned by Zambonin and Jordan[247] in 1967 when they postulated the following as a result of potentiometric measurements and oxidation–reduction reactions in the pure melt:

$$NO_3^- + O^{2-} \rightleftharpoons NO_2^- + O_2^{2-} \quad (94)$$

$$2NO_3^- + O_2^{2-} \rightleftharpoons 2NO_2^- + 2O_2^- \quad (95)$$

and in the presence of oxygen

$$O_2^{2-} + O_2 \rightleftharpoons 2O_2^- \quad (96)$$

This has been followed by more than 20 publications reporting further measurement of these and related equilibria. It became clear that reasonable values for the equilibrium constants were $K_{94} \sim 3$ (the subscript indicating the appropriate equilibrium), $K_{95} \sim 6.7 \times 10^{-11}$, and $K_{96} \sim 3.5 \times 10^5$ (all in $NaNO_3/KNO_3$ at 229°C) if the value for the equilibrium constant of the reaction,

$$2O^{2-} + 2O_2^- \rightleftharpoons 3O_2^{2-} \quad (97)$$

which was measured in a hydroxide melt,[248] could be assumed to be equally applicable to the nitrate melt.

For some five years, Eqs. 94–96 seemed contrary to the experiences of other electrochemists, such as Shams El Din and El Hosary[244–246] or Temple and co-workers (e.g., Fredericks *et al.*[249]), using low concentrations of basic species (typically 10^{-4} *m*); and of inorganic chemists using much higher concentrations (up to the molal range) where oxide, peroxide, and hydroxide anions appeared to be distinct species and displayed different reactions, incidentally suggesting that the equilibrium

$$O^{2-} + H_2O \rightleftharpoons 2OH^- \quad (98)$$

lies to the right. Most electrochemists, who used glass or silica vessels, found the gradient of potential–oxide concentration plots to indicate a two-electron reaction and the value of K_{98} to be low (e.g., 2×10^{-2}; Fredericks *et al.*[249]). But Zambonin and co-workers, who used platinum vessels, found the slope equivalent to one electron and the value of K_{98} to be large (e.g., 10^{18}; Zambonin[250]).

Resolution of these differences become possible when Eq. 96 in a nitrate melt was determined manometrically[251] allowing a value for Eq. 95 to be obtained independent of any electrochemical measurements. This was in good agreement with the voltametric value,[252] suggesting that the assumption of the applicability of the hydroxide melt for K_{96} was justified. The major discrepancies expressed on the importance of Eq. 94 could not be resolved on the basis of anion solvation of oxide to form ortho- or pyronitrate, but could be explained on the assumption of the formation of a silicate species

$$x\mathrm{SiO_2} + \mathrm{O^{2-}} \rightleftharpoons (\mathrm{SiO_2})_x\mathrm{O^{2-}} \quad (99)$$

when silica or glass containers were used. Values for K_{90} and for K_{98} had been obtained both in siliceous containers and in platinum, and enabled two independent values of K_{99} to be calculated. These were in very satisfactory agreement (10^{16} and 2×10^{15}, respectively), suggesting that the basic species in a dilute solution in glass or silica was in fact a silicate anion.[252] However, reaction rate measurements show that more concentrated solutions do not reach equilibrium with the silica container surfaces within the normal experimental time scale, thus explaining why oxide, peroxide, and hydroxide anions show different reactions.[252] In addition, there are at present no reports of the formation of higher concentrations of superoxide in basic melts held in glass or silica vessels and in contact with oxygen, or reactions attributable to it (Schlegel and Priore[251] used a platinum crucible). This would seem initially to be attributed to the relative slowness of Eq. 96 and after longer times to the increasing stabilization of "oxide" as silicate (Eq. 99).

D. Thermal Decomposition of Molten Nitrates

A further important equilibrium occurring in nitrate melts is the solvent decomposition or reduction

$$\mathrm{NO_3^-} \rightleftharpoons \mathrm{NO_2^-} + \tfrac{1}{2}\mathrm{O_2} \quad (100)$$

The value of the equilibrium constant has been the subject of several measurements, at higher temperatures (550°–750°C). The older measurements of Freeman[253,254] were reasonably confirmed by Bartholomew,[255] who reported values for potassium nitrate. A recent measurement at lower temperatures (272°–413°C) also gave values close to those of Freeman when extrapolated.[256]

The chemical importance of the relatively low concentration of nitrite present (e.g., $2 \times 10^{-4}\,m$ at 300°C in $NaNO_3/KNO_3$) is now being demonstrated in an increasingly wide variety of reactions (see Section III,H). Not only can nitrite act directly as a reducing and as an oxidizing agent (Section VI,5), but the self-ionization analogous to Eq. 90

$$\mathrm{NO_2^-} \rightleftharpoons \mathrm{NO^+} + \mathrm{O^{2-}} \quad (101)$$

has a constant some 10^{10} larger, e.g., 1.3×10^{-9} at 300°C in $NaNO_3/KNO_3$.[257] Thus, oxide ions in basic reactions may very well result from the nitrite impurity much more often than is usually realized. In the same way the nitrosonium cation may be the significant acid species in many reactions. Certainly, as a result of careful analysis of kinetic measurements, this cation has been shown to have a role in the oxidation of formate in a nitrate melt and probably also in the oxidation of oxalate and acetate. In addition, a still unidentified nitrogen oxide derived from nitrite seems involved in another route for the oxidation of formate (Section III,J).

Because of the importance of the nitrite impurity, differences in the previous thermal history of the melt can often explain apparent discrepancies between different investigators. For it has been shown that the forward reaction of Eq. 100 is relatively fast (e.g., nitrite was detectable in 10 min, and reached the equilibrium value after 7 hr at 340°C in $NaNO_3/KNO_3$)[257] but the reverse reaction appears to be much slower (e.g., the kinetic experiments of Paniccia and Zambonin[256] were carried out for several months). It has also been demonstrated that these low nitrite concentrations can be removed, for example, by reaction with nitrogen dioxide

$$NO_2^- + NO_2 \rightleftharpoons NO_3^- + NO \tag{102}$$

so that the previous presence of acids (i.e., Eqs. 91 and 101) can also be significant.

E. Solubility and Reactivity of Nonmetallic Elements

The presence of so many potentially reactive species in a nitrate melt makes it readily apparent that fewer elements can be expected to dissolve without reaction. Among those that do so are the rare gases whose solubilities have been the subject of several measurements. Those of Cleaver and Mather[258] indicated values inversely proportional to the radius of the solute atom and only a little less (e.g., He 1.87×10^{-4} *m*, Ar 0.84×10^{-4} *m* in $NaNO_3$ at 331°C) than the solubilities in molten halides and consistent with thc values earlier obtained for xenon.[259] Field and Green[260] later obtained rather similar values, though Paniccia and Zambonin[261] reported values one-third smaller with the temperature coefficient (related to the heat of solution) varying little with the solute gas and thus indicating a relatively small solute–solvent interaction.

The three heavier common halogens have been reported to be rather more soluble, with solubilities increasing with size, (e.g., Cl_2 4.0×10^{-2} *m*, Br_2 5.1×10^{-2} *m*, and I_2 6.5×10^{-2} *m* in $LiNO_3/NaNO_3/KNO_3$ at 550°C)[262] though chlorine was also found to oxidize nitrate anions (in $LiNO_3/KNO_3$ at 180°C).[263] In contrast, iodine is formed by the oxidation of iodide above 147°C (see Section III,H).

The solubility of oxygen has also been the subject of two recent reports by Paniccia and Zambonin[261,264] who found somewhat smaller values than they reported for the rare gases (e.g., O_2 5.3×10^{-6} m at 260°C in $NaNO_3/KNO_3$). Similarly, Cleaver and Mather[258] found nitrogen to be less soluble than the rare gases (N_2 2.5×10^{-5} m in $NaNO_3$ at 331°C, a value one quarter of that reported by Field and Green)[260] though whether the solubility is genuinely larger than that of oxygen, a point of some significance in the calculation of the equilibrium constants of Eqs. 96 and 100, awaits comparative measurements by the same techniques and groups of workers.

Of the other nonmetallic elements so far studied, both carbon and sulfur have been found to react, being oxidized at 250°C to carbon dioxide and sulfate, respectively, with nitrous oxide and nitric oxide or a trace of nitrogen dioxide as the nitrate reduction products.[265]

F. Reactions of the Metallic Elements

In contrast to the molten halides, the known behavior of metallic elements in nitrate melts in every case involves reactions, either at the surface to form insoluble oxide layers, or oxidation to cations (in one case anions) in solutions. The latter reactions occur more frequently in acidic melts.

Of the alkali metals and alkaline earth metals, sodium (as the amalgam), magnesium, and calcium reacted readily to form insoluble metal oxides, while the nitrate solvent anions were reduced to nitrite, nitrogen, nitrous oxide, and nitrogen dioxide. The proportions of the product gases are given in Table VIII. In the case of calcium, a trace of hyponitrite was also detected, which gave additional support to the suggested series of reactions involving successive extraction of oxygen from the nitrogen anions by the very electropositive metals[265] i.e.

$$Ca + NO_3^- \rightarrow CaO + NO_2^- \tag{103}$$

$$Ca + NO_2^- \rightarrow [NO^-] \rightarrow \tfrac{1}{2}N_2O_2^{2-} \tag{104}$$

TABLE VIII

REACTION TEMPERATURES AND GASEOUS REACTION PRODUCTS OF METALS IN NEUTRAL AND ACIDIC $LiNO_3/KNO_3$ MELTS

Reactant metal	Gaseous reaction products (%) from neutral melts			Reaction temperature (°C)		
	N_2	N_2O	NO_2	Neutral melt	0.8 M $K_2Cr_2O_7$	0.1 M $K_2S_2O_7$
Na/Hg	24	76	0	250	—	—
Mg	7	67	26	420	310	160
Ca	70	30	<1	230	220	160

Sodium hyponitrite is known to decompose thermally to nitrite, oxide, nitrogen, and nitrous oxide.[266]

At the higher temperatures of reaction required for magnesium, the proportion of nitrogen dioxide increased (Table VIII) in accordance with the hypothesis of more electron transfer directly to the higher concentrations of nitronium ion via Eq. 90. In the same way, as the nitronium concentration was increased by the presence of weakly acidic dichromate (0.8 *M*) or more strongly acidic pyrosulfate (0.1 *M*), the temperature of the reaction was found to decrease rapidly.

A similar variation in reaction temperatures has been reported for aluminum, a surface reaction in neutral $LiNO_3/KNO_3$ occurring at 450°C, but at 400°C in 0.8 *M* potassium dichromate, while soluble cations were formed in 0.1 *M* potassium pyrosulfate at 250°C.[265] However a very slow reaction at 300°C was detected by x-ray diffraction to form α and γ alumina in sodium nitrate–potassium nitrate.[267] Chromium metal reacted at the same temperature in the pyrosulfate solution but in the neutral ($LiNO_3/KNO_3$) melt reacted at 300°C to form chromate via the initial formation of dichromate[268]

$$2Cr + 7NO_3^- \rightarrow Cr_2O_7^{2-} + 5NO_2^- + 2NO_2 \rightarrow 2CrO_4^{2-} + 3NO_2^- + 3NO_2 + NO \quad (105)$$

Dichromate reacts in a pure ($LiNO_3/KNO_3$) melt at 400°C, but nitrite anions are more reactive (see Section III,D).

The details of the reaction of iron are at present less clear, since it has been reported to be corroded (at 250°C in $NaNO_3/KNO_3$) to form a surface oxide layer,[269] but also not to form an oxide film under the same conditions.[270] In lithium nitrate–potassium nitrate, visible reaction of iron filings (B.D.H.) commenced at 400°C, but analysis of the evolved gases (N_2 10%, N_2O 8%, NO_2 < 1%, CO_2 82%) suggested the carbon impurity was responsible. A purer sample of metal (Johnson–Matthey spectroscopically pure) only formed a surface film above 500°C.[265] A similar reaction temperature was observed in nitrate melts containing chloride.[271]

G. Other Acid–Base Reactions

Metallic cations can act as effective Lux–Flood acids, as can nonmetallic oxides which are acid anhydrides. Among the first class of reactants, beryllium(II) cations acted as a moderately strong acid (in $NaNO_3/KNO_3$ at 240°C) partially decomposing the solvent and evolving nitrogen dioxide.[272]

$$Be^{2+} + 2NO_3^- \rightarrow BeO + 2NO_2 + \tfrac{1}{2}O_2 \quad (106)$$

Aluminum(III) cations were even stronger acids under the same conditions,[272] but as expected the acidity decreased down the group, gallium(III) "foaming" and forming a cloudy solution, while indium(III) sulfate was merely reported as insoluble (both in $LiNO_3/NaNO_3/KNO_3$ at 160°C).[273]

A similar trend in acidity has been observed with the highest oxides of group VA. Phosphorus(V) oxide was early reported to react in potassium nitrate at 350°C, initially to form a metaphosphate which then underwent further reaction on addition of sodium peroxide, first to pyrophosphate and then orthophosphate.[274] In sodium nitrate the first reaction to form trimetaphosphate occurred at 325°C and at 420°C in lithium nitrate[275]

$$P_4O_{10} + NO_3^- \rightarrow (PO_3^-)_n \xrightarrow{Na_2O_2} P_2O_7^{4-} \xrightarrow{Na_2O_2} PO_4^{3-} \qquad (107)$$

Sodium tripolyphosphate and trimetaphosphate have also been potentiometrically titrated with sodium peroxide in sodium nitrate at 350°C yielding products as shown in Eq. 107.[276] These and the tetrametaphosphate will also react with this melt at temperatures above 400°C.[275] As with the titration of boric acid with carbonates in alkali metal chloride melts (see Table III), a series of products has been reported following the titration of sodium metaphosphate with various bases in molten potassium nitrate[277,278] (Table IX), as has a less extensive series when potassium dichromate was neutralized (Table X).[277,279] Unexpectedly, in the case of the metaphosphate, different products were found for "forward" and "backward" reactions. Some products are otherwise unknown, including a phosphorus linked to five oxygens which has been previously found in organic compounds but hitherto not in an inorganic system.

In contrast to this strongly acidic behavior of phosphorus(V) arsenic(V) oxide reacted less vigorously

$$As_2O_5 + NO_3^- \rightarrow AsO_3^- \xrightarrow{Na_2O_2} As_2O_7^{4-} \xrightarrow{Na_2O_2} AsO_4^{3-} \qquad (108)$$

and antimony(V) oxide only very slowly even with added sodium peroxide.[280] The reaction of bismuth(V) oxide has not been reported but bismuth(III) cations act as strong Lux–Flood acids, releasing nitrogen dioxide.[272,273]

Turning to compounds of the transition metals, titanium(IV) in the form of the potassium hexafluorotitanate reacted as a Lux–Flood acid, forming anatase (commencing at 240°C in $LiNO_3/KNO_3$, maximum reaction rate at 360°C)

$$K_2TiF_6 + 4NO_3^- \rightarrow TiO_2 + 4NO_2 + O_2 + 2K^+ + 6F^- \qquad (109)$$

The oxide produced reacted further at higher temperatures (from 420°C) in a pure melt, or at lower temperatures in a basic melt (i.e., containing sodium peroxide, monoxide, or hydroxide) to form three different titanates.[281] Zirconium(IV) reacted in a similar way to form zirconium dioxide and zirconates.[282]

Vanadium(V) oxide, like the lower group VA oxides, reacted with potassium nitrate melt at 350°C, but unlike metaphosphate or metaarsenate, the metavanadate also reacted with the pure melt forming pyrovanadate which in turn could be neutralized with sodium peroxide[283]

$$V_2O_5 + NO_3^- \rightarrow NO_3^- \xrightarrow{NO_3^-} V_2O_7^{4-} \xrightarrow{Na_2O_2} VO_4^{3-} \tag{110}$$

The neutralization of chromium(VI) as dichromate was less straightforward[277,279] (see Table X).

H. Other Oxidation–Reduction Reactions

Several pseudohalides have been found to be soluble in molten nitrates. Potassium cyanide was very soluble (e.g., 3.2 *m* in $LiNO_3/KNO_3$ at 160°C)[284] and reacted in this melt from 310°C (maximum rate at 420°C)

$$CN^- + NO_3^- \rightarrow CO_3^{2-} + N_2 + N_2O + NO_2 + CO_2 \tag{111}$$

TABLE IX

NEUTRALIZATION OF SODIUM METAPHOSPHATE WITH VARIOUS BASES IN MOLTEN POTASSIUM NITRATE AT 350°C

Base	Forward titration: Acid : base ratio	Forward titration: Product	Backward titration: Acid : base ratio	Backward titration: Product
NaOH	2 : 1	$P_2O_7^{4-}$	1 : 2	$(PO_4^{3-} \cdot O^{2-})^a$
Na_2O_2	↓		↓	
Electrolytically produced O^{2-}	1 : 1	PO_4^{3-}	2 : 1	$P_2O_7^{4-}$
			↓	
			3 : 1	$P_3O_{10}^{5-}$
HCOONa	3 : 1	$P_3O_{10}^{5-}$	1 : 1	PO_4^{3-}
$(COONa)_2$	↓		↓	
Na_2CO_3	2 : 1	$P_2O_7^{4-}$	2 : 1	$P_2O_7^{4-}$
K_2CO_3	↓		↓	
$NaHCO_3$	1 : 1	PO_4^{3-}	3 : 1	$P_3O_{10}^{5-}$
CH_3COONa			2 : 1	$P_2O_7^{4-}$
			↓	
			3 : 1	$P_3O_{10}^{5-}$
Li_2CO_3	1 : 1	PO_4^{3-}	1 : 1	PO_4^{3-}
			↓	
			3 : 2	$(PO_4^{3-} \cdot P_2O_7^{4-})^a$
$CaCO_3$	2 : 3	$(2PO_4^{3-} \cdot O^{2-})^a$	1 : 2	$(PO_4^{3-} \cdot O^{2-})^a$
	↓		↓	
	4 : 9	$[4(PO_4^{3-} \cdot O^{2-}) \cdot CO_3^{2-}]^a$	3 : 4	$(3PO_4^{3-} \cdot O^{2-})^a$
$SrCO_3$	4 : 3	$(2PO_4^{3-} \cdot P_2O_7^{4-})^a$	2 : 3	$(2PO_4^{3-} \cdot O^2)^a$
	↓		↓	
	2 : 3	$(2PO_4^{3-} \cdot O^{2-})^a$	3 : 4	$(2PO_4^{3-} \cdot P_2O_7^{4-})^a$
$BaCO_3$	2 : 1	$P_2O_7^{4-}$	1 : 1	PO_4^{3-}
	↓		↓	
	1 : 1	PO_4^{3-}	5 : 1	$(2PO_2O_7^{4-} \cdot PO_3^-)^a$

[a] Indicates a compound not previously reported.

TABLE X

NEUTRALIZATION OF POTASSIUM DICHROMATE WITH VARIOUS BASES IN MOLTEN POTASSIUM NITRATE AT 350°C

Base	Forward titration		Backward titration	
	Acid : base ratio	Product	Acid : base ratio	Product
NaOH			1 : 2	$(CrO_4^{2-} \cdot O^{2-})$[a]
Na_2O_2	1 : 1	CrO_4^{2-}	↓	
Electrolytically produced O^{2-}			1 : 1	CrO_4^{2-}
Na_2CO_3				
$NaHCO_3$				
$(NaOOC)_2$				
NaOOCH	1 : 1	CrO_4^{2-}	1 : 1	CrO_4^{2-}
Li_2CO_3				
K_2CO_3				
$CaCO_3$				
$SrCO_3$				
$BaCO_3$	1 : 1	CrO_4^{2-}	2 : 3	$(2BaCrO_4 \cdot BaCO_3)$[a]
$PbCO_3$	1 : 1	$PbCrO_4 + K_2CrO_4$	2 : 1	$K_2CrO_4 + (PbCrO_4 \cdot PbCO_3)$[a] + PbO
	↓		↓	
	1 : 2	$(PbCrO_4 \cdot PbCO_3)$[a]	3 : 2	$K_2CrO_4 + 2PbCrO_4 + PbO$

[a] Indicates a compound not previously reported.

the first two products being the major ones, the latter three decreasing to trace quantities.[285] Potassium cyanate was less soluble (0.65 *m* at 160°C in $LiNO_3/KNO_3$)[284] and reacted from 300°C (maximum rate 420°C)

$$CNO^- + NO_3^- \rightarrow CO_3^{2-} + N_2 + N_2O + NO_2 + CO_2 \tag{112}$$

The proportions of the products varied with the temperature of reaction.[285] Potassium thiocyanate was the most soluble pseudohalide (9.1 *m* at 160°C),[284] reacting at 300°C (maximum rate at 350°C), and the stoichiometry was approximately

$$2SCN^- + 7NO_3^- \rightarrow CO_3^{2-} + 3NO_2^- + 2SO_4^{2-} + CO_2 + N_2 + 2N_2O \tag{113}$$

though some carbon dioxide and cyanate were also formed. The reaction sometimes became explosive if nickel cations were present.[285] In sodium nitrate–potassium nitrate, reaction occurred from 230°C the products including nitric oxide, nitrogen dioxide, carbon dioxide, cyanogen, and carbonyl sulfide as well as those listed in Eq. 113.[286]

Finally, azide was soluble and stable, though addition of transition metal cations precipitated insoluble azides which were thermally unstable. However even these were more stable than those precipitated from aqueous solutions, the decomposition temperature being some 30°C higher.[287]

Other sulfur-containing anions are also oxidized in nitrate melts, polysulfide explosively at 300°C,[288] and potassium monosulfide at 350°C

$$K_2S + 4KNO_3 \rightarrow K_2SO_4 + 4KNO_2 \tag{114}$$

If lead nitrate was present, black lead sulfide initially precipitated and was then redissolved with a final precipitation of lead sulfate.[238] With the sodium salts, sulfite was also detected but this may well have been due to incomplete oxidation, since the potassium salts were observed to react rapidly at 350°C.[238]

$$K_2SO_3 + KNO_3 \rightarrow K_2SO_4 + KNO_2 \tag{115}$$

The nitrite produced may account for a larger value of the self-dissociation constant as found from emf measurements with the oxygen electrode ($K_{116} = 1.2 \times 10^{-8}$ at 300°C in $NaNO_3/KNO_3$)

$$SO_3^{2-} \rightleftharpoons O^{2-} + SO_2 \text{ (solvated)} \tag{116}$$

than was calculated from the free energy values. The nitrite would then form oxide (Eq. 101) in accordance with its relatively larger dissociation constant. However, a similarly larger constant was also derived from the self-dissociation of sulfate ($K_{117} = 3.6 \times 10^{-10}$)[289]

$$SO_4^{2-} \rightleftharpoons O^{2-} + SO_3 \text{ (solvated)} \tag{117}$$

so that formation of silicate (Eq. 99) from the pyrex containers used may have been responsible for both unexpected experimental values. Sulfur dioxide was also oxidized (at 310°–400°C in $NaNO_3$), in addition to the stoichiometry stated

$$SO_2 + 2NaNO_3 \rightarrow Na_2SO_4 + 2NO_2 \tag{118}$$

traces of nitric oxide and nitrous oxide were also observed.[238]

The potassium halides are stable at low temperatures, the solubilities decreasing with size of the halide anion (KCl 1.0 *m*, KBr 0.36 *m*, and KI 0.01 *m* at 160°C in $LiNO_3/KNO_3$).[284] Iodide is visibly oxidized at 220°C in this melt[239]

$$2I^- + NO_3^- \rightarrow I_2 + NO_2^- + O^{2-} \tag{119}$$

a reaction detected electrochemically from 147°C. Potassium chloride and bromide appear to catalyze the nitrate decomposition (Eq. 100), reaction becoming appreciable from 470° and 300°C, respectively.[239] The halide anions have been found to be oxidized in the presence of nitrogen dioxide and of a number of Lux–Flood acids, dichromate, metaphosphate, and pyrosulfate[75,236–238]

$$2Cl^- + 2NO_2 \rightarrow Cl_2 + 2NO_2^- \tag{120}$$

The halate anions decomposed to liberate oxygen. In the case of iodate this has been shown to be probably via the nitrite ion (formed by Eq. 100), the reaction at 320°C (in $LiNO_3/KNO_3$) being

$$KIO_3 + 3NO_2^- \rightarrow KI + 3NO_3^- \tag{121}$$

In the presence of halide anions reduction took place at lower temperatures. The kinetics of the oxidation of iodide by chlorate[290] or bromate[291] have suggested the transfer of the three oxygen atoms simultaneously in a type of Walden inversion

$$ClO_3^- + I^- \rightarrow Cl^- + IO_3^- \tag{122}$$

The reaction of bromate with bromide has been studied on several occasions and shown to need the presence of a Lux–Flood acid, which may either be a transition metal cation[292,293] (relative acidities in $NaNO_3/KNO_3$: $Co^{2+} > Cu^{2+} > Ni^{2+}, Zn^{2+} > Cd^{2+}, Hg^{2+}, Pb^{2+}, Y^{3+}, La^{3+}$), dichromate,[294] nitrogen dioxide,[239] or even the lithium cation,[295]

$$BrO_3^- + 5Br^- + 6Li^+ \rightarrow 3Li_2O + 3Br_2 \tag{123}$$

a reaction analogous to that found in the aqueous system. A similar stoichiometry was found with relatively high concentrations of dichromate

$$BrO_3^- + 5Br^- + 3Cr_2O_7^{2-} \rightarrow 6CrO_4^{2-} + 3Br_2 \tag{124}$$

though with lower concentrations oxygen was also formed[296]

$$BrO_3^- + Br^- + Cr_2O_7^{2-} \rightarrow 2CrO_4^{2-} + Br_2 + O_2 \tag{125}$$

(the kinetics indicating reaction via BrO_2^+ and BrO_2^- ions).

Perchlorate anions have been reported to be stable (at 300°C in $NaNO_3/KNO_3$)[289] but periodate decomposed to liberate oxygen (at 230°C in $LiNO_3/KNO_3$) though it evolved no gas in the presence of nitrite[297]

$$IO_4^- + NO_2^- \rightarrow IO_3^- + NO_3^- \tag{126}$$

Turning to transition metal compounds, chromium(III) is oxidized to chromium(VI), e.g., at 400°C[298]

$$2CrCl_3 + 8KNO_3 \rightarrow K_2Cr_2O_7 + 6KCl + 8NO_2 + \tfrac{1}{2}O_2 \tag{127}$$

though the assertion that nitrosyl chloride and nitryl chloride were also evolved if chloride ions were added was supported by little evidence.[299] Chromium(III) oxide in lithium nitrate/potassium nitrate was, however, observed to form the chromate(VI) anion even though dichromate anions would have been stable at the temperature of reaction (250°C)[268]

$$Cr_2O_3 + 5NO_3^- \rightarrow 2CrO_4^{2-} + NO_2^- + 4NO_2 \tag{128}$$

Manganese(VII) is much less stable in nitrate melts than is chromium(VI). Permanganate normally decomposes within 5 min at 210°C (in $LiNO_3/KNO_3$) according to the reaction[300]

$$2KMnO_4 + 2Li^+ \rightarrow (Li_{0.92}/K_{0.08})_2Mn_2O_5\downarrow + \tfrac{3}{2}O_2 + 2(Li_{0.08}/K_{0.92})^+ \tag{129}$$

although in the presence of bromate anions the permanganate anion can persist for several months (Fig. 4). Periodate is also effective in the stabilisation of permanganate, as may be shown spectroscopically (Fig. 5). In both cases it was thought that nitrite, produced via Eq. 100, reacted preferentially with halate or perhalate, e.g., Eq. 126, rather than with permanganate[297]

$$2MnO_4^- + 3NO_2^- \rightarrow Mn_2O_5^{2-} + 3NO_3^- \tag{130}$$

Manganate(VI), which normally also decomposed rapidly,[300]

$$K_2MnO_4 + NO_3^- \rightarrow MnO_4^- + K_2MnO_3 + K_2O \rightarrow (Li_{0.96}/K_{0.04})_2MnO_3 + \tfrac{1}{2}O_2 \tag{131}$$

could also be stabilized as green solutions (in $NaNO_3/KNO_3$ at 260°C), by addition of alkali metal hydroxide ($>0.1\ m$)[301] somewhat analogously to the stabilization found in aqueous alkali, the characteristic absorptions being found at 22,800 and 16,200 cm^{-1}. Manganate(V) (absorption band 15,000 cm^{-1}) could also be stabilized as a blue solution, in this case by the presence of sodium monoxide or peroxide (in concentrations $>0.1\ m$). With the lithium nitrate–potassium nitrate melt, manganate(V) was produced with each of the bases but was insoluble since it formed a blue suspension.

The oxidation–reduction equilibria

$$Hg + Hg^{2+} \rightleftharpoons Hg_2^{2+} \tag{132}$$

has been shown to lie largely to the right ($K_{132} = 224$ at 177°C in

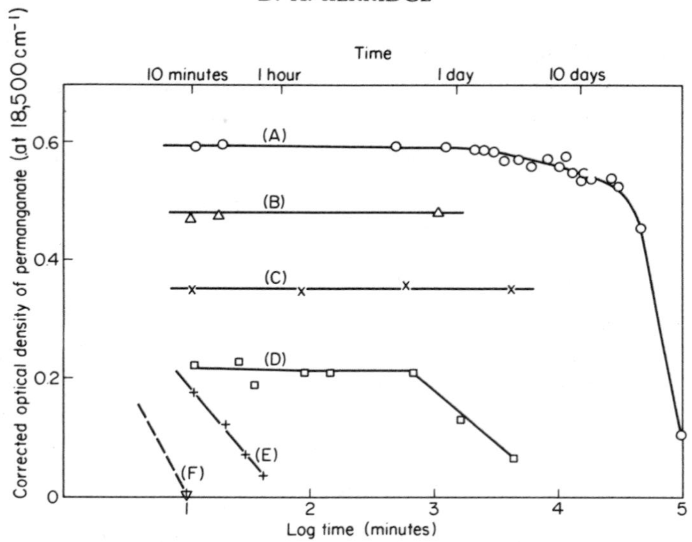

FIG. 4. Stabilization of dissolved permanganate in $LiNO_3/KNO_3$ melts in presence of dissolved halates and perhalates. (A) 0.1 *M* $KBrO_3$; (B) 0.4 *M* $KBrO_3$; (C) 0.4 *M* KIO_4; (D) 0.1 *M* KIO_4; (E) 0.4 *M* $KClO_3$; (F) 0.4 *M* KIO_3: 0.4 *M* $KClO_4$. [From B. J. Brough, D. A. Habboush, and D. H. Kerridge, *Inorg. Chim. Acta* **6**, 259 (1972).]

$LiNO_3/KNO_3$[302] and 150 at 250°C in $NaNO_3/KNO_3$[303]), though acid–base reactions can also occur with mercury(I)

$$3Hg_2SO_4 + 6NO_3^- \rightarrow 2(HgSO_4 \cdot 2HgO)\downarrow + SO_4^{2-} + 6NO_2 + 2O^{2-} \rightarrow 6HgO + 3SO_4^{2-} + 6NO_2 \quad (133)$$

where the insoluble green basic sulphate, schuettite, was first formed; and with mercury(II)

$$3HgSO_4 + 4NO_3^- \rightarrow (HgSO_42HgO)\downarrow + 2SO_4^{2-} + 4NO_2 + O_2 \xrightarrow{2NO_3^-} 3HgO + 3SO_4^{2-} + 6NO_2 + \tfrac{3}{2}O_2 \quad (134)$$

where some yellow-white sublimate [containing Hg(I), Hg(II), and NO_3^-] was formed before the green schuetteite precipitated.[304]

Finally, the new oxidation state neptunium(VII) has been found to be formed rapidly in sodium nitrate–potassium nitrate containing bromate and hydroxide[305]

$$NpO_2^{2+} + BrO_3^- + OH^- \rightarrow Np(VII) \text{ (probably } NpO_6^{5-}) \quad (135)$$

I. Complex Formation and Spectroscopy

A number of cations have been found to form coordination complexes, frequently with halide ions as ligands, in molten nitrates.

For example, an infrared investigation of alkaline earth cations in sodium or potassium nitrates has shown splitting of the ν_3 vibrational band due to

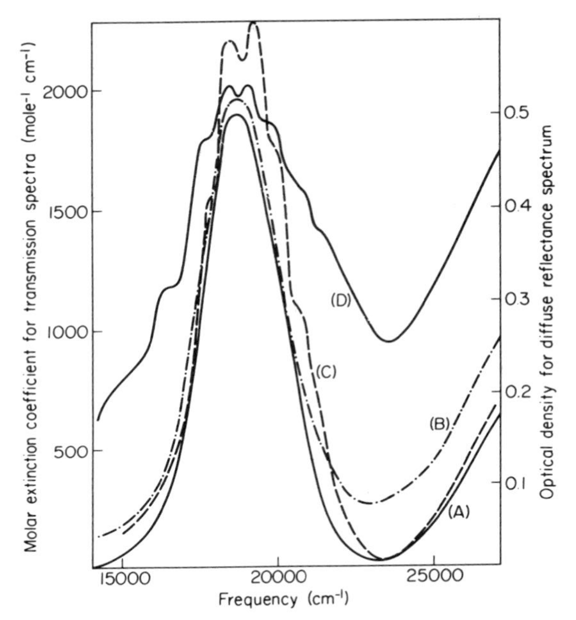

FIG. 5. Spectra of permanganate solutions. (A) 0.1 *M* $KBrO_3$ in $LiNO_3/KNO_3$ at 210°C; (B) 0.1 *M* KIO_4 in $LiNO_3/KNO_3$ at 210°C; (C) in aqueous solution at 20°C; (D) reflectance spectrum of 0.1 *M* KIO_4 in $LiNO_3/KNO_3$ [from B. J. Brough, D. A. Habboush, and D. H. Kerridge, *Inorg. Chim. Acta* **6**, 259 (1972)].

an asymmetric perturbation of the nitrate anion. Addition of potassium chloride, however, restored the original spectrum in the case of calcium, indicating the calcium–chloride complex was stronger than the calcium–nitrate complex, though in the case of magnesium the reverse was true since addition of chloride had no effect on the vibrational spectrum.[306]

The use of such spectroscopic data to deduce the presence of complexes has been the occasion of some controversy since it has been pointed out that the time for the spectroscopic transition is very much less than for thermal motion. Indeed the precise meaning of the term "complex" in an ionic liquid, in which juxtaposition of appropriately charged ions must in any case take place, has caused much debate. Nevertheless, despite uncertainty about the lifetime of a complex, the term is useful as indicating some additional interaction, or local ordering, about some ions rather than others, as in the case above.

Evidence for the existence of halide complexes of calcium and magnesium, as well as of lead and cadmium, has been found from their influence on the equilibrium

$$Cr_2O_7^{2-} + 2NO_3^- \rightleftharpoons 2CrO_4^{2-} + 2NO_2 + \tfrac{1}{2}O_2 \quad (136)$$

the cations producing more product gases by forming less soluble chromates, an effect diminished by the presence of halide. Complex stability constants have been calculated.[307,308] The constants for lead–halide complexes have also been determined cryoscopically[309] and polarographically,[310] as they have for the complexes with sulfate,[311] fluoride,[312] and EDTA[313] by cryoscopic, polarographic, and amperometric techniques, respectively. The group IIB cations, particularly cadmium(II), have been used in many investigations which have yielded information on complex formation. The melts, temperatures, and experimental methods have been varied (Table XI)[307–328] as in some cases have been the reported values of the stability constants. Selection of definitive values awaits measurements on the same solution by more than one technique.

Complexes with silver(I) cations have also been the subject of a number of investigations using a variety of techniques, but frequently making use of the low solubilities of the silver halides.[329–332] Emf, solvent extraction,[333] and chromatographic[334] methods have also been used. The results have in several cases been interpreted in terms of a number of polynuclear as well as mononuclear complexes (e.g., Ag_2X^+, Ag_3X^{2+}, AgX_2^-, and AgX_3^{2-} in $NaNO_3/KNO_3$ at 275°–300°C[330]) though this has been challenged by different views on the soundness of the computer programs involved (e.g., see Inman *et al.*[335] and Braunstein[336]). Binuclear chloro and bromo complexes are not now considered likely in sodium nitrate–barium nitrate eutectic at 350°–400°C,[337] though complicated polynuclear silver species with iodide (in $NaNO_3/KNO_3$ at 280°C) are still being reported.[338] Cyanide anions give more stable complexes[329,339,340] and fluoride less stable[341] than chloride and bromide ions. Ammonia forms complexes with one and two ammonia molecules below 207°C (in $LiNO_3/KNO_3$, $K_2 > K_1$); above 227°C "instability" occurs,[342] presumably an oxidation reaction.

The formation of complexes also shows effects in reaction chemistry, where, for example, addition of excess chloride raised the temperature of the maximum reaction rate of zinc(II) cations considerably (from 405° to 530°C in $LiNO_3/KNO_3$)[343] while the tetrachloroferrate(III) anion was also shown to lead to increased thermal stability.[344] In the case of titanium, the chloro complex was found to be sufficiently stable that it volatilized at 130°C as the dinitrosyl hexachlorotitanate(IV) (from the reaction of $TiCl_4$ or $TiCl_3$ in $LiNO_3/KNO_3$).[345] Electronic spectra have been used on several occasions to indicate the formation of complexes. An early classic investigation by Gruen and McBeth showed the gradual change from an octahedrally coordinated cobalt(II) cation in a pure nitrate melt (a simple absorption band, $[^4T_{1g}(F) \rightarrow {}^4A_{2g}]$ to a tetrahedrally coordinated chloro complex $(^4A_2 \rightarrow {}^4T_1(P))$ split by spin-orbit coupling, of much higher molar absorptivity (Fig. 6)[156] as the chloride/cobalt ratio was increased, though the stability constants were sufficiently small that a ratio a thousand times larger

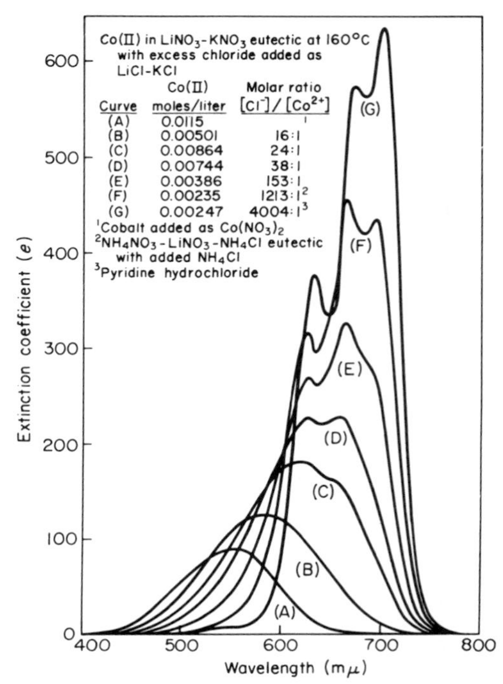

FIG. 6. Spectra of cobalt(II) in $LiNO_3/KNO_3$ at 160°C and in melts with added chloride [from D. M. Gruen, S. Fried, P. Graf, and R. L. McBeth, *Proc. U.N. Int. Conf. Peaceful Uses At. Energy, 2nd, 1958*, Vol. 28, p. 112 (1958)].

than stoichiometric was required for complete conversion (similar results were reported in Tanaev and Dzhurinskii).[346] The same variation in the electronic absorption was found with bromide and with iodide[347] and also with nickel(II) and chloride, where recently numerous detailed measurements have been subjected to computer analysis and assigned to a series of chloro–nitrato complexes of both octahedral and tetrahedral geometries.[348] Complex formation has also been detected by potentiometric,[349] conductometric,[350] and polarographic[310] methods. Rhodium(III) chloride, which formed a yellow solution in sodium nitrate–potassium nitrate (absorption 28,200 cm^{-1}, molar absorbtivity 2000 liter $mole^{-1}$ cm^{-1} at 300°C), changed to a rose color on addition of chloride (24,000 cm^{-1}, $\varepsilon = 100$, 18,800 cm^{-1}, $\varepsilon = 100$) which was interpreted as the formation of the hexachloro complex.[351] The formation of a tetrachloroferrate(III) complex gave little spectroscopic information as would be expected for a d^5 cation with a low energy charge transfer band.[344]

Several ions have been the subject of spectroscopic investigations though not all gave information on complex formation. Copper(II) gave a single band (11,800 cm^{-1} at 184°C in $LiNO_3/KNO_3$).[156] Chromate(VI) anions

TABLE XI

COMPLEX FORMATION WITH ZINC, CADMIUM, AND MERCURY CATIONS IN MOLTEN NITRATES

Complex					Stability constants				
Cation	Ligand	Nitrate solvent cations	Temperature (°C)	Experimental method	K_1	K_2	K_3	K_4	Reference
Zn	Cl	Na	307	Cryoscopy	380	1940	—	—	309
	EDTA	Li/K	160–210	Polarography	—	—	—	—	313
Cd	Cl	Li/Na/K	145	Polarography	155	30	—	—	310
		Li/K	160	emf	158	64	—	—	314
			160	Distribution	156	75	—	—	315
			180	Polarography	200	15	40	5	316
			250	Solubility	74	—	—	—	317
		Na/K	230	Ion exchange	32	20	10^{-2}	—	318
			254	emf	79	24	10	—	319
			263	Chronopotentiometry	100	7	35	—	320
			320	Raman spectroscopy	—	—	3800	—	321
		Na	250	Solubility	20	5	—	—	307
			275	emf	53	—	—	—	322
			307	Cryoscopy	320	1410	—	—	309
			307	Cryoscopy	—	—	—	—	323
		Li	260	Solubility	48	—	—	—	317
	Br	Li/Na/K	145	Polarography	640	140	80	15	310
		Li/Na	240	emf	208	—	—	—	324

		Li/K	240	emf	270	—	—	—	324
		Na/K	254	emf	185	54	21	7	319
			256	emf	108	51	10	—	308
			258	emf	140	—	—	—	325
			263	Chronopotentiometry	100	65	8	0	320
			300	emf	93	—	—	—	324
			320	Raman spectroscopy	—	—	—	9300	321
		Na	307	Cryoscopy	2900	—	—	—	309
			331	emf	53	—	—	—	325
			338	emf	53	—	—	—	324
		K	358	emf	66	—	—	—	325
			358	emf	66	—	—	—	324
Cd	I^-	Li/Na/K	145	Polarography	unstable		—	—	310
		Na/K	263	Chronopotentiometry	500	60	53	78	320
			270	Polarography	—	—	—	—	326
	F^-	Li/Na/K	145	Polarography	0	—	—	—	310
		Na/K		Polarography	14	36	350	—	312
		Li	254	Cryoscopy	0	—	—	—	311
	CN^-	Li/Na/K	145	Polarography	unstable		—	—	310
	SO_4^{2-}	Li	254	Cryoscopy	5.3	—	—	—	311
	EDTA	Li/K	160–210	Polarography	—	—	—	—	313
Hg^{2+}	Cl^-	Na/K	240	emf	—	—	—	—	327
		Li/K	150–200	Amperometric	—	—	—	—	328
Hg_2^{2+}	Cl^-	Li/K	150–200	Amperometric	—	—	—	—	328

showed similar charge transfer bands to those found in other solvents (K_2CrO_4 27,100 cm^{-1}, $\varepsilon = 3620$; $K_2Cr_2O_7$ 26,900 cm^{-1}, $\varepsilon = 2350$; $K_2Cr_3O_{10}$ 26,700 cm^{-1}, $\varepsilon = 5550$ in $LiNO_3/KNO_3$ at 160°C[268]). Praseodymium(III) and neodymium(III) chloride gave spectra similar to those of aqueous solutions (in $LiNO_3/KNO_3$ at 184°C[352]) though the latter compound showed a change in spectrum above 1 *M* concentration (in $NaNO_3/KNO_3$ at 286°–326°C) paralleled by a discontinuity in density and emf which was interpreted as a change in coordination.[353] A change in the absorption spectrum was also observed on addition of chloride.[354] The lanthanide cations have been the subjects of many investigations of solvent extraction demonstrating the formation of complexes with the organic ligands, often various derivatives of phosphine oxides (see, for example, Foos *et al.*[355–357]). Among the actinide elements uranyl(VI) chloride dissolved to give a yellow solution (23,200 cm^{-1}, $\varepsilon = 30$ in $LiNO_3/KNO_3$ at 160°C) which was ascribed to formation of an uranate anion. However, in the presence of ammonium chloride, the absorption band showed the fine structure ascribed to a chloro complex.[156] Uranium(IV) chloride was oxidized to the yellow uranate in the pure nitrate melt.[358] In contrast, neptunyl(V) chloride formed a green solution (10,200 cm^{-1} in $LiNO_3/KNO_3$ at 182°C) which was oxidized with ammonium nitrate, or with ozone, to a yellow-brown solution of neptunium(VI) (8620 cm^{-1}) which slowly reverted over the course of several days, or on heating at 220°C, to green neptunium(V), the most stable neptunium oxidation state.[359]

J. Reactions with Organic Compounds

Certain organic compounds appear to be unreactive in nitrate melts; for example, methane showed only a low solubility (10^{-6} *m* up to 340°C in $NaNO_3/KNO_3$).[261] Similarly no reaction and a low solubility ($< 10^{-4}$ *m*) were reported for chloroform and acetone.[360] Alkylammonium halides have been found surface active (in $LiNO_3/KNO_3$ at 166°C)[361] and phase diagrams have been determined with urea (with $LiNO_3/KNO_3$ and $LiNO_3/NaNO_3$ eutectics) which were stated to be stable up to 160°–170°C when decomposition became rapid.[362]

Phenolphthalein was also found to have some stability up to 260°C (in $LiNO_3/KNO_3$) but perhaps more unexpectedly to function as an indicator, dissolving in a melt containing sodium monoxide to give a purple solution [absorption bands 27,200 cm^{-1} (medium), and 17,800 cm^{-1} (strong), as compared to aqueous alkali 26,600 cm^{-1} (medium) and 18,050 cm^{-1} (strong)]. The indicator was precipitated on addition of pyrosulfate, but could be redissolved with sodium peroxide or hydroxide. Sodium hydrogen sulfate and ammonium chloride also functioned as acids.[363]

Oxalate has often been used as a base in acid–base titrations in molten nitrates, where it yielded the equivalent of one oxide per oxalate. The decomposition reaction was originally considered to be

$$C_2O_4^{2-} \rightarrow CO_2 + CO + O^{2-} \tag{137}$$

(both in KNO_3 at 350°C[364] and in $NaNO_3/KNO_3$ at 250°–300°C with Pt or Ag as a catalyst[365]) but later nitrite was recognized as a product and the reaction was therefore an oxidation with the stoichiometry

$$C_2O_4^{2-} + NO_3^- \rightarrow NO_2^- + CO_3^{2-} + CO_2 \tag{138}$$

(in KNO_3 at 350°C[366,367] and in $NaNO_3/KNO_3$ at 334°–352°C.[368]) A study of the kinetics has shown that nitrite anions also take part in the oxidation and that both processes were acid–base dependant,[369] suggesting the participation of nitronium and nitrosonium cations (Eqs. 90 and 101).

Turning to other organic anions it is now clear that, despite the determination of many phase diagrams of alkali metal nitrates with carboxylates (the reports are usefully summarized in Frazosini *et al.*[370]), these are also systems in which reactions can occur readily. Some disagreement about the products of the oxidation of formate was originally apparent but it now seems certain that at low formate concentrations and high temperatures the reaction approaches the stoichiometry

$$2HCOO^- + 2NO_3^- \rightarrow 2NO_2^- + CO_3^{2-} + CO_2 + H_2O \tag{139}$$

(both in $NaNO_3/KNO_3$[371] and as originally suggested in $NaNO_3$ at 320°–350°C[372]). However, under other conditions nitrite takes part in the oxidation. A detailed kinetic study showed that this took place via both an acid–base dependant and an independent process, and nitrite also acted as a catalyst.[371]

A similar disagreement was originally registered about the oxidation of acetate, the reactions

$$H_3CCOO^- + 4NO_3^- \rightarrow 4NO_2^- + \tfrac{1}{2}CO_3^{2-} + \tfrac{3}{2}CO_2 + \tfrac{3}{2}H_2O \tag{140}$$

(in $NaNO_3$ at 320°C[372]) and

$$4H_3CCOO^- + 5NO_3^- \rightarrow NO_2^- + 4CO_3^{2-} + 4H_3CNO_2 \tag{141}$$

(in KNO_3 at 350°C[373]) being reported. More recently, ammonia has been found as a major reaction product (in $NaNO_3/KNO_3$ at 300°C) as well as nitrite and carbonate. The nitroacetate anion was also found as an intermediate, suggesting nitromethane was indeed an initial product of the reaction (as in Eq. 141, but not as Eq. 140). In addition, cyanide was found among the hydrolysis products, again showing the formation of a C–N bond. It proved impossible to represent all the reactions taking place by a single stoichiometry but all the evidence pointed to the occurrence of the hitherto unknown nitrative decarboxylation reaction.[374]

Oxidation of potassium benzoate or phthallate indicated similar reactions while all three anion reactions were base catalyzed by potassium hydroxide in nitrate and in nitrite melts, giving further evidence for free radicals being produced on decarboxylation.[374]

Nitration reactions have also been sought in acidic nitrate melts. Benzene and a number of derivatives were efficiently nitrated in the presence of pyrosulfate (in $LiNO_3/NaNO_3/KNO_3$ at 250°–300°C), being carried on a stream of nitrogen, and the nitronium cation (Eq. 90) was suggested as the reactant.[375] Later work showed[376] that similar nitrations could also take place in the gas phase with nitrogen dioxide or even nitric acid. However, such observations did not disprove the possibility of a reactant from the liquid phase.

IV. Conclusions

It will by now be apparent that the chemistry of these two molten salt systems is both extensive and complex. Thus this general class of liquids forms a significant part of the chemistry of nonaqueous solvents and indeed of solvent chemistry as a whole. Though up to the present time the greatest emphasis has been placed on investigations of the molten halides, and in particular the molten alkali metal chlorides, in view of the greater complications and possibilities of oxyanion systems, it would seem that future workers will tend to concentrate on this latter area. This might mean increased interest in molten nitrites, acetates, cyanates, hydroxides, perchlorates, and borates, but probably most of all much more emphasis on molten sulfates, carbonates, and silicates because of their greater industrial importance.

It seems likely that if the chemistry of molten nitrates is a good guide to that of other oxyanions, the crucial importance of a number of species, not deliberately introduced into the melt and often only present in low concentrations, will soon be recognized (in the case of molten nitrates these have been shown to include NO_2^+, O^{2-}, SiO_x^{2-}, O_2^{2-}, O_2^-, NO_2^-, NO^+, etc.). Thus the need for much careful study using a variety of techniques, including electrochemical methods and kinetic measurements, will become apparent in order to identify the reactive species so that the chemistry of these interesting and novel liquids may be properly understood.

References

1. G. Charlot and B. Tremillon, "Chemical Reactions in Solvents and Salts." Pergamon, Oxford, 1969.
2. T. C. Waddington, "Non-Aqueous Solvents." Nelson, London, 1969.
3. J. E. Huheey, "Inorganic Chemistry, Principles of Structure and Reactivity." Harper, New York, 1972.

4. B. R. Sundheim, ed., "Fused Salts." McGraw-Hill, New York, 1964.
5. M. Blander, ed., "Molten Salt Chemistry." Wiley (Interscience), New York, 1964.
6. G. Mamantov, ed., "Molten Salts, Characterization and Analysis." Dekker, New York, 1969.
7. H. Bloom, "The Chemistry of Molten Salts." Benjamin, New York, 1967.
8. G. J. Janz, "Molten Salts Handbook." Academic Press, New York, 1967.
9. J. Braunstein, G. Mamantov, and G. P. Smith, eds., "Advances in Molten Salt Chemistry," Vols. 1, 2, and 3. Plenum, New York, 1971, 1973, and 1975 resp.
10. H. Bloom and J. O. M. Bochris, *in* "Fused Salts" (B. R. Sundheim, ed.), p. 1. McGraw-Hill, New York, 1964.
11. T. Førland, *in* "Fused Salts" (B. R. Sundheim, ed.), p. 73. McGraw-Hill, New York, 1964.
12. L. V. Woodcock, *Adv. Molten Salt Chem.* **3**, 1 (1975).
13. B. R. Sundheim and L. V. Woodcock, *Chem. Phys. Lett.* **15**, 191 (1972).
14. V. A. Maroni, E. J. Hathaway, and E. J. Cairns, *J. Phys. Chem.* **75**, 171 (1971).
15. J. Greenberg and B. R. Sundheim, *J. Chem. Phys.* **29**, 1029 (1958).
16. U. M. Ryabukin, *Zh. Neorg. Khim.* **7**, 1101 (1962).
17. U. M. Ryabukin and N. G. Bukin, *Zh. Neorg. Khim.* **13**, 1141 (1968).
18. L. S. Leonova, U. M. Ryabukin, and E. A. Ukshe, *Elektrokhimiya* **5**, 464 (1969).
19. H. v. Wartenberg, *Z. Elektrochem.* **32**, 330 (1926).
20. L. S. Leonova and E. A. Ukshe, *Elektrokhimya* **6**, 892 (1970).
21. T. Nakajima, H. Imoto, and K. Nakanishi, *Denki Kagaku* **42**, 85 (1974).
22. L. P. Kostin, U. G. Prasolov, and A. N. Ketov, *Tr. Estestvennonauchn. Inst. Permsk. Gos. Univ.* **13**, 205 (1972); *Ref. Zh. Khim.* 1973, abst. **12B**, 1296.
23. D. R. Olander and J. L. Camahart, *AIChE J.* **12**, 693 (1966).
24. J. D. Van Norman and R. J. Tivers, *in* "Molten Salts" (G. Mamantov, ed.), p. 509. Dekker, New York, 1969.
25. M. Kowalski and G. W. Harrington, *Inorg. Nucl. Chem. Lett.* **3**, 121 (1967).
26. M. Leroy, *Bull. Soc. Chim. Fr.* p. 968 (1962).
27. F. G. Bodewig and J. A. Plambeck, *J. Electrochem. Soc.* **116**, 607 (1969).
28. F. G. Bodewig and J. A. Plambeck, *J. Electrochem. Soc.* **117**, 904 (1970).
29. J. Greenberg, B. R. Sundheim, and D. M. Gruen, *J. Chem. Phys.* **29**, 461 (1958).
30. W. Giggenbach, *Inorg. Chem.* **10**, 1308 (1971).
31. D. M. Gruen, R. L. McBeth, and A. J. Zielen, *J. Am. Chem. Soc.* **93**, 6691 (1972).
32. J. P. Bernard, A. de Haan, and H. Van der Poorten, *C. R. Hebd. Seances Acad. Sci., Ser. C* **276**, 587 (1973).
33. G. Delarue, *Bull. Soc. Chim. Fr.* p. 906 (1960).
34. A. L. Novozhilov, V. N. Deryatkin, and E. I. Pcheluna, *Fiz. Khim. Elektrolitov* **1**, 118 (1973); *Ref. Zh., Khim.* 1973, abstr. **19B**, 1082.
35. M. A. Bredig, J. W. Johnson, and W. T. Smith, *J. Am. Chem. Soc.* **77**, 307 (1955).
36. J. W. Johnson and M. A. Bredig, *J. Phys. Chem.* **62**, 604 (1958).
37. M. Bettman, *J. Chem. Phys.* **44**, 3254 (1966).
38. R. H. Arendt and N. H. Nachtrieb, *J. Chem. Phys.* **53**, 3085 (1970).
39. D. M. Gruen, M. Krumpelt, and I. Johnson, *in* "Molten Salts" (G. Mamantov, ed.), p. 169. Dekker, New York, 1969.
40. N. H. Nachtrieb, *Adv. Chem. Phys.* **31**, 465 (1975).
41. E. J. Adaev, G. Morachevskii, and A. F. Alabshev, *Fiz. Khim. Electrokhim. Rasplavl. Solei Shlakov. Tr. Vses. Soveshch., 3rd, 1966* p. 104 (1968); V. F. Grishchenko and I. Zarabitskaya, *Zh. Prikl. Khim.* **46**, 2325 (1973).
42. S. A. Rice, *Discuss. Faraday Soc.* **32**, 181 (1961).
43. J. F. Rounsaville and J. J. Lagowski, *J. Phys. Chem.* **72**, 1111 (1968).
44. W. Sundermeyer, unpublished work, quoted in W. Sundermeyer, *Angew. Chem., Int. Ed. Engl.* **4**, 222 (1965).

45. K. E. Johnson and J. R. Mackenzie, *J. Electrochem. Soc.* **110**, 1697 (1969).
46. A. V. Volkovich, B. I. Lyazgin, and O. G. Potapenko, *Izv. Vyssh. Uchebn. Zaved., Tsvetn. Metall.* **16**, 101 (1973).
47. M. S. Foster, C. E. Crouthomel, D. M. Gruen, and R. L. McBeth, *J. Phys. Chem.* **67**, 980 (1964).
48. L. M. Ferris, M. A. Bredig, and F. J. Smith, *J. Phys. Chem* **77**, 2351 (1973).
49. M. Okada, R. A. Guidotti, and J. D. Corbett, *Inorg. Chem.* **7**, 2118 (1968).
50. J. P. Zezyanov and V. A. Il'ichev, *Zh. Neorg. Khim.* **17**, 2541 (1972).
51. A. B. Bezukladnikov, V. N. Devyatkin, and O. N. Il'icheva, *Zh. Fiz. Khim.* **44**, 253 (1970).
52. E. P. Mignonsin and G. Duyckaerts, *Anal. Chim. Acta* **47**, 71 (1969).
53. R. Combes, J. Vedel, and B. Tremillon, *C. R. Hebd. Seances Acad. Sci., Ser. C* **275**, 199 (1972).
54. W. Fischer and H. J. Abendrath, *Z. Anorg. Chem.* **308**, 98 (1961).
55. I. N. Sheiko, E. B. Gitman, and V. Y. Loichenko, *Ukr. Khim. Zh.* **38**, 305 (1972).
56. R. L. Martin and J. B. West, *J. Inorg. Nucl. Chem.* **24**, 105 (1962).
57. R. Molina, *Bull. Soc. Chim. Fr.* p. 301 (1961).
58. F. Colom and M. De La Cruz, *Electrochim. Acta* **15**, 749 (1970).
59. G. Delarue, *J. Electroanal. Chem.* **1**, 285 (1960).
60. A. A. El Hosary, H. D. Taki El Din, and A. M. Shams El Din, *Phys. Chem. Glasses* **12**, 111 (1971).
61. C. O. T. Weiss, *Diss. Abstr. B* **27**, 3451 (1967).
62. R. S. Drago and K. W. Whitten, *Inorg. Chem.* **5**, 677 (1966).
63. V. I. Shapoval and V. F. Grischenko, *Ukr. Khim. Zh.* **39**, 847 (1973).
64. L. Martinot and G. Duyckaerts, *Inorg. Nucl. Chem. Lett.* **5**, 909 (1969).
65. G. Landresse and G. Duyckaerts, *Anal. Chim. Acta* **65**, 245 (1973).
66. G. Landresse and G. Duyckaerts, *Anal. Chim. Acta* **57**, 214 (1971).
67. V. N. Goprindashvili and N. V. Kanashvili, *Soobshch. Akad. Nauk Gruz SSR* **58**, 357 (1970).
68. V. I. Shapoval and V. A. Vasilenko, *Ukr. Khim. Zh.* **40**, 868 (1974).
69. V. I. Shapoval, Y. K. Delimarski, and D. G. Tsiklauri, *Ukr. Khim. Zh.* **40**, 734 (1974).
70. Y. Kaneko and H. Kojima, *Denki Kagaku* **41**, 347 (1973).
71. N. S. Wrench and D. Inman, *J. Electroanal. Chem.* **17**, 319 (1968).
72. K. H. Stern, *J. Phys. Chem.* **66**, 1311 (1962).
73. E. P. Mignonsin, L. Martinot, and G. Duyckaerts, *Inorg. Nucl. Chem. Lett.* **3**, 511 (1967).
74. V. I. Shapoval and V. A. Vasilenko, *Ukr. Khim. Zh.* **40**, 986 (1974).
75. J. D. Van Norman and R. A. Osteryoung, *Anal. Chem.* **32**, 398 (1960).
76. E. Franks and D. Inman, *J. Electroanal. Chem.* **26**, 13 (1970).
77. A. I. Teterevkov, V. V. Pechkovskii, and N. V. Borisova, *Dokl. Akad. Nauk SSSR* **188**, 1069 (1969).
78. P. Pacak, J. Novak, and I. Slama, *Collect. Czech. Chem. Commun.* **38**, 3589 (1973).
79. N. M. Ryutina, N. V. Moksina, and N. A. Finkel'shtein, *Zh. Neorg. Khim.* **18**, 1101 (1973).
80. M. V. Smirnov, Y. V. Yurniv, and Y. V. Nasonov, *Tr. Inst. Elektrokhim., Ural. Nauchn. Tsentr. Akad. Nauk. SSSR* **20**, 23 (1973); *Ref. Zh., Khim.* 1973, abstr. **20B**, 162.
81. N. S. Wrench and D. Inman, *Electrochim. Acta* **12**, 1601 (1967).
82. P. Pacak, I. Slama, and I. Horsok, *Collect. Czech. Chem. Commun.* **38**, 2347 (1973).
83. P. Pacak and I. Slama, *Collect. Czech. Chem. Commun.* **38**, 2355 (1973).
84. R. B. Chessmore, *Diss. Abstr. Int. B* **34**, 568 (1973).
85. G. P. Smith and C. R. Boston, *Ann. N. Y. Acad. Sci.* **79**, 930 (1960).
86. H. A. Laitenen and L. R. Lieta, *Croat. Chem. Acta* **44**, 275 (1972).
87. K. W. Hanck and H. A. Laitenen, *J. Electrochem. Soc.* **118**, 1123 (1971).

88. V. I. Shapoval, Y. K. Delimarski, and D. G. Tsiklauri, *Ukr. Khim. Zh.* **40**, 941 (1974).
89. I. A. Korshunov, A. I. Kryukova, and A. V. Kolysh, *Tr. Khim. Khim. Tekhnol.* **69**, 21 (1969).
90. I. A. Korshunov, A. I. Kryukova, and C. N. Belova, *Tr. Khim. Khim. Tekhnol.* **71**, 3 (1971).
91. A. I. Kryukova, I. A. Korshunov, and Y. I. Zhuravok, *Tr. Khim. Khim. Tekhnol.* **3**, 20 (1973).
92. A. Packter and B. N. Roy, *J. Cryst. Growth* **18**, 86 (1973).
93. I. A. Korshunov, A. I. Kryukova, A. V. Kolysh, and G. E. Didenko, *Tr. Khim. Khim. Tekhnol.* **71**, 25 (1971).
94. H. A. Mal'tsev, V. A. Krokhin, A. S. Buturlov, and I. S. Morozov, *Tsvetn. Met.* **43**, 58 (1970).
95. C. H. Liu, A. J. Zielen, and D. M. Gruen, *J. Electrochem. Soc.* **120**, 67 (1973).
96. C. B. Alcock, K. T. Jacob, and O. P. Mohoptra, *Metall. Trans.* **4**, 1755 (1973).
97. B. Cleaver, A. J. Davies, and D. J. Schiffrin, *Electrochim. Acta* **18**, 747 (1973).
98. J. Mala, J. Novak, and I. Slama, *Collect. Czech. Chem. Commun.* **38**, 3032 (1973).
99. A. L. Novozhilov, V. N. Deryatkin, and E. I. Gribova, *Zh. Fiz. Khim.* **46**, 1856 (1972).
100. A. L. Novozhilov, E. I. Gribova, and V. N. Deryatkin, *Zh. Neorg. Khim.* **17**, 2078 (1972).
101. C. E. Johnson, S. E. Wood, and C. E. Crouthamel, *Inorg. Chem.* **3**, 1487 (1964).
102. W. A. Barber and C. L. Sloan, *J. Phys. Chem.* **65**, 2026 (1961).
103. G. Bienvenu, C. Gentaz, and A. Boussiba, *Rev. Int. Hautes Temp. Refract.* **11**, 61 (1974).
104. A. Bonomi, M. Parodi, and C. Gentaz, *J. Appl. Electrochem.* **6**, 59 (1976).
105. M. Fild, Diploma Thesis, University of Göttingen (1964) (quoted in Sundermeyer[44]).
106. L. M. Litz, German Abstr. 1,096,884 (1961) (quoted in Sundermeyer[44]).
107. G. Tschirne and D. Naumann, *Z. Chem.* **7**, 200 (1967).
108. W. Sundermeyer, *Z. Anorg. Allg. Chem.* **313**, 290 (1961).
109. C. Jackl and W. Sundermeyer, *Chem. Ber.* **106**, 1752 (1973).
110. G. Torsi and G. Mamantov, *Inorg. Chem.* **11**, 1439 (1972).
111. J. H. R. Clarke and R. E. Hester, *Inorg. Chem.* **8**, 1113 (1969).
112. A. J. Easteal and S. J. Heron, *Aust. J. Chem.* **22**, 681 (1969).
113. R. Oyamada, *J. Phys. Soc. Jpn.* **36**, 903 (1974).
114. J. Braunstein, *J. Phys. Chem.* **73**, 754 (1969).
115. J. H. R. Clarke and C. Solomon, *J. Chem. Phys.* **47**, 1823 (1967).
116. K. W. Fung, G. M. Begun, and G. Mamantov, *Inorg. Chem.* **12**, 53 (1973).
117. V. V. Safonov, A. V. Konov, and B. G. Korshunov, *Zh. Neorg. Chem.* **14**, 2880 (1969).
118. F. W. Poulsen and N. J. Bjerrum, *J. Phys. Chem.* **79**, 1610 (1975).
119. D. M. Gruen, *in* "Fused Salts" (B. R. Sundheim, ed.), p. 341. McGraw-Hill, New York, 1964.
120. D. M. Gruen and R. L. McBeth, *Pure. Appl. Chem.* **6**, 23 (1963).
121. D. M. Gruen and R. L. McBeth, *J. Phys. Chem.* **66**, 57 (1962).
122. H. C. Brookes and S. N. Flengas, *Can. J. Chem.* **48**, 55 (1970).
123. D. M. Gruen and R. L. McBeth, *Nature (London)* **194**, 468 (1962).
124. B. R. Sundheim and M. Kukk, *Discuss. Faraday Soc.* **32**, 49 (1961).
125. R. A. Bailey and J. A. McIntyre, *Inorg. Chem.* **5**, 964 and 1940 (1966).
126. G. W. Harrington and B. R. Sundheim, *Ann. N. Y. Acad. Sci.* **79**, 950 (1960).
127. S. Suzuki and K. Tanaka, *J. Jpn. Inst. Metals, Sendai* **34**, 461 (1970).
128. J. R. Dickinson and K. E. Johnson, *Mol. Phys.* **19**, 19 (1970).
129. J. R. Dickinson and K. E. Johnson, *Can. J. Chem.* **45**, 1631 (1967).
130. D. M. Gruen and R. L. McBeth, *J. Phys. Chem.* **63**, 393 (1959).
131. G. P. Smith, C. R. Boston, and J. Brynestad, *J. Chem. Phys.* **45**, 829 (1965).
132. G. P. Smith, C. R. Boston, and J. Brynestad, *J. Chem. Phys.* **47**, 3179 (1967).

133. R. A. Bailey and J. A. McIntyre, *Inorg. Chem.* **5**, 1824 (1966).
134. G. N. Papatheodorou and G. P. Smith, *J. Inorg. Nucl. Chem.* **35**, 799 (1973).
135. J. Brynestad and G. P. Smith, *J. Chem. Phys.* **47**, 3190 (1967).
136. C. A. Angell and D. M. Gruen, *J. Phys. Chem.* **70**, 1601 (1966).
137. W. E. Smith, J. Brynestad, and G. P. Smith, *J. Chem. Phys.* **52**, 3890 (1970).
138. H. A. Øye and D. M. Gruen, *Inorg. Chem.* **3**, 836 (1964).
139. G. P. Smith, J. Brynestad, C. R. Boston, and W. E. Smith, *in* "Molten Salts" (G. Mamantov, ed.), p. 143. Dekker, New York, 1969.
140. J. P. Young, *Inorg. Chem.* **8**, 825 (1969).
141. C. R. Boston, C. H. Liu, and G. P. Smith, *Inorg. Chem.* **7**, 1938 (1968).
142. D. Inman, B. Jones, and S. H. White, *J. Inorg. Nucl. Chem.* **32**, 927 (1970).
143. K. E. Johnson and J. N. Sandoe, *Can. J. Chem.* **46**, 3457 (1968).
144. R. A. Bailey and J. A. McIntyre, *Rev. Sci. Instrum.* **36**, 968 (1965).
145. K. E. Johnson, J. R. Mackenzie, and J. N. Sandoe, *J. Chem. Soc. A* p. 2644 (1968).
146. K. E. Johnson and J. R. Mackenzie, *J. Electrochem. Soc.* **116**, 1697 (1964).
147. R. Foethmann, G. Vogel, and A. Schneider, *Z. Anorg. Allg. Chem.* **367**, 19 (1969).
148. R. Foethmann and A. Schneider, *Z. Anorg. Allg. Chem.* **369**, 27 (1969).
149. G. Vogel and A. Schneider, *Z. Anorg. Allg. Chem.* **388**, 97 (1972).
150. Original references listed in Foethmann and Schneider.[148]
151. V. N. Desyatnik, N. M. Emel'yanov, and S. P. Raspopin, *Zh. Fiz. Khim.* **48**, 1880 (1974).
152. R. Oyamada, *J. Phys. Soc. Jpn.* **32**, 1044 (1972).
153. J. Yoshida, R. Oyamada, and T. Kuroda, *J. Electrochem. Soc. Jpn.* **35**, 183 (1967).
154. J. Yoshida, T. Iyana, and T. Kuroda, *Denki Kagaku* **41**, 217 (1973).
155. V. N. Desyatnik, S. P. Raspopin, and K. I. Trifona, *Zh. Fiz. Khim.* **48**, 1881 (1974).
156. D. M. Gruen, S. Fried, P. Graf, and R. L. McBeth, *Proc. U.N. Int. Conf. Peaceful Uses At. Energy, 2nd, 1958* Vol. 28, p. 112 (1958).
157. V. N. Desyatnik, S. F. Katyshev, and S. P. Raspopin, *Izv. Vyssh. Uchebn. Zaved., Khim. Khim. Tekhnol.* **17**, 362 (1974).
158. R. Lysy, G. Landresse, and G. Duyckaerts, *Bull. Soc. Chim. Belg.* **83**, 227 (1974).
159. R. Lysy, G. Landresse, and G. Duyckaerts, *Anal. Chim. Acta* **72**, 307 (1974).
160. G. Landresse and G. Duyckaerts, *Anal. Chim. Acta* **73**, 121 (1974).
161. U. A. Barbonel, V. P. Kotlin, and V. R. Klokman, *Radiokhimiya* **15**, 366 (1973).
162. M. V. Smirnov and N. Y. Chukreev, *Zh. Neorg. Khim.* **4**, 2536 (1959).
163. K. Ohinae and J. Kuroda, *J. Electrochem. Soc. Jpn.* **36**, 163 (1968).
164. J. M. Shafir and J. A. Plambeck, *Can. J. Chem.* **48**, 2131 (1970).
165. M. V. Smirnov, T. A. Puzanova, and N. A. Loginov, *Elektrokhimiya* **7**, 369 (1971).
166. M. V. Smirnov, T. A. Puzanova, and N. A. Loginov, *Tr. Inst. Electrokhim., Ural. Nauchn. Tsentr, Akad. Nauk SSSR* **18**, 21 (1972); *Ref. Zh., Khim.* 1973, abstr. **7B**, 804.
167. M. V. Smirnov, A. V. Pokpovskii, and N. A. Loginov, *Zh. Neorg. Khim.* **15**, 3154 (1970).
168. M. Okada, K. Yoshida, and Y. Hisomatsu, *J. Electrochem. Soc. Jpn.* **32**, 99 (1964).
169. M. V. Smirnov and L. E. Ivanovskii, *Zh. Fiz. Khim.* **31**, 801 (1967).
170. D. M. Gruen, *Ann. N. Y. Acad. Sci.* **79**, 941 (1960).
171. M. V. Smirnov, V. Y. Kudyakov, U. V. Posokhin, and U. V. Krasnov, *At. Energ.* **28**, 419 (1970).
172. M. V. Smirnov, V. Y. Kudyakov, and U. V. Posokhin, *Tr. Inst. Elektrokhim., Ural. Nauchn. Tsentr, Akad. Nauk SSSR* **18**, 27 (1972); *Ref. Zh., Khim.* 1970, abstr. **8L**, 343.
173. J. D. Corbett, *Inorg. Chem.* **7**, 198 (1968).
174. T. C. F. Munday and J. D. Corbett, *Inorg. Chem.* **5**, 1263 (1966).
175. B. J. Prince, J. D. Corbett, and B. Rabisch, *Inorg. Chem.* **9**, 2731 (1970).
176. R. Fehrmann, N. J. Bjerrum, and H. A. Andreasen, *Inorg. Chem.* **14**, 2259 (1975).

177. N. J. Bjerrum, *Inorg. Chem.* **10**, 2578 (1971).
178. N. J. Bjerrum and H. A. Andreasen, *Inorg. Chem.* **15**, 2187 (1976).
179. U. Anders and J. A. Plambeck, *Can. J. Chem.* **47**, 3055 (1969).
180. G. Torsi and G. Mamantov, *Inorg. Nucl. Chem. Lett.* **6**, 843 (1970).
181. A. N. Popov, I. F. Nichkov, and S. P. Raspopin, *Izv. Vyssh. Uchebn. Zaved., Tsvetn, Metall.* **16**, 70 (1973).
182. A. N. Popov, I. F. Nichkov, and S. P. Raspopin, *Izv. Vyssh. Uchebn. Zaved., Tsvetn. Metall.* **14**, 88 (1971).
183. E. P. Aleksandrov, B. D. Vasin, N. G. Zhezler, and I. F. Nichkov, *Izv. Vyssh. Uchebn. Zaved., Tsvetn. Metall.* **16**, 79 (1973).
184. E. P. Aleksandrov, B. D. Vasin, N. G. Zhezler, and I. F. Nichkov, *Izv. Vyssh. Uchebn. Zaved., Tsvetn. Metall.* **12**, 63 (1969).
185. A. N. Popov, I. F. Nichkov, and S. A. Raspopin, *Izv. Vyssh. Uchebn. Zaved., Tsvetn. Metall.* **15**, 67 (1972).
186. A. N. Popov, I. F. Nichkov, and S. A. Raspopin, *Izv. Vyssh. Uchebn. Zaved., Tsvetn. Metall.* **16**, 98 (1973).
187. B. D. Vasin, I. F. Nichkov, and S. A. Raspopin, *Izv. Vyssh. Uchebn. Zaved., Tsvetn. Metall.* **15**, 104 and 122 (1972).
188. L. M. Ferris, F. J. Smith, and J. C. Mailen, *J. Inorg. Nucl. Chem.* **34**, 2921 (1972); J. C. Mailer and M. J. Bell, *ibid.* p. 313.
189. B. D. Vasin, *Tr. Ural. Politekh. Inst.* **220**, 45 (1973).
190. S. Senderoff and G. W. Mellors, *J. Electrochem. Soc.* **114**, 556 (1967).
191. "Statistical Year Book 1975," Dept. Econ. & Soc. Affairs, United Nations, New York, 1976.
192. K. Huber and E. Jost, German Patent 1,092,217 (1961).
193. P. V. Polyakov and V. V. Buruakin, *Elektrokhimiya* **8**, 26 (1972).
194. S. Singh and J. Balachandra, *J. Electrochem. Soc. India* **22**, 222 (1973).
195. V. Danek and K. Matiasovsky, *Koroze. Ochr. Mater.* **17**, 89 (1973).
196. C. Decroly, A. Mukhtar, and R. Winand, *J. Electrochem. Soc.* **115**, 905 (1968).
197. V. S. Balikhin, *Zashch. Met.* **10**, 459 (1974).
198. F. R. Clayton, G. Mamantov, and D. L. Manning, *J. Electrochem. Soc.* **121**, 86 (1974).
199. F. Jesnitze-Endmann and P. Paschen, *Arch. Eisenhuettenwes.* **43**, 423 (1972).
200. H. Monk and D. J. Fray, *Inst. Min. Metall., Trans., Sect. C* **82**, 161 and 240 (1973).
201. D. R. Holmes, *Corros. Sci.* **13**, 627 (1973).
202. A. N. Zelikman, S. C. Stefayuk, C. L. Chang, and L. M. Leonova, *Izv. Vyssh. Uchebn. Zaved., Tsvetn. Metall.* **13**, 78 (1970).
203. A. I. Teterevkov, V. V. Pechkovskii, and V. V. Tamanov, *Izv. Vyssh. Uchebn. Zaved., Khim. Khim. Tekhnol.* **16**, 911 (1973).
204. F. Colom and A. Bodalo, *Corros. Sci.* **12**, 731 (1972).
205. F. Colom and A. Bodalo, *Collect. Czech. Chem. Commun.* **36**, 674 (1971).
206. L. G. Levina, E. I. Storchai, and N. S. Buranov, *Zashch. Met.* **8**, 726 (1972).
207. D. C. Namby and A. R. Hahimen, *J. Electrochem. Soc.* **121**, 104 (1974).
208. A. V. Volkovitch, B. I. Lyazgin, and O. G. Potapenko, *Izv. Vyssh. Uchebn. Zaved., Tsvetn. Metall.* **76**, 101 (1973).
209. P. M. Usov and G. N. Saratova, *Izv. Vyssh. Uchebn. Zaved., Khim. Khim. Tekhnol.* **17**, 1026 (1976).
210. V. F. Pimenov, *Izv. Vyssh. Uchebn. Zaved., Tsvetn. Metall.* **12**, 90 (1969).
211. T. Suzuki, *Electrochim. Acta* **15**, 127 and 303 (1970).
212. A. A. Nobile, *Diss. Abstr. Int. B* **32**, 805 (1971).
213. D. Inman, R. S. Sethi, and R. Spencer, *J. Electroanal. Chem.* **29**, 137 (1971).

214. S. N. Shkol'nikov and M. I. Manenkov, *Zh. Prikl. Khim.* (*Leningrad*) **46**, 1918 (1973).
215. M. I. Manenkov and S. N. Shkol'nikov, *Izv. Vyssh. Uchebn. Zaved., Tsvetn. Metall.* **17**, 65 (1974).
216. R. A. Bailey and A. A. Nobile, *Electrochim. Acta* **17**, 1139 (1972).
217. M. Henderson, J. Lewis, D. J. Machin, and A. Thompson, *Nature* (*London*) **211**, 966 (1966).
218. W. Sundermeyer, *Chem. Ber.* **96**, 1293 (1963).
219. J. Stenzel and W. Sundermeyer, *Chem. Ber.* **100**, 3368 (1967).
220. D. I. Packham and F. A. Rackley, *Chem. Ind.* (*London*) p. 899 (1966).
221. F. Peter, J. Hitzke, and J. Guion, *Bull. Soc. Chim. Fr.* p. 742 (1972).
222. R. Baker, D. H. Kerridge, and A. Onions, private communication.
223. J. E. Gordon, *Tech. Methods Org. Organomet. Chem.* **1**, 51 (1969).
224. H. L. Jones and R. A. Osteryoung, *Adv. Molten Salt Chem.* **3**, 121 (1975).
225. T. Kunugi, H. Tominaga, A. Sawanabari, and M. Nushi, *J. Chem. Soc. Jpn., Ind. Chem. Sect.* **72**, 2385 (1969).
226. T. Kunugi, H. Tominaga, and T. Nishimura, *J. Chem. Soc. Jpn., Ind. Chem. Sect.* **73**, 170 (1970).
227. S. Kikkawa, T. Hayashi, T. Toni, and H. Oshima, *J. Chem. Soc. Jpn., Ind. Chem. Sect.* **73**, 964 (1970).
228. S. Kikkawa, T. Hayashi, and N. Shibahara, *Nippon Kagaku Kaishi* p. 625 (1972).
229. R. A. Bailey and S. F. Prest, *Can. J. Chem.* **49**, 1 (1971).
230. H. Lux, *Z. Elektrochem.* **45**, 303 (1939).
231. D. H. Kerridge, *Inorg. Chem., Ser. One* **2**, 29 (1972).
232. F. R. Duke and M. L. Iverson, *J. Am. Chem. Soc.* **80**, 5061 (1958).
233. F. R. Duke and S. Yamamoto, *J. Am. Chem. Soc.* **81**, 6378 (1959).
234. J. A. Luthy and F. R. Duke, *U.S.A.E.C., Rep.* **IS-742** (1963).
235. R. N. Kust and F. R. Duke, *J. Am. Chem. Soc.* **85**, 3338 (1963).
236. L. E. Topol, R. A. Osteryoung, and J. H. Christie, *J. Phys. Chem.* **70**, 2857 (1966).
237. R. F. Bartholomew and D. W. Donigan, *J. Phys. Chem.* **72**, 3545 (1968).
238. T. R. Kozlowski, R. F. Bartholomew, and H. M. Garfinkel, *J. Inorg. Nucl. Chem.* **32**, 401 (1970).
239. D. H. Kerridge and D. A. Habboush, *Inorg. Chim. Acta* **4**, 81 (1970).
240. B. J. Brough, D. A. Habboush, and D. H. Kerridge, *J. Inorg. Nucl. Chem.* **30**, 2870 (1968).
241. E. Zintl and W. Morowietz, *Z. Anorg. Allg. Chem.* **236**, 372 (1938).
242. R. Kohlmuller, *Ann. Chim.* **4**, 1183 (1959).
243. R. N. Kust, *Inorg. Chem.* **3**, 1035 (1964).
244. A. M. Shams El Din and A. A. El Hosary, *J. Inorg. Nucl. Chem.* **28**, 3043 (1966).
245. A. M. Shams El Din and A. A. El Hosary, *Electrochim. Acta* **12**, 1665 (1967).
246. A. A. El Hosary and A. M. Shams El Din, *Electrochim. Acta* **16**, 143 (1971).
247. P. G. Zambonin and J. Jordan, *J. Am. Chem. Soc.* **89**, 6365 (1967).
248. J. Goret and B. Tremillon, *Bull. Soc. Chim. Fr.* p. 97 (1966).
249. M. Fredericks, R. B. Temple, and G. W. Thickett, *J. Electroanal. Chem.* **38**, App. 5 (1972).
250. P. G. Zambonin, *J. Electroanal. Chem.* **33**, 243 (1971).
251. J. M. Schlegel and D. Priore, *J. Phys. Chem.* **76**, 2841 (1972).
252. J. D. Burke and D. H. Kerridge, *Electrochim. Acta* **19**, 251 (1974).
253. E. S. Freeman, *J. Phys. Chem.* **60**, 1487 (1956).
254. E. S. Freeman, *J. Am. Chem. Soc.* **79**, 838 (1957).
255. R. F. Bartholomew, *J. Phys. Chem.* **70**, 3442 (1966).
256. F. Paniccia and P. G. Zambonin, *J. Phys. Chem.* **77**, 1810 (1973).
257. R. N. Kust and J. D. Burke, *Inorg. Nucl. Chem. Lett.* **6**, 333 (1970).
258. B. Cleaver and D. E. Mather, *Trans. Faraday Soc.* **66**, 2469 (1970).

259. P. I. Protsenko and A. G. Bergman, *Zh. Obshch. Khim.* **20**, 1365 (1950).
260. P. E. Field and W. J. Green, *J. Phys. Chem.* **75**, 821 (1971).
261. F. Paniccia and P. G. Zambonin, *J. Chem. Soc., Faraday Trans. I* **68**, 2083 (1972).
262. Y. K. Delimarsky and G. V. Shilina, *Electrochim. Acta* **10**, 971 (1965).
263. G. G. Bombi, G. A. Sacchetto, and G.-A. Mazzochin, *J. Electroanal. Chem.* **24**, 23 (1970).
264. E. Desimoni, F. Paniccia, and P. G. Zambonin, *J. Electroanal. Chem.* **38**, 373 (1972).
265. B. J. Brough and D. H. Kerridge, *Inorg. Chem.* **4**, 1353 (1965).
266. T. M. Oza and V. T. Oza, *J. Phys. Chem.* **60**, 192 (1956).
267. L. Campanella and A. Conte, *J. Electrochem. Soc.* **114**, 144 (1969).
268. B. J. Brough, D. H. Kerridge, and S. A. Tariq, *Inorg. Chim. Acta* **1**, 267 (1967).
269. T. Notoya, T. Ishikawa, and K. Midorikawa, *Denki Kagaku Oyobi Kogyo Butsuri Kagaku* **40**, 62 (1972).
270. A. Conte and S. Casadio, *Ric. Sci.* **36**, 433 (1966).
271. V. P. Kochergin, E. A. Drzhinina, G. V. Men'shenina, and E. P. Asanova, *Zh. Prikl. Khim.* **33**, 1580 (1960).
272. G. G. Bombi and M. Fiorani, *Talanta* **12**, 1053 (1965).
273. M. Steinberg and N. H. Nachtrieb, *J. Am. Chem. Soc.* **72**, 3558 (1950).
274. A. M. Shams El Din, A. A. El Hosary, and A. A. A. Gerges, *J. Electroanal. Chem.* **6**, 131 (1963).
275. J. L. Copeland and L. Gutierez, *J. Phys. Chem.* **77**, 20 (1973).
276. N. Coumert, M. Porthault, and J-C. Merlin, *Bull. Soc. Chim. Fr.* p. 90 (1965).
277. A. M. Shams El Din and A. A. El Hosary, *Electrochim. Acta* **13**, 135 (1968).
278. A. M. Shams El Din, H. D. Taki El Din, and A. A. El Hosary, *Electrochim. Acta* **13**, 407 (1968).
279. A. M. Shams El Din and A. A. El Hosary, *J. Electroanal. Chem.* **16**, 557 (1968).
280. A. M. Shams El Din, A. A. El Hosary, and A. A. A. Gerges, *J. Electroanal. Chem.* **8**, 312 (1964).
281. D. H. Kerridge and J. Cancela Rey, *J. Inorg. Nucl. Chem.* **37**, 2257 (1975).
282. D. H. Kerridge and J. Cancela Rey, *J. Inorg. Nucl. Chem.* **39**, 405 (1977).
283. A. M. Shams El Din and A. A. El Hosary, *J. Electroanal. Chem.* **7**, 464 (1964).
284. B. J. Brough and D. H. Kerridge, *J. Chem. Eng. Data* **11**, 260 (1966).
285. B. J. Brough, Ph.D. Thesis, Southampton University (1965).
286. M. E. Martins, A. J. Calandria, and A. J. Arvia, *J. Inorg. Nucl. Chem.* **36**, 1705 (1974).
287. H. C. Egghart, *Inorg. Chem.* **6**, 2121 (1967).
288. B. Cleaver and A. J. Davies, *Electrochim. Acta* **18**, 727 (1973).
289. M. Fredericks and R. B. Temple, *Aust. J. Chem.* **25**, 2319 (1972).
290. F. R. Duke and G. Franke, *in* "Fused Salts" (B. R. Sundheim, ed.), p. 417. McGraw-Hill, New York, 1964.
291. J. M. Schlegel and J. Perrin, *J. Inorg. Nucl. Chem.* **34**, 2087 (1972).
292. F. R. Duke and W. L. Lawrence, *J. Am. Chem. Soc.* **83**, 1269 (1961).
293. F. R. Duke and W. L. Lawrence, *J. Am. Chem. Soc.* **83**, 1271 (1961).
294. J. M. Schlegel, *J. Phys. Chem.* **69**, 3638 (1965).
295. F. R. Duke and E. A. Shute, *J. Phys. Chem.* **66**, 2114 (1962).
296. J. M. Schlegel, *J. Phys. Chem.* **73**, 4152 (1969).
297. B. J. Brough, D. A. Habboush, and D. H. Kerridge, *Inorg. Chim. Acta* **6**, 259 (1972).
298. M. W. Y. Spink, *Diss. Abstr.* **26**, 4274 (1966).
299. M. W. Y. Spink, Ph.D. Thesis, Pennsylvania State University, University Park (1965).
300. D. H. Kerridge and S. A. Tariq, *Inorg. Chim. Acta* **2**, 371 (1968).
301. B. J. Brough, D. A. Habboush, and D. H. Kerridge, *Inorg. Chim. Acta* **6**, 366 (1972).
302. G.-A. Mazzochin, G. G. Bombi, and M. Fiorani, *J. Electroanal. Chem.* **17**, 95 (1968).

303. H. S. Swofford and J. Dietz, *Anal. Chem.* **44**, 3232 (1972).
304. D. H. Kerridge and J. Cancela Rey, *J. Inorg. Nucl. Chem.* **39**, 297 (1977).
305. N. N. Krot, M. P. Mefod'eva, C. P. Shilov, and A. D. Gelman, *Radiokhimiya* **12**, 471 (1970).
306. R. E. Hester and K. Krishnan, *J. Chem. Phys.* **46**, 3405 (1966); **47**, 1747 (1967).
307. F. R. Duke and M. L. Iverson, *J. Phys. Chem.* **62**, 417 (1958).
308. H. M. Garfinkel and F. R. Duke, *J. Phys. Chem.* **65**, 1627 and 1629 (1961).
309. E. R. Van Artsdalen, *J. Phys. Chem.* **60**, 172 (1956).
310. D. Inman, D. G. Lovering, and R. Narayan, *Trans. Faraday Soc.* **64**, 2476 (1968).
311. R. E. Isbell, E. W. Wilson, and D. F. Smith, *J. Phys. Chem.* **70**, 2493 (1966).
312. K. M. Boika, *Ukr. Khim. Zh.* **35**, 596 (1969).
313. M. V. Susic, D. A. Markovic, and N. N. Hecigonja, *J. Electroanal. Chem.* **41**, 119 (1973).
314. G. G. Bombi, G.-A. Mazzochin, and M. Fiorani, *Ric. Sci.* **36**, 573 (1966).
315. J. D. Liljenzin, H. Reinhardt, H. Wirries, and R. Lindner, *Z. Naturforsch., Teil A* **18**, 840 (1963).
316. J. H. Christie and R. A. Osteryoung, *J. Am. Chem. Soc.* **82**, 1841 (1960).
317. T. P. Flaherty and J. Braunstein, *Inorg. Chim. Acta* **1**, 335 (1967).
318. M. Liquornik and J. W. Irvine, *Inorg. Chem.* **9**, 1330 (1970).
319. D. Inman, *Electrochim. Acta* **10**, 11 (1965).
320. D. Inman and J. O.M. Bockris, *Trans. Faraday Soc.* **57**, 2308 (1961).
321. J. H. R. Clarke, P. J. Hartley, and Y. Kuroda, *Inorg. Chem.* **11**, 29 (1972).
322. C. Sinistri, *Ric. Sci.* **32**, 638 (1962).
323. E. L. Heric, A. J. Moon, and W. H. Waggoner, *J. Chem. Eng. Data* **14**, 318 (1969).
324. J. Braunstein and A. S. Minao, *Inorg. Chem.* **5**, 942 (1966).
325. H. Braunstein, J. Braunstein, and D. Inman, *J. Phys. Chem.* **70**, 2726 (1964).
326. R. M. Novik and Y. S. Lyalikov, *Zh. Anal. Chem.* **13**, 691 (1958).
327. J. L. Videl Garcia, *Acta Cient. Compostelana* **6**, 129 (1969).
328. G.-A. Mazzochin, G. A. Sacchetto, and G. G. Bombi, *J. Electroanal. Chem.* **24**, 31 (1970).
329. J. L. Videl Garcia, *Acta Cient. Compostelana* **6**, 47 (1969).
330. C. Sinistri and E. Pazzoti, *Gazz.* **97**, 1116 (1967).
331. S. Hill and F. E. W. Whitmore, *Can. J. Chem.* **32**, 864 (1954).
332. H. C. Gaur and N. P. Bansal, *J. Chem. Soc., Faraday Trans. I* **68**, 1368 (1972).
333. I. J. Gal, J. Mendez, and J. W. Irvine, *Inorg. Chem.* **7**, 985 (1968).
334. G. Alberti and G. Grassini, *J. Chromatogr.* **4**, 425 (1960).
335. D. Inman, S. H. White, I. Wilmot, and B. Jones, *J. Electroanal. Chem.* **33**, 225 (1971).
336. J. Braunstein, *J. Electroanal. Chem.* **33**, 235 (1971).
337. H. C. Gaur and N. P. Bansal, *Rev. Roum. Chim.* **18**, 1903 (1973).
338. B. Holmberg, *Acta Chem. Scand.* **27**, 875, 3550 (1973).
339. G. G. Bombi, M. Fiorani, and G.-A. Mazzochin, *J. Electroanal. Chem.* **9**, 457 (1965).
340. J. Pendergast, *Diss. Abstr.* **23**, 2304 (1963).
341. G. G. Bombi, G. A. Sacchetto, and C. Macca, *J. Electroanal. Chem.* **42**, 373 (1973).
342. C. Macca, G. G. Bombi, and G. A. Sacchetto, *J. Electroanal. Chem.* **51**, 425 (1974).
343. D. H. Kerridge and J. Cancela Rey, *J. Inorg. Nucl. Chem.* **37**, 975 (1975).
344. D. H. Kerridge and A. Y. Khudhari, *J. Inorg. Nucl. Chem.* **37**, 1893 (1975).
345. D. H. Kerridge and J. Cancela Rey, *J. Inorg. Nucl. Chem.* **37**, 2257 (1975).
346. I. Y. Tanaev and B. F. Dzhurinskii, *Dokl. Akad. Nauk SSSR* **134**, 1374 (1960).
347. I. Y. Tanaev and B. F. Dzhurinskii, *Dokl. Akad. Nauk SSSR* **135**, 94 (1960).
348. T. R. Griffiths and P. J. Potts, *Inorg. Chem.* **14**, 1039 (1975).
349. P. Pacak and I. Slama, *Collect. Czech. Chem. Commun.* **36**, 2988 (1971).
350. P. C. Papaioannou and G. W. Harrington, *J. Phys. Chem.* **68**, 2424 (1964).
351. F. B. Ogilvie and O. G. Holmes, *Can. J. Chem.* **44**, 447 (1966).

352. D. M. Gruen, *J. Inorg. Nucl. Chem.* **4**, 74 (1957).
353. J. Padova, M. Peleg, and J. Soriano, *J. Inorg. Nucl. Chem.* **29**, 1895 (1967).
354. J. Foos, A. S. Kertes, and M. Peleg, *J. Inorg. Nucl. Chem.* **36**, 837 (1974).
355. J. Foos and R. Guilloumont, *Radiochem. Radioanal. Lett.* **10**, 201 (1972).
356. J. Foos and J. Mesplede, *J. Inorg. Nucl. Chem.* **34**, 2051 (1972).
357. J. Torghetta, J. Mesplede, and M. Porthault, *J. Inorg. Nucl. Chem.* **36**, 445 (1974).
358. D. M. Gruen, R. L. McBeth, J. Kooi, and W. T. Cornell, *Ann. N. Y. Acad. Sci.* **79**, 941 (1960).
359. D. Cohen, *J. Am. Chem. Soc.* **83**, 4094 (1961).
360. S. Allulli, *J. Phys. Chem.* **73**, 1084 (1969).
361. K. F. Guenther, *J. Phys. Chem.* **67**, 2851 (1963).
362. N. A. Shul'ga and A. G. Bergman, *Russ. J. Inorg. Chem.* (*Engl. Transl.*) **14**, 1034 (1969).
363. B. J. Brough, D. H. Kerridge, and M. Mosley, *J. Chem. Soc. A* p. 1556 (1966).
364. N. Coumert, M. Porthault, and J.-C. Merlin, *Bull. Soc. Chim. Fr.* p. 332 (1967).
365. P. G. McCormick and H. S. Swofford, *Anal. Chem.* **41**, 146 (1969).
366. A. A. El Hosary and A. M. Shams El Din, *J. Electroanal. Chem.* **30**, 33 (1971).
367. A. G. Keenan and C. G. Fernandez, *J. Electrochem. Soc.* **120**, 1697 (1973).
368. J. M. Schlegel and C. A. Pitak, *J. Inorg. Nucl. Chem.* **32**, 2088 (1970).
369. J. D. Burke, Ph.D. Thesis, University of Utah, Salt Lake City (1972).
370. P. Franzosini, P. Ferloni, and G. Spinolo, " Molten Salts with Organic Anions. An Atlas of Phase Diagrams." University of Pavia, Italy, 1973.
371. D. H. Kerridge and J. D. Burke, *J. Inorg. Nucl. Chem.* **38**, 1307 (1976).
372. T. R. Kozlowski and R. F. Bartholomew, *Inorg. Chem.* **7**, 2247 (1968).
373. A. M. Shams El Din and A. A. El Hosary, *Electrochem. Acta* **13**, 135 (1968).
374. J. D. Burke and D. H. Kerridge, *J. Inorg. Nucl. Chem.* **37**, 751 (1975).
375. R. B. Temple, C. Fay, and J. Williamson, *Chem. Commun.* p. 966 (1967).
376. R. B. Temple and G. W. Thickett, *Aust. J. Chem.* **26**, 667 (1973).

Author Index

Numbers in parentheses are reference numbers and indicate that an author's work is referred to although his name is not cited in the text. Numbers in italics show the page on which the complete reference is listed.

B

D

E

H

I

J

K

L

N

T

U

V

W

Y

Z

Subject Index

B

C

A 8
B 9
C 0
D 1
E 2
F 3
G 4
H 5
I 6
J 7